RNA Technologies in Cardiovascular Medicine and Research

Volker A. Erdmann • Wolfgang Poller
Jan Barciszewski
Editors

RNA Technologies in Cardiovascular Medicine and Research

Springer

Prof. Dr. rer.nat Volker A. Erdmann
Institute for Chemistry/Biochemistry
Free University of Berlin,
Thielallee 63
14195 Berlin, Germany
erdmann@chemie.fu-berlin.de

Prof. Dr. med Wolfgang Poller
Clinic for Cardiology and Pneumology
Campus Benjamin Franklin
Charite University Medicine Berlin
Hindenburgdamm 30
12200 Berlin, Germany
wolfgang.poller@charite.de

Prof. Dr. hab. Jan Barciszewski
Institute of Bioorganic Chemistry of the
Polish Academy of Sciences
Noskowskiego 12
61-704 Poznan
Poland
jan.barciszewski@ibch.poznan.pl

ISBN 978-3-642-09746-1 e-ISBN 978-3-540-78709-9

DOI: 10.1007/978-3-540-78709-9

Cover Design: WMX Design GmbH, Heidelberg, Germany

Printed on acid-free paper

5 4 3 2 1

springer.com

Preface

The heart may respond to chronic and acute injury by hypertrophic growth and pathological remodeling. Cardiomyocyte hypertrophy is a dominant cellular response to all kinds of hemodynamic overload, inherited mutations in many structural and contractile proteins, and other factors and may be compensatory or maladaptive. In the latter case pathological remodeling may result from diverse molecular pathomechanisms which are still incompletely understood.

Small deviations in a mechanism controlling cardiac morphology and function may lead to enormous negative effects including loss of function and, in severe cases, even death. Despite great progress in understanding various aspects of heart development, cardiovascular diseases remain a major problem for medicine. Therefore, there is a need for new diagnostic and therapeutic strategies to detect, classify and cure heart diseases.

Ribonucleic acid (RNA) in its many facets of structure and function is more and more understood, and therefore it is possible to design and use RNAs as valuable tools in molecular biology and medicine. An understanding of the role of RNAs within the cell has changed dramatically in recent years (Fig. 1). Its status expanded with reports on catalytic RNAs (ribozymes) 25 years ago, of endogenous RNA interference 15 years later, and other noncoding RNA very recently. Today, it is obvious that RNAs are not merely the intermediary molecules between DNA and proteins, but that they can also be functional end products. Large stretches of genomic DNA do not contain protein-coding sequences and have, therefore, been considered as 'junk'. However, a significant fraction of this noncoding DNA have actually been found to hold the information for some of these functional noncoding RNAs. Diverse eukaryotic organisms harbor a class of noncoding small RNAs which are thought to function as regulators of gene expression. Thus, RNAs can be the transmitters (mRNAs) of genetic information to the ribosome for proteins to be synthesized, and also the regulators in protein synthesis. The conclusion to be drawn is that RNA is much more than solely a messenger RNA, and therefore this class of molecules are truly renaissance molecules. Most of the noncoding DNA is occupied by various units of repeats, satellite sequences and transposons. These sequences have been sought to be epigenetic elements that control stability of gene expression programs, and organize heterochromatic domains at centromers and telomers. Their role appears to be mainly regulatory. Although the effects of antisense RNAs on the corresponding

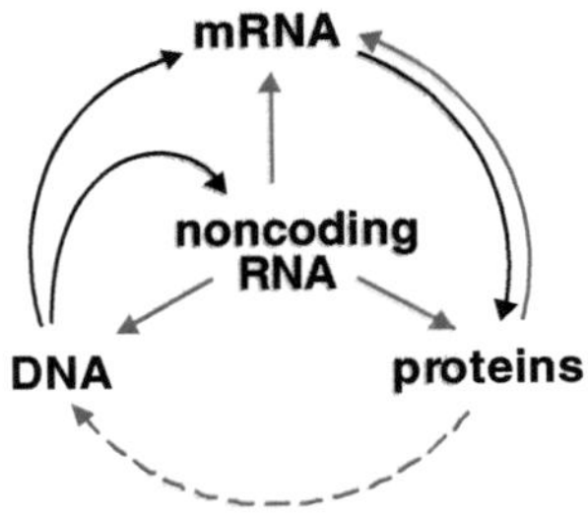

Fig. 1 Genetic information from DNA is transcribed into mRNA containing instruction for protein synthesis and regulatory RNAs which take part in systems controlling expression of genes

sense RNAs have not been clearly established, a number of examples indicate that they may exert control at various levels of gene expression, such as transcription, mRNA processing, splicing, stability, transport, and translation.

RNA has become a focus of investigations into novel therapeutic schemes. Ribozymes, antisense RNAs, RNA decoys, aptamers, micro RNAs and small interfering (siRNAs) have been used to down regulate undesired gene expression (Fig. 2). Multiple challenges, such as optimization of selectivity, stability, delivery and long term safety, have to be addressed in order for RNA drugs to become successful therapeutic agents. Not all RNA classes (e.g., ribozymes or RNA decoys) have been so far successfully developed as drugs. The recognition of the biological roles of small molecular weight RNAs have been one of the most significant discoveries in molecular biology. These RNA molecules influence the translation of messenger RNAs (mRNAs) in post-transcriptional manner that makes the regulation of RNAs even more complex.

The use of RNA-mediated interference (RNAi) for gene silencing has provided a powerful tool for loss-of-function studies in a variety of metazoans. SiRNA mediated gene silencing by degradation of target messenger RNAs have been widely used in gene function characterizations.

Compared with the laborious, time-consuming, and very costly gene knockout models, siRNA provides an efficient, specific and cheap solution for inhibiting expression of target genes. Efficient siRNA delivery is essential for the success of specific gene silencing. As the popularity of RNAi technology grows, so does the frustration it still causes for many researchers. Direct measurement at the mRNA level is always needed for direct verification that RNA interference is decreasing the amount of mRNA.

Because high doses of siRNAs may provoke an altered expression of many other genes, selections of optimal conditions are essential to minimize potential side effects. The most informative experiments in understanding the specificity of a siRNA would consider acquiring global gene expression of relevant genes, which unfortunately is lacking in many siRNA studies. These small RNAs of about 15–49 nucleotides in length guide the RNA-induced silencing complex (RISC). The beauty of the system lies in the application of short RNAs, which can be synthesized at reasonable cost and can evolve quickly, to regulate a large and complex protein synthesis.

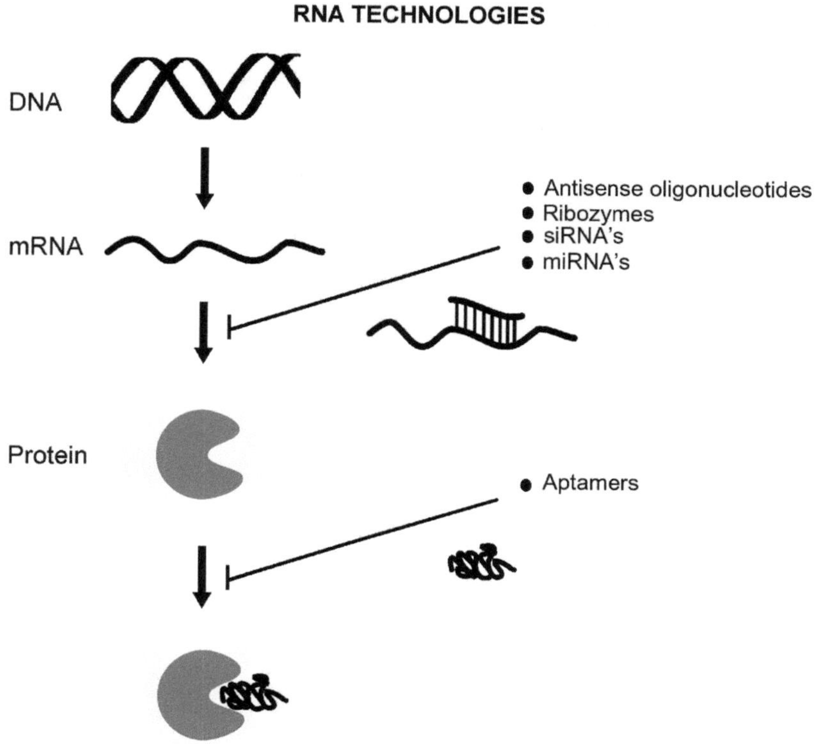

Fig. 2 The potentials of the different RNA technologies which can be applied to inhibit protein biosynthesis on RNA levels (*antisense oligonucleotides*, *ribozymes*, *short interference RNAs* and *microRNAs*) or protein functional level (*aptamers*)

In several recent studies, microarray analyses were performed to determine whether miRNAs are deregulated in hyperthrophic and failing hearts. The results implicate that miRNAs function as negative regulators of cell growth or as regulators of prosurvival pathways such that their downregulation predisposes the heart to pathological remodeling. A major challenge for the future will be to identify the mRNA targets of RNAs that participate in cardiac remodeling and to understand the functions of their target mRNAs.

Finally, the recent application of advanced vector technologies developed initially in the gene therapy field has had an enormous impact on the efficacy by which RNAi and microRNAs can be employed for therapeutic purposes in vivo. These most recent developments have brought clinical translation of certain RNA-based therapies within reach.

Berlin, Germany — *Volker A. Erdmann*
Berlin, Germany — *Wolfgang Poller*
Poznan, Poland — *Jan Barciszewski*
February 2008

Contents

Contributors

Maha Abdellatif
Department of Cell Biology and Molecular Medicine, Cardiovascular Research Institute, University of Medicine and Dentistry of New Jersey, 185 S Orange Ave, Newark, NJ 07103, USA
abdellma@umdnj.edu

Jan Barciszewski
Institute of Bioorganic Chemistry of the Polish Academy of Sciences, Noskowskiego 12, 61704 Poznan, Poland
Jan.Barciszewski@ibch.poznan.pl

Charlene M. Blake
Duke Institute for Genome Sciences and Policy, Duke University Medical Center, Durham, NC 27710, USA
charlene.blake@duke.edu

Thomas E. Callis
Carolina Cardiovascular Biology Center, Department of Cell and Developmental Biology, 8340C Medical Biomolecular Research Building (MBRB), CB #7126, University of North Carolina, Chapel Hill, NC 27599-7126, USA

Daniele Catalucci
Division of Cardiology, Department of Medicine, University of California San Diego, 9500 Gilman Drive, BSB 5022, LaJolla, San Diego, CA 92093-0613, USA Scientific and Technology Pole, IRCCS MultiMedica, Via G. Fantoli 16/15, 20138 Milan, Italy
dcatalucci@ucsd.edu

Gianluigi Condorelli
Division of Cardiology, Department of Medicine, University of California San Diego, 9500 Gilman Drive, BSB 5022, LaJolla, San Diego, CA 92093-0613, USA Scientific and Technology Pole, IRCCS MultiMedica, Via G. Fantoli 16/15, 20138 Milan, Italy
gcondorelli@ucsd.edu

Volker A. Erdmann
Institut für Chemie/Biochemie, Freie Universität Berlin,
Thielallee 63, 14195 Berlin, Germany
erdmann@chemie.fu-berlin.de

Henry Fechner
Department of Cardiology and Pneumology, Campus-Benjamin Franklin,
Charite University Medicine, Berlin, Germany

Noboru Fukuda
Division of Nephrology and Endocrinology, Department of Medicine,
Nihon University School of Medicine, Tokyo, Japan
Advanced Research Institute of Science and Humanities,
Nihon University School of Medicine, Tokyo, Japan
fukudan@med.nihon-u.ac.jp

Gabriele Grassi
Department of Internal Medicine, University Hospital of Trieste,
Cattinara 34149, Trieste, Italy
Department of Molecular Pathology, University Hospital
of Tübingen, Liebermeisterstr. 8, 72076 Tübingen, Germany
ggrassi@units.it

Mario Grassi
Department of Chemical Engineering (DICAMP), University of Trieste,
Piazzale Europa 1, Trieste 34127, Italy

Hugo A. Katus
Innere Medizin III, Universitatsklinikum Heidelberg, Im Neuenheimer
Feld 410, 69120 Heidelberg, Germany

Nicholas Kipshidze
Lenox Hill Hospital, Department of Interventional Cardiac & Vascular
Services, 30 East 77th Street, New York, NY 10021, USA
NKipshidze@Lenoxhill.net

Michael V.G. Latronico
Scientific and Technology Pole, IRCCS MultiMedica, Via G. Fantoli 16/15,
20138 Milan, Italy

Osako Kiomy Mariana
Division of Clinical Gene Therapy, Graduate School of Medicine, Osaka
University, Suita 565-0871, Japan

Sabine Merl
Deutsches Herzzentrum and 1 Medizinische Klinik, Technische Universität,
Lazarettstrasse 36, 80636 München, Germany

Ryuichi Morishita
Division of Clinical Gene Therapy, Graduate School of Medicine, Osaka
University, Suita 565-0871, Japan

Oliver J. Müller
Innere Medizin III, Universitatsklinikum Heidelberg, Im Neuenheimer Feld 410, 69120 Heidelberg, Germany
oliver.mueller@med.uni-heidelberg.de

Hironori Nakagami
Division of Clinical Gene Therapy, Graduate School of Medicine, Osaka University, Suita 565-0871, Japan

Division of Gene Therapy Science, Graduate School of Medicine, Osaka University, Suita 565-0871, Japan
nakagami@gts.med.osaka-u.ac.jp

Dobrawa Napierala
Department of Molecular and Human Genetics, Baylor College of Medicine, One Baylor Plaza, BCM-225, Houston, TX 77030, USA
dobrawan@bcm.tmc.edu

Marek Napierala
University of Texas M.D. Anderson Cancer Center, 1515 Holcombe Blvd., Houston, TX 77030, USA
mnapiera@mdanderson.org

Shahid M. Nimjee
Duke Institute for Genome Sciences and Policy, Duke University Medical Center, Durham, NC 27710, USA
shahid.nimjee@mc.duke.edu

Sabah Oney
Duke Institute for Genome Sciences and Policy, Duke University Medical Center, Durham, NC 27710, USA
sabah.oney@duke.edu

Vladimir Pekarik
Center for Anatomy, Institute for Cell- and Neurobiology, Charité University Hospital, Phillippstrasse 12, 10115 Berlin, Germany
Centre de Medicina Regenerativa de Barcelona (CMR[B]), Parc de Recerca Biomèdica de Barcelona (PRBB), c/ Dr. Aiguader 88, 7 planta, 08003 Barcelona
v.pekarik@cmrb.eu

Shannon Pendergrast
Archemix Corp. 300 Third St. 2nd Floor, Cambridge, MA 02139, USA
pendergrast@archemix.com

Sandra Pinkert
Department of Cardiology and Pneumology, Campus-Benjamin Franklin, Charite University Medicine, Berlin, Germany

Wolfgang Poller
Department of Cardiology and Pneumology, Campus-Benjamin Franklin, Charite University Medicine, Berlin, Germany
wolfgang.poller@charite.de

Robert G. Schaub
Archemix Corp. 300 Third St. 2nd Floor, Cambridge, MA 02139, USA

Lennart Suckau
Department of Cardiology and Pneumology, Campus-Benjamin Franklin, Charite University Medicine, Berlin, Germany

Bruce Sullenger
Duke Institute for Genome Sciences and Policy, Duke University Medical Center, Durham, NC 27710, USA
b.sullenger@cgct.duke.edu

Maciej Szymański
Institute of Bioorganic Chemistry of the Polish Academy of Sciences, Noskowskiego 12, 61-704 Poznan, Poland
mszyman@ibch.poznan.pl

Mariko Tatsuguchi
Carolina Cardiovascular Biology Center, Department of Cell and Developmental Biology, 8340C Medical Biomolecular Research Building (MBRB), CB #7126, University of North Carolina, Chapel Hill, NC 27599-7126, USA

Kristin M. Thompson
Archemix Corp. 300 Third St. 2nd Floor, Cambridge, MA 02139, USA

Naruya Tomita
Division of Nephrology, Department of Internal Medicine, Kawasaki Medical School, 577 Matsushima, Kurashiki 701-0192, Japan

Da-Zhi Wang
Carolina Cardiovascular Biology Center, Department of Cell and Developmental Biology, 8340C Medical Biomolecular Research Building (MBRB), CB #7126, University of North Carolina, Chapel Hill, NC 27599-7126, USA
dawang@med.unc.edu

Edwin Wang
Biotechnology Research Institute, National Research Council of Canada, 6100 Royalmount Avenue, Montreal, QC, Canada

Center for Bioinformatics, McGill University, Montreal, QC, Canada
Edwin.Wang@cnrc-nrc.gc.ca

Rainer Wessely
Deutsches Herzzentrum and 1 Medizinische Klinik, Technische Universität, Lazarettstrasse 36, 80636 München, Germany
rwessely@web.de

En-Hui Yao
Division of Nephrology and Endocrinology, Department of Medicine, Nihon University School of Medicine, Tokyo, Japan

Part I
MicroRNA

An Overview of MicroRNA

E. Wang

Contents

Abstract Recently, microRNAs (miRNAs) have emerged as central posttranscriptional regulators of gene expression. miRNAs regulate many key biological processes, including cell growth, death, development and differentiation. This discovery is challenging the central dogma of molecular biology. miRNAs have been known to be involved in development, cell proliferation and apoptosis. Several reports have recently shown that miRNAs might also be involved in filtering out gene expression noise by regulating positive regulatory loops in cells. Loss- or gain-of-function of specific miRNAs appears to be a key event in the genesis of many diverse diseases. Recent studies have shown that miRNAs are important during heart development and adult cardiac physiology, and modulate a diverse spectrum of cardiovascular functions in vivo. miRNAs have been shown to regulate pathways controlled by genes like p53, MYC and RAS, which are closely related to cancer. Single-nucleotide polymorphisms (SNPs) of miRNA binding sites are associated with gene expression

E. Wang
Biotechnology Research Institute, National Research Council of Canada, 6100 Royalmount Avenue, Montreal, Quebec, Canada, and Center for Bioinformatics, McGill University, Montreal, Quebec, Canada
e-mail: Edwin.Wang@cnrc-nrc.gc.ca

V.A. Erdmann et al. (eds.) *RNA Technologies in Cardiovascular Medicine and Research*,
doi: 10.1007/978-3-540-78709-9,

levels of the target alleles and cancer. Finally, these miRNA studies also have implications for understanding complex pathways, e.g., interactions between miRNAs, cell signaling and transcription factors, involved in human diseases, and can lead to potential opportunities in manipulating miRNAs as therapeutic targets.

1 Introduction

According to the central dogma of molecular biology, RNAs are passive messengers and only take charge of transferring genetic information, or carrying out DNA instructions, or code, for protein production in cells. However, this central dogma is getting challenged by the recent findings that tiny fragments of noncoding RNA typically ~22 nucleotides in length, namely microRNA (miRNA), are able to negatively regulate protein-coding genes by interfering with mRNA's original instructions. Recent studies indicate that miRNAs have emerged as central posttranscriptional repressors of gene expression. miRNAs suppress gene expression via imperfect base pairing to the 3′ untranslated region (3′UTR) of target mRNAs, leading to repression of protein production or mRNA degradation (Bartel 2004; Carthew 2006; Valencia-Sanchez et al. 2006). These noncoding regulatory RNA molecules have been found in diverse plants, animals, some viruses and even algae species, and it now seems likely that all multicellular eukaryotes, and perhaps some unicellular eukaryotes, utilize these RNAs to regulate gene expression.

Some researchers claimed that the human genome might encode more than 1,000 miRNAs (Bentwich et al. 2005). However, a recent sequencing survey of miRNA expression across 26 distinct organ systems and cell types of human and rodents validated that only 300+ miRNAs are present in humans and/or rodents (Landgraf et al. 2007). Computational predictions indicate that thousands of genes could be targeted by miRNAs in mammals (John et al. 2004; Krek et al. 2005; Lewis et al. 2005; Rajewsky 2006). Experimental analysis revealed that 100–200 target mRNAs are repressed and destabilized by a single miRNA (Krutzfeldt et al. 2005; Lim et al. 2005; Yu et al. 2007a). It is estimated that more than one-third of human genes are potentially regulated by miRNAs. These findings suggest that miRNAs play an integral role in genome-wide regulation of gene expression.

miRNAs regulate many key biological processes, including cell growth, death, development and differentiation, by determining how and when genes turn on and off. Animals that fail to produce certain mature miRNAs are unable to survive or reproduce (Bernstein et al. 2003; Cao et al. 2006; Forstemann et al. 2005; Ketting et al. 2001; Plasterk 2006; Shivdasani 2006; Wienholds et al. 2003). Thus, a single, malfunctioning microRNA can be sufficient to cause cancer in mice (Costinean et al. 2006). These discoveries offer new insights into another layer of gene regulation and at the same time underscore the powerful role that these tiny snippets of noncoding RNA play in cells. These discoveries indicate that it is no longer the genes, nor mRNAs themselves, that hold the most mystery, but the miRNAs that influence their behavior and the result that such gene regulation process produces. Thus, miRNA has become an important force in biology.

In this chapter, I will summarize the basic knowledge of miRNAs, including their biogenesis and functions, especially highlighting recently discovered miRNA functions, such as filtering out gene expression noise by miRNAs, and the potential function of maintaining tissue identity by miRNAs. The recent progress of miRNA involvements in heart development, heart physiology and heart diseases will be discussed. Finally, I will review the relationship of miRNAs to cancer.

2 miRNA Biogenesis

miRNAs as posttranscriptional regulatory molecules were first discovered to regulate expression of partially complementary mRNAs in *Caenorhabditis elegans* (Lee et al. 1993; Moss et al. 1997; Wightman et al. 1993). miRNAs are encoded in either intergenic regions of genomes or within introns of known protein-coding genes. miRNAs are transcribed by RNA polymerase II as long precursor transcripts, which are called primary miRNAs (pri-miRNAs). The pri-miRNAs are capped and polyadenylated, and can reach several kilobases in length (Cullen 2005; Kim 2005). A single pri-miRNA might contain one, or up to several, miRNAs. Several sequential steps of transcript processing are required to produce mature miRNAs from pri-miRNAs. In the nucleus, there is a microprocessor complex in which the major components are the RNase-III enzyme Drosha and its partner DGCR8/Pasha (Denli et al. 2004; Gregory et al. 2004; Landthaler et al. 2004), which initially recognize pri-miRNAs and then excise the stem–loop hairpin structure that contains the miRNA, a 60–80 nucleotide intermediate known as precursor miRNA (pre-miRNA) with pri-miRNAs. Exportin-5, a nuclear export factor, recognizes and transports the pre-miRNAs to cytoplasm (Bohnsack et al. 2004; Lund et al. 2004; Yi et al. 2003). In the cytoplasm, Dicer, a second RNase-III enzyme, cleavages the pre-miRNAs to generate double-stranded 18–24 nucleotide-long RNA molecules – miRNAs (Bernstein et al. 2003; Grishok et al. 2001; Hutvagner et al. 2001; Ketting et al. 2001; Knight and Bass 2001). RNA-induced silencing complex (RISC), the core component of which is the Argonaute protein (Kim 2005), incorporates one of these two strands – the guide strand of miRNAs. Finally, the miRNA guides the RISC complex to the target mRNA to suppress gene expression via imperfect base pairing to the 3′UTR of target mRNAs, leading to repression of protein production and, in some cases, mRNA degradation (Bartel 2004; Carthew 2006; Valencia-Sanchez et al. 2006).

3 Biological Functions of miRNAs

3.1 miRNA Emerges as a Central Regulator for Development

Several reports indicated that miRNAs repress a large set of targets so that the targets are expressed at low levels in the miRNA-expressing cells (Krutzfeldt and Stoffel 2006; Lim et al. 2005; Yu et al. 2007a). This might offer a second layer of regulation to reinforce transcriptional controls at posttranscriptional level. A number of lines of evidence suggested that miRNAs are involved in regulating developmental processes. We conducted a genome-wide survey of transcription factor binding sites in the promoter regions of human genes and found that developmental genes are significantly regulated by more transcription factors (Cui et al. 2007a). Furthermore, we showed that the more transcription factors a gene is regulated by, the more miRNAs that gene is regulated by (Cui et al. 2007a). Certain miRNAs have been suggested to be essential regulators for developmental programs (Giraldez et al. 2005). For instance, without miR-430, zebrafish embryos develop defects, which can be rescued and complemented by supplying miR-430 (Giraldez et al. 2005). Genes in this process by miR-430 seem to be direct miR-430 targets based on miRNA seed matches, and are misregulated in the absence of miR-430 (Giraldez et al. 2006). Another example comes from the study of *C. elegans* miRNAs, lin-4 and let-7. Without lin-4, *C. elegans* is unable to make the transition from the first to the second larval stage due to a differentiation defect, which is caused by a failure to posttranscriptionally repress the lin-14 gene, which is the target gene of lin-4 (Lee et al. 1993; Wightman et al. 1993). Similarly, without let-7, a failure of larval-to-adult transition was observed (Reinhart et al. 2000). It is known that lin-41, hbl-1, daf-12 and the forkhead transcription factor pha-4 are the direct targets of let-7 during this transition (Abrahante et al. 2003; Grosshans et al. 2005; Slack et al. 2000).

3.2 miRNAs are Involved in Cell Proliferation and Apoptosis

miRNAs have been shown to regulate key genes for tumorigenesis and cancer progression, which coordinately controls cell proliferation and apoptosis. For instance, miRNA let-7 promotes tumorigenesis by regulation KRAS and NRAS transcripts (Johnson et al. 2005). miRNAs are known to regulate pathways controlled by genes such as p53, MYC and RAS. Furthermore, miR17-92 cluster has been shown to be able to act as a functional switch between cell proliferation and apoptosis.

3.3 miRNAs Act as Regulators for Noise Filtering and Buffering

Eukaryotic cells are noisy environments in which transcription often occurs in a bursting manner, causing the number of mRNAs per cell to fluctuate significantly (Blake et al. 2006; Golding et al. 2005; Raj et al. 2006). Moreover, such fluctuations can propagate through the network, e.g., fluctuations in the level of an upstream transcription factor can significantly induce the expression fluctuations of downstream genes (Pedraza and van Oudenaarden 2005; Rosenfeld et al. 2005). In positive regulatory loops, noise or stochastic fluctuations of gene transcripts and protein molecules leads to randomly switching cell phenotypes in yeast, while a negative regulator adding in the positive regulatory loops often helps in reducing such noise in biological systems and making a robust decision for cell development

(Acar et al. 2005). Because miRNAs can tune target protein levels more rapidly at the posttranscriptional level, they might significantly shorten the response delay and, in turn, provide more effective noise buffering. The miRNA miR-17 might play a role in preventing noise-driven transition from apoptosis to cell proliferation. c-Myc and E2F1 are known to reciprocally activate transcription of one another, establishing a positive feedback circuit (Fernandez et al. 2003; Leone et al. 1997). This architectural structure of the circuit makes it possible for miRNAs to support a shift from apoptosis toward proliferation by repressing E2F1. Expression of E2F1 promotes G1 to S phase progression by activating genes involved in cell cycle (Bracken et al. 2004). High expression of E2F1, however, is sufficient to induce apoptosis (Johnson et al. 1994a,b; Matsumura et al. 2003). In the absence of additional regulatory mechanisms, this circuit might be expected to overactivate E2F1, leading to apoptosis, when c-Myc simultaneously activates E2F1 transcription and miR17-92 cluster, which in turn negatively regulates E2F1. This might promote a proliferative signal but not an apoptotic signal. Another example is the fly miR-9a, which is suggested to set up a "threshold" for signals in a positive feedback loop, so that it can filter out noise (Li et al. 2006). Without miR-9a, flies produce extra sense organs (Li et al. 2006). During fly sensory organ development, a fly gene, *senseless* expression is activated by proneural proteins and feedbacks positively to reinforce proneural gene expression. If *senseless,* the target of miR-9a is highly expressed and the defects mentioned above occur. miR-9a has been suggested to set a threshold that *senseless* expression has to overcome to induce the normal developmental program. In agreement with these findings, we found that cross-species expression divergences of miRNA target genes are significantly smaller than those of other genes (Cui et al. 2007b). Similar observations have been found between human and chimpanzee, between human and mouse, between *Drosophila* species, and between *D. melanogaster* and *D. simulans* (Cui et al. 2007b). These results suggest that miRNAs might provide a genetic buffer to constrain gene expression divergence. We showed that miRNAs preferentially regulate positive regulatory loops (Cui et al. 2006). It is possible that miRNAs serve to buffer stochastically fluctuating expression of genes in positive regulatory loop. Given that positive feedback circuits are abundant in genomes (Brandman et al. 2005; Ferrell 2002), we surmise that miRNAs provide a common mechanism in buffering gene expression noise by frequently regulating positive regulatory loops (Fig. 1).

Buffering by miRNAs decreases the detrimental effects of errors in gene regulation. miRNA buffering might also provide a way for silence-accumulating mutations without being subjected to selective forces and thus might contribute to evolvability.

3.4 miRNAs Might Contribute to Maintaining Tissue Identity

We conducted a genome-wide analysis of the expression profiles of mRNA targets in human, mouse and *Drosophila* (Yu et al. 2007a). We found that the expression levels of miRNA targets are significantly lower in all mouse mature tissues and *Drosophila*

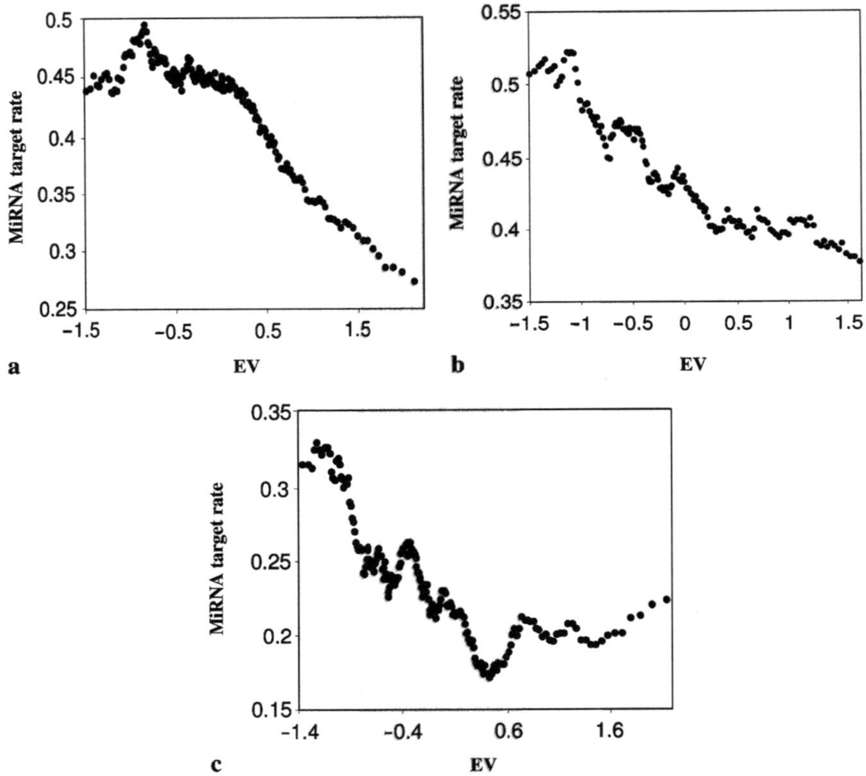

Fig. 1 Distribution of miRNA targets among the genes with different cross-species gene expression variations (*EVs*). Genes were grouped according to their EV values. The ratio of miRNA targets to total genes in each group (miRNA target rate) was calculated. (**a**) Human and chimpanzee. (**b**) Human and mouse. (**c**) *Drosophila melanogaster* and *D. simulans*

later life stages than in the embryos. These results indicate that miRNAs might play roles in determining the timing of tissue differentiation during the larva period of *Drosophila* development and maintaining the tissue identity during adulthood.

4 miRNAs in Human Disease

4.1 miRNAs and Heart Diseases

Loss- or gain-of-function of specific miRNAs appears to be a key event in the genesis of many diverse diseases. Recently, an interesting question how miRNAs influence heart development and disease has been addressed. Four recent papers highlighted the role of miRNAs in the heart. These reports showed that miRNAs are essential for heart development and regulating the expression of genes which take part in cardiac function in vivo: the conductance of electrical signals, heart muscle contraction, and heart growth and morphogenesis.

It has been reported that there is a function for miR-1 in heart conductivity. miR-1 levels were positively correlated with coronary artery disease and rats after cardiac infarction (Yang et al. 2007). Loss-of-function of miR-1 prevented heart arrhythmia, whereas miR-1 overexpression caused heart arrhythmia in normal and infarcted hearts. They further showed that both gain- and loss-of-function of miR-1 affect conductivity through affecting potassium channels. These results suggest that miR-1 has a prominent effect on the development of cardiac arrhythmia, irregular electrical activity in the heart. A separate study also focused on miR-1 by creating mice that are deficient in a muscle-specific miRNA, miR-1-2 (Zhao et al. 2007). They showed that the miRNA-deficient embryos have cardiac failure and a variety of developmental defects, including pericardial edema and underdevelopment of the ventricular myocardium, an increase in cardiomyocyte proliferation and electrophysiological defects, a reduction in heart rate and the prolonging of ventricular depolarization. Interestingly, these phenotypes are similar to the defects during heart development in zebrafish embryos, when miRNAs are non-functional (Giraldez et al. 2005). Both studies identify miR-1 targets that might, at least in part, account for the manifestation of the associated diseases.

In another study, it was found that the muscle-specific miR-133 is a negative regulator of cardiac hypertrophy, which is an essential adaptive physiological response to mechanical and hormonal stress and heart size (Care et al. 2007). To understand the molecular mechanism by which miR-133 controls heart size, they showed that Rhoa, Cdc42 and Whsc2 are the direct targets of miR-133. Moreover, the heart-specific miRNA miR-208 also modulates the genes that are controlling the hypertrophic response (van Rooij et al. 2007). The main function of miR-208 seems to be mediating the switch from expression of the heavy chain of α-myosin to that of β-myosin during stress or thyroid-hormone-induced cardiac growth (van Rooij et al. 2007). These results suggest that miR-208 is an important regulator for cardiac growth and gene expression in response to stress and hypothyroidism.

Taken together, it seems clear that miRNAs have an important role in regulating gene expression in the heart. These studies indicate that miRNAs are important during heart development and adult cardiac physiology, and modulate a diverse spectrum of cardiovascular functions in vivo. The findings revealed a level of molecular control of heart physiology that is beyond the well-accepted regulatory role of signaling and transcription factor complexes in the heart. Furthermore, these studies also have implications for understanding complex pathways, e.g., interactions between miRNAs, cell signaling and transcription factors, involved in heart diseases, and can lead to potential opportunities in manipulating miRNAs as therapeutic targets.

4.2 miRNAs and Cancer

Human cancer studies are always the hotspots in life science research. Much significant progress in miRNAs and cancer has been made in the past few years. Genome-wide studies of miRNA expression profiling have shown that miRNA expression levels are altered in primary human tumors (Calin et al. 2004; Calin and Croce 2006; Lu et al. 2005). Significant signatures of miRNA expression profiles can be linked to various types of tumors, suggesting that miRNA profiling has diagnostic and perhaps prognostic potential (Lu et al. 2005; Calin and Croce 2006). Certain miRNAs could be tumor suppressors, because loss of these miRNAs is often associated with cancers. miR-15a and miR-16-1 genes are deleted in most cases of chronic lymphocytic leukemia (Calin et al. 2004). Loss of miRNA let-7 in lung tumors correlates with high RAS protein expression, suggesting that let-7 promotes tumorigenesis by regulation of popular oncogenes, KRAS and NRAS transcripts (Johnson et al. 2005). miR-372 and miR-373 have been shown to be able to overcome oncogenic Ras-mediated arrest and, therefore, induced tumorigenesis (Voorhoeve et al. 2006). miR-21 was demonstrated to be consistently upregulated in human glioblastoma tumor tissues, primary tumor cultures and established glioblastoma cell lines relative to normal fetal and adult brain tissue (Chan et al. 2005). Knockdown of miR-21 in glioblastoma cell lines led to activation of caspases and a corresponding induction of apoptotic cell death. Furthermore, two recent studies have placed the miR-24 family into the p53 tumor suppressor network. miR-24, regulating apoptosis and cell proliferation, has become an essential component of the p53 network (He et al. 2007; Raver-Shapira et al. 2007), which is closely associated with cancer.

4.3 Single-Nucleotide Polymorphisms (SNPs) of miRNA Binding Sites and Human Diseases

Mapping human SNP genotype data (25,000 SNPs) generated in the HapMap and Perlegen projects onto the 3′-UTR regions of human gene transcripts (Chen and Rajewsky 2006), uncovered that SNP density in conserved miRNA sites was lower than in conserved control sites. These results indicate that a large class of computationally predicted conserved miRNA target sites is under significant negative selection. Similarly, we showed the same trend when mining NCBI's dbSNP database (Yu et al. 2007b). These results have implications that SNPs located at miRNA-binding sites are likely to affect the expression of the miRNA target and might contribute to the susceptibility of humans to common diseases. Indeed, naturally occurring polymorphisms in miRNA binding sites have been documented in Tourette's syndrome in humans and muscularity in sheep (Abelson et al. 2005; Clop et al. 2006).

Motivated by this concept, we explored the effects of miRNA-binding SNPs on cancer susceptibility by genome-wide analysis of the data deposited in NCBI's dbSNP database and human dbEST database (Yu et al. 2007b). Interestingly, we found that the frequencies of the minor alleles (non-target alleles) of the miRNA-binding site SNPs are extremely low. Furthermore, we showed that the average expression level of the non-target alleles of miRNA-binding site SNPs is significantly higher than that of the target alleles. Moreover, we identified a set of potential candidates for miRNA-binding site SNPs with an aberrant allele frequency present in the human cancer EST database. Finally, we experimentally validated them by sequencing clinical tumor samples.

Although the miRNA inducing disease studies are still in their infancy, miRNAs are known to regulate pathways controlled by genes such as p53, MYC and RAS. These findings emphasize the need to integrate the study of miRNA expression and function into other cellular processes, such as signaling, gene regulation, and others, in order to achieve a complete understanding of this group of disorders. Unraveling miRNA regulatory circuits, even miRNA regulation of cellular networks that are involved in disease development, is challenging, but is essential to gain a comprehensive understanding of the molecular mechanisms of the diseases. Luckily, there have been recent developments in technologies such as microarray and systemic delivery of small RNA systems that allow high-throughput studies of the function of miRNAs (Krutzfeldt et al. 2005; Soutschek et al. 2004). These approaches provide promise for understanding miRNA function at systems-level and eventually developing therapeutic strategies based on miRNA overexpression or inhibition.

5 Summary

miRNA research has been conducted for only a few years. There are still lots of unknown but exciting knowledge to be revealed about miRNAs. Here, we highlighted the recently discovered new functions of miRNAs. miRNAs might act as regulators for filtering out gene expression noise and letting cells make right decisions for normal development. In addition, miRNAs might maintain tissue identity, although further experimental evidence is needed to validate this hypothesis.

Although miRNAs have been thought to take part in many kinds of human diseases, it was still exciting to find that miRNAs become important players in heart development, affecting adult cardiac physiology, and modulating a diverse spectrum of cardiovascular functions in vivo. Many miRNAs have been implicated in tumorigenesis and cancer progression. A few studies in genome-wide microarray profiling of miRNAs of tumor samples suggested that miRNAs could be used as tumor biomarkers. More and more miRNAs have been documented in causing cancer, but it is still unclear whether and how miRNAs could cooperate to take part in cancer development.

Finally, most of the miRNA studies revealed a level of molecular control of human diseases that is beyond the well-accepted regulatory systems such as regulation by transcription factors and signaling proteins. Furthermore, these studies also have implications that, in a cell, a more complex regulatory system, e.g., interactions between miRNAs, cell signaling and transcription factors, is involved in many cellular activities and human diseases. Therefore, it is essential to study the underlying interactions between miRNAs and cellular regulatory systems and pathways. Thus, it leads to understanding gene regulation in a more comprehensive manner and opening up new opportunities in manipulating miRNAs as diagnostic and therapeutic targets.

Acknowledgements This work is partially supported by Genome and Health Imitative. I thank our team members, Dr. Y. Deng and Mrs. M. Mistry, for comments.

References

Abelson JF, Kwan KY, O'Roak BJ et al (2005) Sequence variants in SLITRK1 are associated with tourette's syndrome. Science 310:317–320

Abrahante JE, Daul AL, Li M et al (2003) The *Caenorhabditis elegans* hunchback-like gene lin-57/hbl-1 controls developmental time and is regulated by microRNAs. Dev Cell 4:625–637

Acar M, Becskei A, van Oudenaarden A (2005) Enhancement of cellular memory by reducing stochastic transitions. Nature 435:228–232

Bartel DP (2004) MicroRNAs: genomics, biogenesis, mechanism, and function. Cell 116:281–297

Bentwich I, Avniel A, Karov Y et al (2005) Identification of hundreds of conserved and nonconserved human microRNAs. Nat Genet 37:766–770

Bernstein E, Kim SY, Carmell MA et al (2003) Dicer is essential for mouse development. Nat Genet 35:215–217

Blake WJ, Balazsi G, Kohanski MA et al (2006) Phenotypic consequences of promoter-mediated transcriptional noise. Mol Cell 24:853–865

Bohnsack MT, Czaplinski K, Gorlich D (2004) Exportin 5 is a RanGTP-dependent dsRNA-binding protein that mediates nuclear export of pre-miRNAs. RNA 10:185–191

Bracken AP, Ciro M, Cocito A et al (2004) E2F target genes: unraveling the biology. Trends Biochem Sci 29:409–417

Brandman O, Ferrell JE Jr, Li R et al (2005) Interlinked fast and slow positive feedback loops drive reliable cell decisions. Science 310:496–498

Calin GA, Croce CM (2006) MicroRNA signatures in human cancers. Nat Rev Cancer 6:857–866

Calin GA, Liu CG, Sevignani C et al (2004) microRNA profiling reveals distinct signatures in B cell chronic lymphocytic leukemias. Proc Natl Acad Sci USA 101:11755–11760

Cao X, Yeo G, Muotri AR et al (2006) Noncoding RNAs in the mammalian central nervous system. Annu Rev Neurosci 29:77–103

Care A, Catalucci D, Felicetti F et al (2007) MicroRNA-133 controls cardiac hypertrophy. Nat Med 13:613–618

Carthew RW (2006) Gene regulation by microRNAs. Curr Opin Genet Dev 16:203–208

Chan JA, Krichevsky AM, Kosik KS (2005) MicroRNA-21 is an antiapoptotic factor in human glioblastoma cells. Cancer Res 65:6029–6033

Chen K, Rajewsky N (2006) Natural selection on human microRNA binding sites inferred from SNP data. Nat Genet 38:1452–1456

Clop A, Marcq F, Takeda H et al (2006) A mutation creating a potential illegitimate microRNA target site in the myostatin gene affects muscularity in sheep. Nat Genet 38:813–818

Costinean S, Zanesi N, Pekarsky Y et al (2006) Pre-B cell proliferation and lymphoblastic leukemia/high-grade lymphoma in E(mu)-miR155 transgenic mice. Proc Natl Acad Sci USA 103:7024–7029

Cui Q, Yu Z, Purisima EO, Wang E (2006) Principles of microRNA regulation of a human cellular signaling network. Mol Syst Biol 2:46
Cui Q, Yu Z, Pan Y et al (2007a) MicroRNAs preferentially target the genes with high transcriptional regulation complexity. Biochem Biophys Res Commun 352:733–738
Cui Q, Yu Z, Purisima EO, Wang E (2007b) MicroRNA regulation and interspecific variation of gene expression. Trends Genet 23:372–375
Cullen BR (2005) Human immunodeficiency virus: nuclear RNA export unwound. Nature 433:26–27
Denli AM, Tops BB, Plasterk RH et al (2004) Processing of primary microRNAs by the microprocessor complex. Nature 432:231–235
Fernandez PC, Frank SR, Wang L et al (2003) Genomic targets of the human c-Myc protein. Genes Dev 17:1115–1129
Ferrell JE (2002) Self-perpetuating states in signal transduction: positive feedback, double-negative feedback and bistability. Curr Opin Cell Biol 14:140–148
Forstemann K, Tomari Y, Du T et al (2005) Normal microRNA maturation and germ-line stem cell maintenance requires Loquacious, a double-stranded RNA-binding domain protein. PLoS Biol 3:e236
Giraldez AJ, Cinalli RM, Glasner ME et al (2005) MicroRNAs regulate brain morphogenesis in zebrafish. Science 308:833–838
Giraldez AJ, Mishima Y, Rihel J et al (2006) Zebrafish MiR-430 promotes deadenylation and clearance of maternal mRNAs. Science 312:75–79
Golding I, Paulsson J, Zawilski SM et al (2005) Real-time kinetics of gene activity in individual bacteria. Cell 123:1025–1036
Gregory RI, Yan KP, Amuthan G et al (2004) The microprocessor complex mediates the genesis of microRNAs. Nature 432:235–240
Grishok A, Pasquinelli AE, Conte D et al (2001) Genes and mechanisms related to RNA interference regulate expression of the small temporal RNAs that control *C. elegans* developmental timing. Cell 106:23–34
Grosshans H, Johnson T, Reinert KL et al (2005) The temporal patterning microRNA let-7 regulates several transcription factors at the larval to adult transition in *C. elegans*. Dev Cell 8:321–330
He L, He X, Lim LP et al (2007) A microRNA component of the p53 tumour suppressor network. Nature 447:1130–1134
Hutvagner G, McLachlan J, Pasquinelli AE et al (2001) A cellular function for the RNA-interference enzyme Dicer in the maturation of the let-7 small temporal RNA. Science 293:834–838
John B, Enright AJ, Aravin A et al (2004) Human microRNA targets. PLoS Biol 2:e363
Johnson DG, Cress WD, Jakoi L et al (1994a) Oncogenic capacity of the E2F1 gene. Proc Natl Acad Sci USA 91:12823–12827
Johnson DG, Ohtani K, Nevins JR (1994b) Autoregulatory control of E2F1 expression in response to positive and negative regulators of cell cycle progression. Genes Dev 8:1514–1525
Johnson SM, Grosshans H, Shingara J et al (2005) RAS is regulated by the let-7 microRNA family. Cell 120:635–647
Ketting RF, Fischer SE, Bernstein E et al (2001) Dicer functions in RNA interference and in synthesis of small RNA involved in developmental timing in *C. elegans*. Genes Dev 15:2654–2659
Kim VN (2005) MicroRNA biogenesis: coordinated cropping and dicing. Nat Rev Mol Cell Biol 6:376–385
Knight SW, Bass BL (2001) A role for the RNase III enzyme DCR-1 in RNA interference and germ line development in *Caenorhabditis elegans*. Science 293:2269–2271
Krek A, Grun D, Poy MN et al (2005) Combinatorial microRNA target predictions. Nat Genet 37:495–500
Krutzfeldt J, Stoffel M (2006) MicroRNAs: a new class of regulatory genes affecting metabolism. Cell Metab 4:9–12
Krutzfeldt J, Rajewsky N, Braich R et al (2005) Silencing of microRNAs in vivo with 'antagomirs'. Nature 438:685–689

Landgraf P, Rusu M, Sheridan R et al (2007) A mammalian microRNA expression atlas based on small RNA library sequencing. Cell 129:1401–1414
Landthaler M, Yalcin A, Tuschl T (2004) The human DiGeorge syndrome critical region gene 8 and Its *D. melanogaster* homolog are required for miRNA biogenesis. Curr Biol 14:2162–2167
Lee RC, Feinbaum RL, Ambros V (1993) The *C. elegans* heterochronic gene lin-4 encodes small RNAs with antisense complementarity to lin-14. Cell 75:843–854
Leone G, DeGregori J, Sears R et al (1997) Myc and Ras collaborate in inducing accumulation of active cyclin E/Cdk2 and E2F. Nature 387:422–426
Lewis BP, Burge CB, Bartel DP (2005) Conserved seed pairing, often flanked by adenosines, indicates that thousands of human genes are microRNA targets. Cell 120:15–20
Li Y, Wang F, Lee JA et al (2006) MicroRNA-9a ensures the precise specification of sensory organ precursors in *Drosophila*. Genes Dev 20:2793–2805
Lim LP, Lau NC, Garrett-Engele P et al (2005) Microarray analysis shows that some microRNAs downregulate large numbers of target mRNAs. Nature 433:769–773
Lu J, Getz G, Miska EA et al (2005) MicroRNA expression profiles classify human cancers. Nature 435:834–838
Lund E, Guttinger S, Calado A et al (2004) Nuclear export of microRNA precursors. Science 303:95–98
Matsumura I, Tanaka H, Kanakura Y (2003) E2F1 and c-Myc in cell growth and death. Cell Cycle 2:333–338
Moss EG, Lee RC, Ambros V (1997) The cold shock domain protein LIN-28 controls developmental timing in *C. elegans* and is regulated by the lin-4 RNA. Cell 88:637–646
Pedraza JM, van Oudenaarden A (2005) Noise propagation in gene networks. Science 307:1965–1969
Plasterk RH (2006) MicroRNAs in animal development. Cell 124:877–881
Raj A, Peskin CS, Tranchina D et al (2006) Stochastic mRNA synthesis in mammalian cells. PLoS Biol 4:e309
Rajewsky N (2006) microRNA target predictions in animals. Nat Genet 38(Suppl):S8–S13
Raver-Shapira N, Marciano E, Meiri E et al (2007) Transcriptional activation of miR-34a contributes to p53-mediated apoptosis. Mol Cell 26:731–743
Reinhart BJ, Slack FJ, Basson M et al (2000) The 21-nucleotide let-7 RNA regulates developmental timing in *Caenorhabditis elegans*. Nature 403:901–906
Rosenfeld N, Young JW, Alon U et al (2005) Gene regulation at the single-cell level. Science 307:1962–1965
Shivdasani RA (2006) MicroRNAs: regulators of gene expression and cell differentiation. Blood 108:3646–3653
Slack FJ, Basson M, Liu Z et al (2000) The lin-41 RBCC gene acts in the *C. elegans* heterochronic pathway between the let-7 regulatory RNA and the LIN-29 transcription factor. Mol Cell 5:659–669
Soutschek J, Akinc A, Bramlage B et al (2004) Therapeutic silencing of an endogenous gene by systemic administration of modified siRNAs. Nature 432:173–178
Valencia-Sanchez MA, Liu J, Hannon GJ et al (2006) Control of translation and mRNA degradation by miRNAs and siRNAs. Genes Dev 20:515–524
van Rooij E, Sutherland LB, Qi X et al (2007) Control of stress-dependent cardiac growth and gene expression by a microRNA. Science 316:575–579
Voorhoeve PM, le Sage C, Schrier M et al (2006) A genetic screen implicates miRNA-372 and miRNA-373 as oncogenes in testicular germ cell tumors. Cell 124:1169–1181
Wienholds E, Koudijs MJ, van Eeden FJ et al (2003) The microRNA-producing enzyme Dicer1 is essential for zebrafish development. Nat Genet 35:217–218
Wightman B, Ha I, Ruvkun G (1993) Posttranscriptional regulation of the heterochronic gene lin-14 by lin-4 mediates temporal pattern formation in *C. elegans*. Cell 75:855–862
Yang B, Lin H, Xiao J et al (2007) The muscle-specific microRNA miR-1 regulates cardiac arrhythmogenic potential by targeting GJA1 and KCNJ2. Nat Med 13:486–491

Yi R, Qin Y, Macara IG et al (2003) Exportin-5 mediates the nuclear export of pre-microRNAs and short hairpin RNAs. Genes Dev 17:3011–3016

Yu Z, Jian Z, Shen SH et al (2007a) Global analysis of microRNA target gene expression reveals that miRNA targets are lower expressed in mature mouse and *Drosophila* tissues than in the embryos. Nucleic Acids Res 35:152–164

Yu Z, Li Z, Jolicoeur N et al (2007b) Aberrant allele frequencies of the SNPs located in microRNA target sites are potentially associated with human cancers. Nucleic Acids Res 35:4535–4541

Zhao Y, Ransom JF, Li A et al (2007) Dysregulation of cardiogenesis, cardiac conduction, and cell cycle in mice lacking miRNA-1-2. Cell 129:303–317

MicroRNAs and Their Potential

M. Abdellatif

Contents

Abstract With the advent of microRNA (miRNA) we are compelled to revise our understanding of the mechanisms underlying gene regulation during health and disease. A miRNA is approximately 21 ribonucleotides long, genetically encoded, with a potential to recognize multiple mRNA targets guided by sequence complementarity and RNA-binding proteins. This class of molecules is functionally versatile, with the capacity to specifically inhibit translation initiation or elongation, as well as induce mRNA degradation, through predominantly targeting the 3′-untranslated regions of mRNA. Early on it was realized that the levels of individual miRNA varied under different developmental, biological, or pathological conditions, thus implicating these molecules in normal and pathological cellular attributes. In this chapter, we will discuss how the functions of miRNA relate to our existing knowledge on post-transcriptional regulation of gene expression that is the underlying mechanism of many diseases, including cardiac hypertrophy and failure, and their potential as biomarkers and therapeutic targets in diseases.

M. Abdellatif
University of Medicine and Dentistry of New Jersey, Cardiovascular Research Institute, Department of Cell Biology and Molecular Medicine, Room I-517, 185 S Orange Ave, Newark, NJ 07103, USA
e-mail: abdellma@umdnj.edu

V.A. Erdmann et al. (eds.) *RNA Technologies in Cardiovascular Medicine and Research*,
doi: 10.1007/978-3-540-78709-9,

1 The Importance of Being miRNA

In computer terms, a macro is a single command that is programmed to replace multiple separate commands that perform a series of actions, thus simplifying, expediting, and minimizing error. Similarly, the cell employs a single miRNA to regulate post-transcriptional expression of an array of genes that are involved in a particular cellular function. This mechanism will ensure synchronization of the regulatory effects and promptness of the response. Moreover, post-transcriptional regulation of multiple genes via a single miRNA circumvents the need for transcriptional regulation of individual genes and is, thus, potentially faster and more energy efficient. This is what Beyer et al. (2004) refer to as "translation on demand," not overlooking the fact that regulating RNA with miRNA is unhindered by the need for a translation product. Thus, it becomes clear that the functions of miRNA complement perfectly our current knowledge of post-transcriptional regulation of expression.

2 Post-Transcriptional Regulatory Mechanisms During Cardiac Hypertrophy

Compensatory cardiac hypertrophy is characterized by a change in the gene expression pattern that recapitulates the neonatal profile (Johnatty et al. 2000). This switch is triggered by regulation of transcriptional and post-transcriptional functions (Fig. 1). Transcription is regulated by selective accessibility of the promoter to the initiation complex and upstream enhancers and/or repressors. Some of the regulators that have been shown to play a role during cardiac hypertrophy include the histone variant, H2A.z (Chen et al. 2006), histone acetylases (Dai and Markham 2001; Gusterson et al. 2003), and deacetylase (Chang et al. 2004; Zhang et al. 2002). Once access to the promoter is granted the availability and/or activity of RNA polymerase II and transcriptional regulators, such as GATA4 (Molkentin et al. 1998), SRF (Zhang et al. 2001), and NKx2-5 (Chen et al. 1996), among others, will determine the level of mRNA produced. But it is well established that translation of mRNA is also a tightly controlled function. It is regulated by: targeted mRNA localization, mRNA stability (half-life), the rate of translation initiation, and, finally, when translation is completed, the rate of protein degradation will determine the extent of the functional outcome. MiRNA are a newly discovered class of post-transcriptional regulators that can inhibit translation and/or induce mRNA degradation (Fig. 2).

3 Post-Transcriptional Regulation by miRNA

Transcription of a gene does not connote automatic translation of the transcript. In other words, we cannot use relative mRNA levels as predictive measures of relative protein levels of a given gene (Gygi et al. 1999), which suggest post-transcriptional

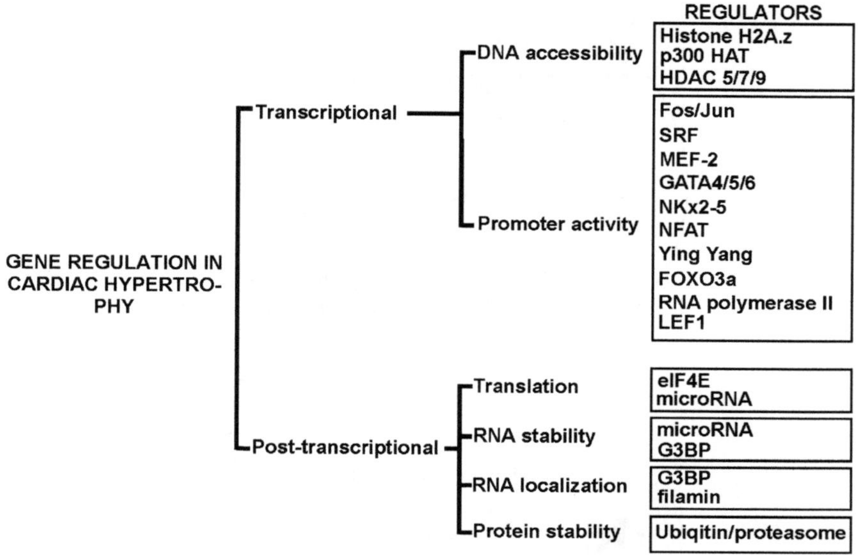

Fig. 1 A schematic showing gene expression regulatory mechanisms involved in cardiac hypertrophy. Gene expression is regulated by transcriptional and post-transcriptional mechanisms. Transcription is restricted by factors that regulate genomic DNA accessibility, while the activity of a promoter is regulated by the availability and activity of various transcription factors, some of which are highlighted in the diagram. Post-transcriptional mechanisms involve regulation of translation initiation and elongation, protein stability, and mRNA stability and localization. MicroRNA plays a role in both translation initiation and elongation, as well as, mRNA stability

regulation of gene expression. Post-transcriptional regulation refers to events that limit the availability or accessibility of mRNA for translation. The study by Beyer et al. (2004) confirms weak or no correlation between mRNA and protein abundance for the whole cell. Similar results were obtained when the calculations were repeated on separate cellular compartments. But, interestingly, a more positive correlation emerged when they grouped functionally related genes. From what we know, an individual miRNA (or a miRNA family) also has the potential to post-transcriptionally regulate a set of specific genes involved in a given cellular function. For example, we observed that miR-1 targets an array of growth-related genes that have been previously implicated in the development of cardiac hypertrophy, which include Ras GTPase-activating protein, cyclin-dependent kinase 9, endothelin, fibronectin, Ras homologue enriched in brain, eukaryotic initiation factor 4E, JunD, quaking, insulin-like growth factor, and Rap1 (Table 1). Concordantly, upon induction of cardiac hypertrophic growth, miR-1 is down-regulated, allowing for the up-regulation of these targets (Sayed et al. 2007). Similarly, when John at al. (2004) analyzed individual miRNA for gene ontology terms they observed that some favored certain terms. For example, miR-208 targets were biased toward "transcription factor," while miR-105 had a preponderance of "small GTPase mediated signal transduction." To the best of our knowledge, there is no other class of molecules identified to

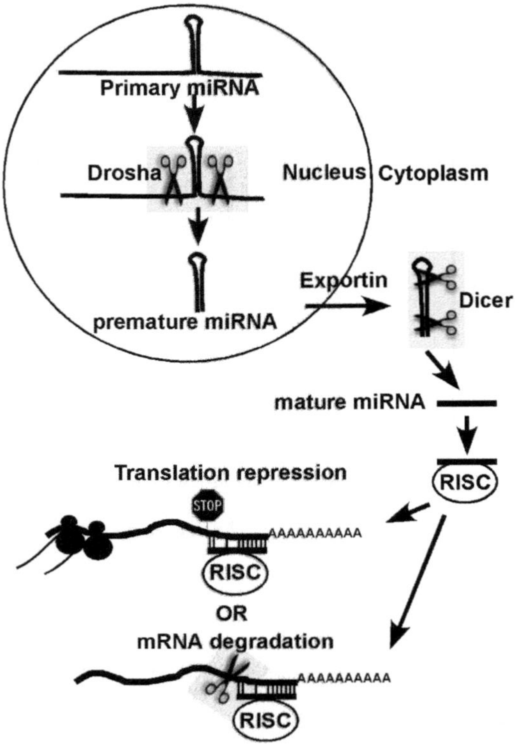

Fig. 2 A diagram showing the processing of microRNA and its fate. MicroRNA are expressed as part of a much longer transcript that matures via a two-step process, one in the nucleus and the second in the cytoplasm, by the enzymes Drosha and Dicer, respectively. Guided by sequence complimentarity and the RISC complex, the mature microRNA targets mRNA and has the potential to induce translation inhibition or mRNA degradation

date that has the potential to fulfill this function as globally and specifically as observed by miRNA. This supports the idea that miRNA are major post-transcriptional regulators.

Transcriptional regulation is the main first step in determining the availability of mRNA, but it immediately follows that the mRNA's half-life will dictate its temporal availability for translation of the open reading frame. Wang et al. (2002) have reported that in yeast the half-lives of mRNA ranged from 3 to 90 min. There was no correlation between the half-life and ribosome density (translational activity), but the half-lives of proteins that form a physical complex were very similar. For example, the 4 core histones have a decay rate of $t_{1/2} = 7 \pm 2$ min, while 131 ribosomal protein mRNA have a $t_{1/2} = 22 \pm 6$ min. Similarly, decay rates of mRNA of functionally related genes were also comparable. In general, genes related to metabolism had longer half-lives than those enrolled in regulatory functions, such

Table 1 Growth-related "TargetscanS" predicted miR-1 targets and their relevant functions in cardiac hypertrophy

TargetscanS predicted miR-1 target genes	Hypertrophy-related functions
Ras GTPase activating protein, RasGAP	Induces RNA polymerase II activity and hypertrophy (Abdellatif et al. 1998)
RasGAP SH3-binding protein-2, G3BP2	A downstream effector of RasGAP necessary for mediating its hypertrophic effects (Lypowy et al. 2005)
Cyclin-dependent kinase 9, Cdk9	An RNA polymerase II kinase involved in mRNA elongation and is necessary for hypertrophy (Sano et al. 2002).
Fibronectin	An extracellular substrate that increases during, and promotes, hypertrophy (Samuel et al. 1991)
Ras homologue enriched in brain, Rheb	An upstream regulator of the mTOR pathway necessary for hypertrophy (Iwata et al. 2005; Nakamura et al. 2005)
Hepatocyte growth factor receptor, MET	A growth factor that enhances angiogenesis and inhibits fibrosis and remodeling in the heart (Iwata et al. 2005; Nakamura et al. 2005)
Endothelin	A growth factor that increases during, and promotes, hypertrophy (Iwata et al. 2005; Nakamura et al. 2005)
Insulin-like growth factor, IGF1	A growth factor that increases during, and promotes, hypertrophy (Ito et al. 1993)
Eukaryotic initiation factor 4E, eIF4E	A factor necessary for initiation of protein synthesis during cardiac hypertrophy (Wada et al. 1996)
JunD	A transcription factor involved in cardiac hypertrophy (Ricci et al. 2005)
Rap1A and Rap1B	A Ras family members involved in cytoskeletal organization and increased during hypertrophy (Johnatty et al. 2000)
Quaking	An RNA-binding protein increased during hypertrophy (Johnatty et al. 2000)
TATA binding protein, TBP	A factor necessary for RNA polymerase II-mediated transcription (Killeen et al. 1992)
Connexin 43	A gap junction protein increased during compensatory hypertrophy (Formigli et al. 2003)

as initiation of translation. These data reconcile well with the fact that mRNA and protein abundance correlate best within the same functional group of genes. Although mammalian miRNAs are commonly known for inhibiting translation versus inducing mRNA degradation, there is substantial evidence to support the latter as well. Farh et al. (2005) showed that predicted mRNA targets of tissue-specific miRNA were lower in the corresponding tissue than they were in others. More convincingly, those targets were, initially, preferentially high in those tissues, before declining, upon the inclining of the targeting miRNA, although they were not completely eliminated. This suggested that miRNA are involved in both mRNA decay, as well as, translational repression. In support, Lim et al. (2005) showed that by expressing the muscle-specific miR-1 or the brain-specific miR-124 in HeLa cells the mRNA expression pattern shifted towards the corresponding tissue.

The mechanism of mRNA degradation is similar to that employed by siRNA, where the endonuclease "Dicer" mediates mRNA cleavage. Alternatively, miRNA may induce deadenylation, which induces mRNA degradation (Wu et al. 2006).

It is well established that, in metazoans, miRNA inhibits mRNA translation. One study showed that miRNA requires both the Cap and poly(A) tail structures to inhibit translation initiation, but the mechanism remains unknown (Humphreys et al. 2005). It was previously shown that both of these structures act in synergy to enhance translation. This may be explained by the fact that eukaryotic initiation factors 4G (eIF4G), a subunit of the 5′cap-binding initiation factor, binds the polyA-binding protein (PABP) and promotes circularization of the mRNA molecule, a structure that is translationally superior to the linear form (Sachs 2000). Thus, one can envision a mechanism whereby the 3′UTR-binding miRISC complex may interfere with the formation of this structure and compromise translation. Alternatively, in an intact circular mRNA, the miRISC complex may come into close proximity to the translation initiation complex eIF4F and physically or biochemically hinder its function. But in case the mRNA is already engaged in translation, miRNA also has the capacity to inhibit post-translation initiation steps (Olsen and Ambros 1999; Seggerson et al. 2002). So, what dictates whether a miRNA will induce mRNA degradation versus translation inhibition? We believe that this will be dictated by duration of exposure of a gene to its targeting miRNA. Chronic high levels of a miRNA in a given tissue will eventually eliminate its target mRNA.

Description of post-transcriptional regulation is incomplete without discussing mRNA localization. The cytoskeletal framework provides mRNA with an anchor, as a mean for targeted localization within the cell. By using in situ hybridization and electron microscopy Bassell et al. (1994) were able to visualize mRNA that was mainly localized to microfilaments (72%), and partially to vimentin filaments and microtubules in human fibroblasts. Both the poly(A) mRNA and the polysomes colocalized to the actin crosslinking proteins, filamin, α-actinin, and actin-binding elongation factor 1α. This feature of mRNA is mediated by the 3′ untranslated region (3′UTR). More than a decade ago, Gottlieb (1990) suggested that mRNA localization evolved by inducing degradation of all un-localized mRNA. In concordance, there is a correlation between translation repression and degradation of some transcripts (Cooperstock and Lipshitz 1997). Interestingly, one example of a molecule that is involved in this process is a non-coding RNA molecule, Yellow Crescent (YC), found in the ascidian eggs (Swalla and Jeffery 1996). Swalla and Jeffery (1996) showed that YC is complimentary to the 3′UTR of proliferating cell nuclear antigen (PCNA), but the two do not exist in the same compartment. This suggested that their association results in degradation of PCNA. This brings to our attention the untapped potential of miRNA to specifically repress un-localized transcripts, earmark them for localization, or maybe aid in regulating the targeting process. Since miRNA have the capacity to both inhibit translation and induce degradation of the mRNA, it is plausible that it exercises both effects on the same transcript in a temporal and spatial manner. For example, if a translationally repressed target does not reach its destination promptly, it must face degradation. More will be revealed when we know if, when, and where miRNA is localized.

4 Transcriptional and Post-Transcriptional Regulation of miRNA

Before discussing the changes of miRNA levels in disease and their potential as diagnostic and therapeutic targets, it is necessary to briefly describe how they themselves are regulated. MiRNA were first discovered by means of large scale cloning and sequencing of the mature forms (Lagos-Quintana et al. 2001; Lau et al. 2001; Lee and Ambros 2001). This group of miRNA was then used as a "training-set" for computer software to generate algorithms for computational prediction of further miRNA (reviewed in (Bentwich 2005). The data from that information was then further examined in an attempt to delineate the transcriptional units of those genes (Rodriguez et al. 2004). Out of 232 mammalian miRNA analyzed, approximately 40% existed within introns of coding RNA, 10% within introns of non-coding RNA (ncRNA), and 13% within exons of ncRNA. In addition, a small portion of miRNA exists within exons (3′UTR) of coding RNA and antisense strands of overlapping genes. The remaining miRNA are intergenic, with as yet undefined transcriptional units. MiRNA contained within host genes are expected to be expressed from the same transcriptional unit, as was indeed confirmed for a subset (Rodriguez et al. 2004). As for intergenic miRNA, Jegga et al. (2007) and Zhou et al. (2007) have developed searchable databases for prediction of miRNA core promoters. From these studies. it was predicted that those genes have features common to RNA polymerase II-dependent promoters, as well as unique sequence motifs. Besides that, there are only a handful of studies that examine the transcriptional regulation of individual miRNA promoters. Examples include regulation of mouse miR-1 promoter by serum response factor, myoD, and muscle enhancer factor 2 (Zhao et al. 2005) and *Drosophila* miR-1 by twist and muscle enhancer factor 2 (Sokol and Ambros 2005) among others (Dews et al. 2006; O'Donnell et al. 2005). Also, recently a fair number of miRNA present within areas of Alu repeats were found to be transcribed from RNA-polymerase III-dependent promoters (Borchert et al. 2006). In addition, it was suggested that miRNA found within CpG islands may be regulated by methylation (Weber et al. 2007).

The dogma is that the primary miRNA transcripts are processed in the nucleus into the premature form by Drosha, which is then exported to the cytoplasm for further processing by Dicer (Lee et al. 2002). But, as with other RNA species, miRNA transcripts are not exempt from post-transcriptional regulation. This was convincingly demonstrated by Thomson et al. (2006), where they compared the mature versus the primary miRNA profile for normal and cancer tissues and found no correlation. We recently found that hypoxia induces down-regulation of miR-199a through specifically inhibiting the processing of the stem-loop precursor of the miRNA (unpublished data). It thus appears that mature miRNA abundance is regulated at the levels of Drosha and/or Dicer processing. This may be a consequence of regulation of the activity or availability of those enzymes or the targeted localization of the primary or premature miRNA. Thus, for diagnostic or prognostic purposes, we must rely on measurements of the mature form.

5 The Mechanism of miRNA-Target Recognition

To evaluate the significance of a change in the levels of a certain miRNA during health or disease, we have to understand how it selects and targets specific mRNA. Very few miRNA targets have been identified or validated experimentally. But there are multiple searchable databases that have computationally predicted miRNA targets in several species, using various algorithms, which continue to be modified as more targets are being validated experimentally. At present, these databases are vital in guiding us in experimentally validating miRNA targets. The miRanda software uses parameters that include: (1) binding energy between miRNA and its target; (2) asymmetric 5′-3′ base-pairing, with more weight on the 5′ end, which can, to some degree, be compensated for by stronger complementation with the target at the 3′ end, also allowing G:U wobble pairing; and (3) the most important criterion is evolutionary conservation of the sequence and its position in 3′UTRs of human, mouse, and rat genes (John et al. 2004). A second database, TargetScanS, uses somewhat different parameters, which includes: (1) exact Watson–Crick (W–C) pairing of nucleotides 2–7 of the miRNA to the target, with no weight given to pairing of its 3′ end; (2) a conserved adenosine at the first position and/or a W–C match at position 8; (3) disregarding the binding energy; and (4) using the human, mouse, rat, dog, and chicken 3′UTR, for determining target site conservation (Lewis et al. 2005). A third searchable database, PicTar, looks for: (1) a perfect ~7 nucleotide seed at position 1–7 or 2–8, but mismatches are allowed if the free binding energy does not increase (Krek et al. 2005); (2) it takes into account the binding energy of the entire miRNA-mRNA duplex; and (3) it includes the genomes of chimpanzee, pufferfish, and zebrafish, in addition to all those listed above, for determining conservation, which further reduces false positive predictions. It should be noted, though, that non-conserved miRNA target sites respond equally well to inhibition by miRNA (Farh et al. 2005). This first suggested that other factors such as secondary structures, RNA binding protein, and other pairing features must contribute to the in vivo specificity of a miRNA. But, upon a closer examination, the investigators established that those messages appeared to have specifically evolved to avoided co-localization with the corresponding miRNA. Those genes were dubbed "antitargets." While the three databases overlap in many predicted targets they diverge in others. Thus, it might be beneficial to search all the databases for potential targets of a miRNA of interest, for experimental validation. But one has to bear in mind that not all predicted targets are genuine, as they may be subject to spatial and temporal restrictions. In addition, binding to the targeting site might be modulated by 3′UTR *cis*-acting sequences or transacting factors. Moreover, a single 3′UTR may be targeted by multiple miRNA. Thus, the level of an mRNA or its translation product is the governed by the combinatorial effect of its targeting miRNA.

While computational analysis predicted the structural basis for miRNA : target pairing, Brennecke et al. (2005) experimentally validated some of the parameters applied above. Essentially, they confirmed that nucleotides 2–8 from the 5′ end of the miRNA are the most critical in establishing base pairing with the target. This criterion could be relaxed to 2–5 nucleotides, under conditions where the nucleotides in the 3′ end had strong complementarity with the target. Also, the position of the base pairing rather than the pairing energy determined the functional efficiency of the miRNA.

It is worth noting that another dimension is added to the mechanism of miRNA targeting by introducing selective miRNA editing. This process converts adenosine to inosine, catalyzed by adenine deaminase in double stranded RNA substrates (reviewed in (Maas and Rich 2000). Luciano et al. (2004) were the first to notice that premature miR-22 is edited at multiple sites. It is estimated that 6% of human miRNA are edited (Blow et al. 2006). Importantly, some of the edited sites are within the "seed" region and result in reassignment of targets (Kawahara et al. 2007).

6 miRNA in Development and Disease

The first miRNA was discovered in 1993 when Lee et al. (1993) found that lin-4, which downregulates the levels of lin-14 during the development of *C. elegans*, expresses a small non-coding 22 nucleotide RNA. Not until 7 years later did Reinhart et al. (2000) report the discovery of the second miRNA, let-7, which induces downregulation of lin-41 protein. This was shortly followed by identification of a multitude of human and fly miRNA that were highly conserved (Lagos-Quintana et al. 2001). Thence, it was noticed that in each of the heart, brain, and liver, there is a distinct tissue-specific miRNA that predominates, which suggested that miRNAs must be involved in tissue specification or differentiation (Lagos-Quintana et al. 2002). MiRNA also play a role in stem cell division (reviewed in Shcherbata et al. 2006). To date, more than 4,000 miRNA sequences have been identified in a wide range of species. These are annotated and catalogued in a searchable web-based data registry by Welcome Trust Sanger Institute (Griffiths-Jones 2004; Griffiths-Jones et al. 2006).

MiR-1 has a dominant expression pattern and function in the heart. First identified as a muscle-specific miRNA (Lagos-Quintana et al. 2002), its expression is detected as early as embryonic day (E) 8.5 in the mouse heart, and increases with the progression of differentiation (Zhao et al. 2005). In *Drosophila,* a DmiR-1^{KO} mutant dies as a small second in-star larvae, 2–7 days after hatching, from apparent paralysis (Sokol and Ambros 2005). Similarly, disrupting miR-1-2 in a homozygous mouse model results in ventricular septal defects and lethality between E15.5 to immediate postnatal (Zhao et al. 2007). Recently, several reports have also implicated miR-1 in development of cardiac hypertrophy (Care et al. 2007, Sayed et al. 2007) and arrhythmias (Yang et al. 2007). While other miRNA are also involved in the progression of hypertrophy, including miR-133 (Care et al. 2007), miR-21 (Cheng et al. 2007), miR-208 (van Rooij et al. 2007), and miR-195 (van Rooij et al. 2006), the emphasis remains on miR-1, which is singularly downregulated within 24 h of triggering hypertrophy. Notably, adult cardiac hypertrophy starts out as a compensatory growth that is characterized by recapitulating the neonatal heart gene expression profile (Johnatty et al. 2000). Concordantly, our results confirm similar changes in miRNA patterns in adult hypertrophied versus sham and neonatal versus adult hearts (unpublished data). Thus, we expected that subsequent progress into heart failure would coincide with a deviation of the miRNA signature from this pattern. To test this idea, we searched for miRNA that were deregulated during the period of transition from hypertrophy to failure in a mouse model. While the results showed an overall continuum of the hypertrophy signature, only one miRNA, miR-451, was sharply upregulated at the onset of contractile dysfunction. Such a molecule would serve as an ideal therapeutic target.

Because of their major roles as oncogenes and tumor suppressors, a group of miRNAs have been dubbed "oncomirs" (Esquela-Kerscher and Slack 2006). In cancer, the discovery of miRNA is offering answers to previously unresolved questions. For example, after years of futile attempts to identify the gene(s) in the deleted 13q14 region that participate in the pathogenesis of chronic lymphocytic leukemia (CLL), Calin et al. (2002) discovered that miR-15 and miR-16 are located within this sequence and that both are downregulated in ~68% of cases. These miRNA molecules are now considered tumor suppressors, one of their targets being Bcl2, the level of which reciprocally correlates with that of miR-15/16 in CCL (Cimmino et al. 2005). Calin et al. (2004b) have then gone on to show that 52.5% of miRNA are located in previously established cancer-associated regions. In contrast, over-expression of miRNA is also associated with cancer. For example, miR-155 is upregulated in B cell lymphoma (Eis et al. 2005). Notably, tissue-specific over-expression of miR-155 in a transgenic mouse model proved it sufficient for induction of lymphoblastic leukemia (Costinean et al. 2006). Overall, unique miRNA expression profiles have been detected in chronic lymphocytic leukemia (Calin et al. 2004a), breast (Iorio et al. 2005), and lung cancers (Yanaihara et al. 2006).

Although we have learned much about the miRNA signatures in cardiac hypertrophy and cancer we have very little knowledge about the targets and functions of individual miRNA deregulated during these diseases. This is indeed quite a challenging task. But until we learn more about function, we cannot use miRNA successfully as therapeutic targets. In an attempt to gain some insight, we examined the differences between the miRNA expression patterns that emerged from studies of various diseases. Accordingly, we tabulated some of the miRNA that were emphasized in cardiac hypertrophy, different cancers, hypoxia, and atherosclerosis (Table 2). We searched for trends, such as an association between certain miRNA and known disease mechanisms. For example, it was immediately noticed that miR-21 is upregulated in all cancer forms, as well as cardiac hypertrophy and atherosclerosis, but not hypoxia. Unlike cancer or atherosclerosis, cardiac hypertrophy is a process of cellular growth in the absence of cell cycling. Thus, by deduction, we may predict that miR-21 may be related to a growth process common to both hypertrophy and proliferation. On the other hand, we could also infer that miR-21 does not play a direct role in the cell cycle or hypoxia. In contrast, miR-15/16 cluster is deleted in chronic lymphocytic leukemia (CLL), but remains high and unchanged in the heart during hypertrophy, which later shows a further increase in miR-15b in

Table 2 Select tissue-specific and disease-related microRNA

Tissue-specific	Cardiac hypertrophy	Cancer	Hypoxia	Athero-sclerosis
miR-1, heart and skeletal muscle (Lagos-Quintana et al. 2002)	miR-1 ↓ (Sayed et al. 2007)	miR-107 ↑, H&N (Tran et al. 2007), colon, and stomach (Volinia et al. 2006), pancreas (Volinia et al. 2006; Bloomston et al. 2007)	miR-107 ↑ (Kulshreshtha et al. 2007)	miR-21 ↑ (Ji et al. 2007)
miR-122, liver (Lagos-Quintana et al. 2002)	miR-107 ↑ (Sayed et al. 2007)	miR-15/16 ↓, CLL (Calin et al. 2002; Fulci et al. 2007)	miR-210 ↑ (Kulshreshtha et al. 2007)	
miR-124, brain (Lagos-Quintana et al. 2002)	miR-133 ↓ (Care et al. 2007)	miR-150 ↑, lung (Yanaihara et al. 2006), CLL (Fulci et al. 2007)	miR-213 ↑ (Kulshreshtha et al. 2007)	
miR-133, heart and skeletal (Lagos-Quintana et al. 2002)	miR-15a/b ↑ (Sayed et al. 2007)	miR-122 ↓, liver (Kutay et al. 2006)	miR-23a ↑ (Kulshreshtha et al. 2007)	
miR-208, heart	miR-150 ↓ (Sayed et al. 2007; Cheng et al. 2007)	miR-125b ↑, H&N (Tran et al. 2007), pancreas (Lee et al. 2006)	miR-24-1 ↑ (Kulshreshtha et al. 2007)	
miR-375, Pancreatic islet (Poy et al. 2004)	miR-155 ↓ (Sayed et al. 2007)	miR-143↓, colon (Michael et al. 2003), pancreas (Bloomston et al. 2007)	miR-26b ↑ (Kulshreshtha et al. 2007)	
	miR-185 ↓ (Sayed et al. 2007; Cheng et al. 2007)	mi-145 ↓, colon (Michael et al. 2003)		
	miR-195 ↑ (van Rooij et al. 2006)	miR-155 ↑, B cell lymphoma (Eis et al. 2005), breast and colon (Volinia et al. 2006), lung (Yanaihara et al. 2006), pancreas (Bloomston et al. 2007), CLL (Fulci et al. 2007)		
	miR-199a/b/a* ↑ (Sayed et al. 2007)	miR-199a ↑, lung (Yanaihara et al. 2006), pancreas (Volinia et al. 2006; Bloomston et al. 2007), prostate (Volinia et al. 2006)		
	miR-208 ↑ (van Rooij et al. 2006)	miR-205 ↑, H&N (Tran et al. 2007), pancreas (Bloomston et al. 2007), lung (Yanaihara et al. 2006)		

(continued)

Table 2 (continued)

Tissue-specific	Cardiac hypertrophy	Cancer	Hypoxia	Athero-sclerosis
	miR-21 ↑ (Sayed et al. 2007; Cheng et al. 2007)	miR-21 ↑, H&N (Tran et al. 2007), breast, colon, lung, prostate, and stomach (Volinia et al. 2006), pancreas (Bloomston et al. 2007), CLL (Fulci et al. 2007), liver (Kutay et al. 2006)		
	miR-214 ↑ (Sayed et al. 2007; Cheng et al. 2007)	miR-210 ↑, pancreas (Bloomston et al. 2007)		
	miR-221 ↑ (Sayed et al. 2007)	miR-213 ↑, pancreas (Bloomston et al. 2007),		
	miR-222 ↑ (Sayed et al. 2007)	miR-214 ↑, pancreas, prostate, stomach, lung (Volinia et al. 2006)		
	miR-23a ↑ (Sayed et al. 2007)	miR-221 ↑, H&N (Tran et al. 2007), colon and smotach (Volinia et al. 2006), pancreas (Bloomston et al. 2007)		
	miR-23b ↑ (Sayed et al. 2007)	miR-222 ↑, pancreas (Bloomston et al. 2007)		
	miR-27a ↑ (Sayed et al. 2007; Cheng et al. 2007)	miR-23a ↑, pancreas (Bloomston et al. 2007), liver (Kutay et al. 2006)		
	miR-27b ↑ (Sayed et al. 2007; Cheng et al. 2007)	miR-23b ↑, H&N (Tran et al. 2007), pancreas (Bloomston et al. 2007), liver (Kutay et al. 2006)		
	miR-29a ↓ (Sayed et al. 2007; Cheng et al. 2007)	miR-27a ↑, H&N (Tran et al. 2007)		
		miR-29a ↑, H&N (Tran et al. 2007)		

Select miRNA were compiled from various reports in an attempt to highlight some of the differences and similarities in disease patterns, as well as, the involvement of tissue specific miRNA in pathogenesis.

Abbreviations: H&N, head and neck; CLL, chronic lymphocytic leukemia; ↑, upregulation; ↓, downregulation late hypertrophy and failure. This agrees with a role for the miRNA cluster in cell cycling, as indeed has been established recently. Another trend involves downregulation of tissue-specific miRNA (miR-1, miR-133, miR-122) in disease states. Overall, such comparisons may also prove useful when evaluating miRNA therapeutic agents and their side effects.

Besides cardiovascular diseases and cancer very little is known about miRNA in other disorders. But there is little doubt that it will soon prove that miRNA have a role in almost every aspect of a cell's function, as seen in its emerging role in regulating metabolism and the lifespan (reviewed in Boehm and Slack (2006).

7 The Diagnostic and Prognostic Potential of miRNA

Within the past 5 years we have seen the rise and fall of cDNA microarrays as diagnostic and prognostic measures in disease (Ioannidis 2005; Michiels et al. 2005; Reis-Filho et al. 2006). While much work is still being done in an attempt to optimize and standardize the methodology, there will remain inconsistencies inherent to the characteristics of mRNA itself: having multiple isoforms and splice variants, and a notoriously unstable nature.

So what does miRNA offer us in the way of a diagnostic and prognostic tool that mRNA does not? While a single mRNA generally dictates the translation of a single protein, a single miRNA molecule has the capacity to regulate the translation of an array of genes governing a certain function (John et al. 2004; Sayed et al. 2007). The significance of which is underscored upon recalling that the underlying apparatus, generally regulating normal cellular functions and pathologies, consist of a consortium of genes. Those genes frequently exhibit deliberate structural and functional redundancy, to ensure a fail-safe operation. More importantly, changes in a gene's mRNA levels can be counterbalanced by post-transcriptional regulation. Thus, we must come to the realization that changes in the levels of mRNA are not necessarily predictive of functional outcome. Using the counter argument, it is expected that miRNA levels are more tightly regulated and less variable, as a single miRNA has the capacity to regulate a cellular function. Moreover, mature miRNA are 18–25 nt long, end products that have developed beyond post-transcriptional regulation and are inherently very stable. Thus, it is expected that miRNA provide us with a more superior predictive parameter in diseases. When Lu et al. (2005) put this idea to the test, they found that miRNA profiling was highly accurate in predicting the differentiation state of tumors and in classification of poorly differentiated tumors, a fact that they could not determine by mRNA profiling. But clearly, more studies are needed to further validate the predictive powers of miRNA in cancer.

We also hope that miRNA might aid in the diagnosis, prognosis, and treatment of heart diseases, which remain the leading cause of death in the United States. Currently, diagnosis and prognosis of heart failure relies mainly on clinical signs and symptoms, aided with only a handful of circulating biomarkers

(Macabasco-O'Connell and Miller 2006). Unfortunately, there "isn't a single patient characteristic that can be used to reliably predict a patient's outcome" (Frankel et al. 2006). This is probably due to the fact that the utility of more direct biomarkers is hindered by the impracticality of routine cardiac biopsies. Alternatively, miRNA profiling of other secondarily affected tissues, in particular the liver, might provide a promising substitute for establishing miRNA biomarkers in heart failure. Our initial results look promising, showing differential regulation of miRNA in liver during heart failure in a mouse model.

8 miRNA as Therapeutic Targets

MiRNA are characterized by targeting a functional group of genes, versus a single mRNA, which renders them a potentially powerful therapeutic target. As an example, it proved more effective to simultaneously, rather than individually, inhibit the cell cycle-related miR-16 targets (Linsley et al. 2007). Thus, it becomes clear that before therapeutically targeting a miRNA, we require knowledge of its gene targets, function, and tissue distribution. Since the discovery of mammalian miRNA is still in its early phase, we have very little information in that regard. In the mean time, however, Krutzfeldt et al. (2005) developed a method for antagonizing miRNA function in vivo. The design involves synthesis of a miRNA-specific, anti-sense, cholesteryl-conjugated, 2′-*O*-methyl oligoribonucleotide, which was dubbed "antagomir." This non-hydrolysable oligoribonucleotide is cell permeable and inhibits the miRNA by inducing its elimination. Hence, it offers a long-lived effect and circumvents the need for a cell delivery vehicle. This approach, though, is not suitable for replenishing depleted miRNA. That is because it would require delivery of the premature miRNA in its native form, so as to allow for proper processing and incorporation into the RISC complex. Thus, other known delivery methods employed in gene therapy, including viral vectors and liposomes, will have to serve as substitutes in this case.

9 Concluding Remarks

As pointed out in the above discussion, we remain in dire need of new biomarkers and therapeutic targets for the two leading causes of death in the United States, cardiovascular diseases and cancer. MiRNA offer us that prospect. But in order to maximally exploit their potential, we must continue to dissect their function in great detail. It is still necessary to identify the specific mode of transcriptional and post-transcriptional regulation of individual miRNA, their cellular localization, targets, and overall function, in parallel with continuing to devise modes for in vivo delivery of miRNA and anti-miRNA.

References

Abdellatif M, Packer SE, Michael LH et al (1998) A Ras-dependent pathway regulates RNA polymerase II phosphorylation in cardiac myocytes: implications for cardiac hypertrophy. Mol Cell Biol 18:6729–6736

Bassell GJ, Powers CM, Taneja KL et al (1994) Single mRNAs visualized by ultrastructural in situ hybridization are principally localized at actin filament intersections in fibroblasts. J Cell Biol 126:863–876

Bentwich I (2005) Prediction and validation of microRNAs and their targets. FEBS Lett 579:5904–5910

Beyer A, Hollunder J, Nasheuer HP et al (2004) Post-transcriptional expression regulation in the yeast Saccharomyces cerevisiae on a genomic scale. Mol Cell Proteomics 3:1083–1092

Bloomston M, Frankel WL, Petrocca F et al (2007) MicroRNA expression patterns to differentiate pancreatic adenocarcinoma from normal pancreas and chronic pancreatitis. JAMA 297:1901–1908

Blow MJ, Grocock RJ, van Dongen S et al (2006) RNA editing of human microRNAs. Genome Biol 7:4

Boehm M, Slack FJ (2006) MicroRNA control of lifespan and metabolism. Cell Cycle 5:837–840

Borchert GM, Lanier W, Davidson BL (2006) RNA polymerase III transcribes human microRNAs. Nat Struct Mol Biol 13:1097–1101

Brennecke J, Stark A, Russell RB et al (2005) Principles of microRNA-target recognition. PLoS Biol 3:0404–0418

Calin GA, Dumitru CD, Shimizu M et al (2002) Frequent deletions and down-regulation of microRNA genes miR15 and miR16 at 13q14 in chronic lymphocytic leukemia. Proc Natl Acad Sci USA 99:15524–15529

Calin GA, Liu CG, Sevignani C et al (2004a) MicroRNA profiling reveals distinct signatures in B cell chronic lymphocytic leukemias. Proc Natl Acad Sci USA 101:11755–11760

Calin GA, Sevignani C, Dumitru CD et al (2004b) Human microRNA genes are frequently located at fragile sites and genomic regions involved in cancers. Proc Natl Acad Sci USA 101:2999–3004

Care A, Catalucci D, Felicetti F et al (2007) MicroRNA-133 controls cardiac hypertrophy. Nat Med 13:613–618

Chang S, McKinsey TA, Zhang CL et al (2004) Histone deacetylases 5 and 9 govern responsiveness of the heart to a subset of stress signals and play redundant roles in heart development. Mol Cell Biol 24:8467–8476

Chen CY, Croissant J, Majesky M et al (1996) Activation of the cardiac alpha-actin promoter depends upon serum response factor, Tinman homologue, Nkx-2.5, and intact serum response elements. Dev Genet 19:119–130

Chen IY, Lypowy J, Pain J et al (2006) Histone H2A.z is essential for cardiac myocyte hypertrophy but opposed by silent information regulator 2alpha. J Biol Chem 281:19369–19377

Cheng Y, Ji R, Yue J et al (2007) MicroRNAs are aberrantly expressed in hypertrophic heart. Do they play a role in cardiac hypertrophy? Am J Pathol 170:1831–1840

Cimmino A, Calin GA, Fabbri M et al (2005) miR-15 and miR-16 induce apoptosis by targeting BCL2. Proc Natl Acad Sci USA 102:13944–13949

Cooperstock RL, Lipshitz HD (1997) Control of mRNA stability and translation during *Drosophila* development. Semin Cell Dev Biol 8:541–549

Costinean S, Zanesi N, Pekarsky Y et al (2006) Pre-B cell proliferation and lymphoblastic leukemia/high-grade lymphoma in E(mu)-miR155 transgenic mice. Proc Natl Acad Sci USA 103:7024–7029

Dai YS, Markham BE (2001) p300 Functions as a coactivator of transcription factor GATA-4. J Biol Chem 276:37178–37185

Dews M, Homayouni A, Yu D et al (2006) Augmentation of tumor angiogenesis by a Myc-activated microRNA cluster. Nat Genet 38:1060–1065

Eis PS, Tam W, Sun L et al (2005) Accumulation of miR-155 and BIC RNA in human B cell lymphomas. Proc Natl Acad Sci USA 102:3627–3632

Esquela-Kerscher A, Slack FJ (2006) Oncomirs – microRNAs with a role in cancer. Nat Rev Cancer 6:259–269

Farh KK, Grimson A, Jan C et al (2005) The widespread impact of mammalian MicroRNAs on mRNA repression and evolution. Science 310:1817–1821

Formigli L, Ibba-Manneschi L, Perna AM et al (2003) Altered Cx43 expression during myocardial adaptation to acute and chronic volume overloading. Histol Histopathol 18:359–369

Frankel DS, Piette JD, Jessup M et al (2006) Validation of prognostic models among patients with advanced heart failure. J Card Fail 12:430–438

Fulci V, Chiaretti S, Goldoni M et al (2007) Quantitative technologies establish a novel microRNA profile of chronic lymphocytic leukemia. Blood 109:4944–4951

Gottlieb E (1990) Messenger RNA transport and localization. Curr Opin Cell Biol 2:1080–1086

Griffiths-Jones S (2004) The microRNA Registry. Nucleic Acids Res 32:D109–D111

Griffiths-Jones S, Grocock RJ, van Dongen S et al (2006) miRBase: microRNA sequences, targets and gene nomenclature. Nucleic Acids Res 34:D140–144

Gusterson RJ, Jazrawi E, Adcock IM et al (2003) The transcriptional co-activators CREB-binding protein (CBP) and p300 play a critical role in cardiac hypertrophy that is dependent on their histone acetyltransferase activity. J Biol Chem 278:6838–6847

Gygi SP, Rochon Y, Franza BR et al (1999) Correlation between protein and mRNA abundance in yeast. Mol Cell Biol 19:1720–1730

Humphreys DT, Westman BJ, Martin DI et al (2005) MicroRNAs control translation initiation by inhibiting eukaryotic initiation factor 4E/cap and poly(A) tail function. Proc Natl Acad Sci USA 102:16961–16966

Ioannidis JP (2005) Microarrays and molecular research: noise discovery? Lancet 365:454–455

Iorio MV, Ferracin M, Liu CG et al (2005) MicroRNA gene expression deregulation in human breast cancer. Cancer Res 65:7065–7070

Ito H, Hiroe M, Hirata Y et al (1993) Insulin-like growth factor-I induces hypertrophy with enhanced expression of muscle specific genes in cultured rat cardiomyocytes. Circulation 87:1715–1721

Iwata K, Sawa Y, Kitagawa-Sakakida S et al (2005) Gene transfection of hepatocyte growth factor attenuates the progression of cardiac remodeling in the hypertrophied heart. J Thorac Cardiovasc Surg 130:719–725

Jegga AG, Chen J, Gowrisankar S et al (2007) GenomeTrafac: a whole genome resource for the detection of transcription factor binding site clusters associated with conventional and microRNA encoding genes conserved between mouse and human gene orthologs. Nucleic Acids Res 35:D116–D121

Ji R, Cheng Y, Yue J et al (2007) MicroRNA expression signature and antisense-mediated depletion reveal an essential role of MicroRNA in vascular neointimal lesion formation. Circ Res 100:1579–1588

John B, Enright AJ, Aravin A et al (2004) Human MicroRNA targets. PLoS Biol 2:1862–1879

Johnatty SE, Dyck JR, Michael LH et al (2000) Identification of genes regulated during mechanical load-induced cardiac hypertrophy. J Mol Cell Cardiol 32:805–815

Kawahara Y, Zinshteyn B, Sethupathy P et al (2007) Redirection of silencing targets by adenosine-to-inosine editing of miRNAs. Science 315:1137–1140

Killeen M, Coulombe B, Greenblatt J (1992) Recombinant TBP, transcription factor IIB, and RAP30 are sufficient for promoter recognition by mammalian RNA polymerase II. J Biol Chem 267:9463–9466

Krek A, Grun D, Poy MN et al (2005) Combinatorial microRNA target predictions. Nat Genet 37:495–500

Krutzfeldt J, Rajewsky N, Braich R et al (2005) Silencing of microRNAs in vivo with 'antagomirs'. Nature 438:685–689

Kulshreshtha R, Ferracin M, Wojcik SE et al (2007) A microRNA signature of hypoxia. Mol Cell Biol 27:1859–1867
Kutay H, Bai S, Datta J et al (2006) Downregulation of miR-122 in the rodent and human hepatocellular carcinomas. J Cell Biochem 99:671–678
Lagos-Quintana M, Rauhut R, Lendeckel W et al (2001) Identification of novel genes coding for small expressed RNAs. Science 294:853–858
Lagos-Quintana M, Rauhut R, Yalcin A et al (2002) Identification of tissue-specific microRNAs from mouse. Curr Biol 12:735–739
Lau NC, Lim LP, Weinstein EG et al (2001) An abundant class of tiny RNAs with probable regulatory roles in *Caenorhabditis elegans*. Science 294:858–862
Lee RC, Ambros V (2001) An extensive class of small RNAs in *Caenorhabditis elegans*. Science 294:862–864
Lee RC, Feinbaum RL, Ambros V (1993) The *C. elegans* heterochronic gene lin-4 encodes small RNAs with antisense complementarity to lin-14. Cell 75:843–854
Lee Y, Jeon K, Lee JT et al (2002) MicroRNA maturation: stepwise processing and subcellular localization. EMBO J 21:4663–4670
Lee EJ, Gusev Y, Jiang J et al (2006) Expression profiling identifies microRNA signature in pancreatic cancer. Int J Cancer 120:1046–1054
Lewis BP, Burge CB, Bartel DP (2005) Conserved seed pairing, often flanked by adenosines, indicates that thousands of human genes are microRNA targets. Cell 120:15–20
Lim LP, Lau NC, Garrett-Engele P et al (2005) Microarray analysis shows that some microRNAs downregulate large numbers of target mRNAs. Nature 433:769–773
Linsley PS, Schelter J, Burchard J et al (2007) Transcripts targeted by the microRNA-16 family cooperatively regulate cell cycle progression. Mol Cell Biol 27:2240–2252
Lu J, Getz G, Miska EA et al (2005) MicroRNA expression profiles classify human cancers. Nature 435:834–838
Luciano DJ, Mirsky H, Vendetti NJ et al (2004) RNA editing of a miRNA precursor. RNA 10:1174–1177
Lypowy J, Chen IY, Abdellatif M (2005) An alliance between RAS GTPASE-activating protein, filamin C, and G3BP regulates myocyte growth. J Biol Chem 280:35717–25728
Maas S, Rich A (2000) Changing genetic information through RNA editing. Bioessays 22:790–802
Macabasco-O'Connell A, Miller PS (2006) Biomarkers for heart failure. Prog Cardiovasc Nurs 21:215–218
Michael MZ, SM OC, van Holst Pellekaan NG et al (2003) Reduced accumulation of specific microRNAs in colorectal neoplasia. Mol Cancer Res 1:882–891
Michiels S, Koscielny S, Hill C (2005) Prediction of cancer outcome with microarrays: a multiple random validation strategy. Lancet 365:488–492
Molkentin JD, Lu JR, Antos CL et al (1998) A calcineurin-dependent transcriptional pathway for cardiac hypertrophy. Cell 93:215–228
Nakamura T, Matsumoto K, Mizuno S et al (2005) Hepatocyte growth factor prevents tissue fibrosis, remodeling, and dysfunction in cardiomyopathic hamster hearts. Am J Physiol Heart Circ Physiol 288:H2131–2139
O'Donnell KA, Wentzel EA, Zeller KI et al (2005) c-Myc-regulated microRNAs modulate E2F1 expression. Nature 435:839–843
Olsen PH, Ambros V (1999) The lin-4 regulatory RNA controls developmental timing in *Caenorhabditis elegans* by blocking LIN-14 protein synthesis after the initiation of translation. Dev Biol 216:671–680
Poy MN, Eliasson L, Krutzfeldt J et al (2004) A pancreatic islet-specific microRNA regulates insulin secretion. Nature 432:226–230
Reinhart BJ, Slack FJ, Basson M et al (2000) The 21-nucleotide let-7 RNA regulates developmental timing in Caenorhabditis elegans. Nature 403:901–906
Reis-Filho JS, Westbury C, Pierga JY (2006) The impact of expression profiling on prognostic and predictive testing in breast cancer. J Clin Pathol 59:225–231

Ricci R, Eriksson U, Oudit GY et al (2005) Distinct functions of junD in cardiac hypertrophy and heart failure. Genes Dev 19:208–213

Rodriguez A, Griffiths-Jones S, Ashurst JL et al (2004) Identification of mammalian microRNA host genes and transcription units. Genome Res 14:1902–1910

Sachs A (2000) Translational control of gene expression. Cold Spring Harbor Laboratory, Cold Spring Harbor

Samuel JL, Barrieux A, Dufour S et al (1991) Accumulation of fetal fibronectin mRNAs during the development of rat cardiac hypertrophy induced by pressure overload. J Clin Invest 88:1737–1746

Sano M, Abdellatif M, Oh H et al (2002) Activation and function of cyclin T-Cdk9 (positive transcription elongation factor-b) in cardiac muscle-cell hypertrophy. Nat Med 8:1310–1317.

Sayed D, Hong C, Chen IY et al (2007) MicroRNAs play an essential role in the development of cardiac hypertrophy. Circ Res 100:416–424

Seggerson K, Tang L, Moss EG (2002) Two genetic circuits repress the Caenorhabditis elegans heterochronic gene lin-28 after translation initiation. Dev Biol 243:215–225

Shcherbata HR, Hatfield S, Ward EJ et al (2006) The MicroRNA pathway plays a regulatory role in stem cell division. Cell Cycle 5:172–175

Sokol NS, Ambros V (2005) Mesodermally expressed *Drosophila* microRNA-1 is regulated by Twist and is required in muscles during larval growth. Genes Dev 19:2343–2354

Swalla BJ, Jeffery WR (1996) PCNA mRNA has a 3′UTR antisense to yellow crescent RNA and is localized in ascidian eggs and embryos. Dev Biol 178:23–34

Thomson JM, Newman M, Parker JS et al (2006) Extensive post-transcriptional regulation of microRNAs and its implications for cancer. Genes Dev 20:2202–2207

Tran N, McLean T, Zhang X et al (2007) MicroRNA expression profiles in head and neck cancer cell lines. Biochem Biophys Res Commun 358:12–17

van Rooij E, Sutherland LB, Liu N et al (2006) A signature pattern of stress-responsive microRNAs that can evoke cardiac hypertrophy and heart failure. Proc Natl Acad Sci USA 103:18255–18260

van Rooij E, Sutherland LB, Qi X et al (2007) Control of stress-dependent cardiac growth and gene expression by a MicroRNA. Science 316:575–579

Volinia S, Calin GA, Liu CG et al (2006) A microRNA expression signature of human solid tumors defines cancer gene targets. Proc Natl Acad Sci USA 103:2257–2261

Wada H, Ivester CT, Carabello BA et al (1996) Translational initiation factor eIF-4E. A link between cardiac load and protein synthesis. J Biol Chem 271:8359–8364

Wang Y, Liu CL, Storey JD et al (2002) Precision and functional specificity in mRNA decay. Proc Natl Acad Sci USA 99:5860–5865

Weber B, Stresemann C, Brueckner B et al (2007) Methylation of human microRNA genes in normal and neoplastic cells. Cell Cycle 6:1001–1005

Wu L, Fan J, Belasco JG (2006) MicroRNAs direct rapid deadenylation of mRNA. Proc Natl Acad Sci USA 103:4034–4039

Yanaihara N, Caplen N, Bowman E et al (2006) Unique microRNA molecular profiles in lung cancer diagnosis and prognosis. Cancer Cell 9:189–198

Yang B, Lin H, Xiao J et al (2007) The muscle-specific microRNA miR-1 regulates cardiac arrhythmogenic potential by targeting GJA1 and KCNJ2. Nat Med 13:486–491

Zhang X, Azhar G, Chai J et al (2001) Cardiomyopathy in transgenic mice with cardiac-specific overexpression of serum response factor. Am J Physiol Heart Circ Physiol 280:H1782–1792

Zhang CL, McKinsey TA, Chang S et al (2002) Class II histone deacetylases act as signal-responsive repressors of cardiac hypertrophy. Cell 110:479–488.

Zhao Y, Samal E, Srivastava D (2005) Serum response factor regulates a muscle-specific microRNA that targets Hand2 during cardiogenesis. Nature 436:214–220

Zhao Y, Ransom JF, Li A et al (2007) Dysregulation of cardiogenesis, cardiac conduction, and cell cycle in mice lacking miRNA-1-2. Cell 129:303–317

Zhou X, Ruan J, Wang G et al (2007) Characterization and identification of MicroRNA core promoters in four model species. PLoS Comput Biol 3:0412–0423

miRNAs and Their Emerging Role in Cardiac Hypertrophy

T.E. Callis, M. Tatsuguchi, and D.-Z. Wang(✉)

Contents

Abstract MicroRNAs (miRNAs) are a class of small non-coding RNAs that negatively regulate gene expression by inhibiting translation or promoting de-gradation of targeted messenger RNAs (mRNAs). More than one-third of human protein-coding mRNAs are predicted as being regulated by miRNAs, which have been implicated in processes as diverse as cancer and muscle biology. Now, recent studies demonstrate that the expression profiles of many miRNAs change during cardiac hypertrophy. Most importantly, both gain- and loss-of-function studies have established a clear functional correlation between miRNAs and cardiac hypertrophy. This previously unrecognized relationship sheds new light onto the regulatory mechanisms underlying the heart's response to pathological stress and suggests the potential of miRNAs as therapeutic targets for cardiovascular disease.

D.-Z. Wang
Carolina Cardiovascular Biology Center, Department of Cell and Developmental Biology, 8340C Medical Biomolecular Research Building (MBRB), CB # 7126, University of North Carolina, Chapel Hill, NC 27599-7126, USA
e-mail: dawang@med.unc.edu

V.A. Erdmann et al. (eds.) *RNA Technologies in Cardiovascular Medicine and Research*, doi: 10.1007/978-3-540-78709-9,

1 Introduction

Cardiac hypertrophy is the primary cellular response of the heart to stress caused by pathological and physiological hemodynamic overload, abnormal hormonal signaling, and certain inherited disorders involving mutations in particular transcription factors or contractile proteins (Hunter and Chien 1999; Frey and Olson 2003). The hypertrophic growth response sustains cardiac output in the face of such stress and involves numerous changes including enhanced protein synthesis, increased sarcomeric density and increased size of cardiomyocytes. Re-activation of several cardiac fetal genes is also a well-established change accompanying hypertrophic growth. Although cardiac hypertrophy induced by pathological stimuli is considered an adaptive mechanism to sustain cardiac output, prolonged hypertrophy has adverse consequences associated with heart failure and sudden death. Much effort has been expended by the cardiovascular research community to map and understand the complex genetic pathways required for hypertrophy, with the ultimate goal of acquiring new therapeutic strategies to improve both cardiac output and heart patient prognosis.

Recent studies have linked a novel class of small regulatory RNAs, known as microRNAs (miRNAs), to cardiac hypertrophy, and offer new insight into the regulation of this disease process. More than 600 human miRNA genes have been identified, many of which are evolutionarily conserved, and perhaps many more miRNA genes await discovery (Griffiths-Jones 2004; Bentwich et al. 2005; Landgraf et al. 2007). miRNAs are now known to play roles in remarkably diverse processes, but given the prevalence of these molecules, relatively few specific miRNAs have any precise biological functions assigned. Interestingly, the expression profiles for a number of miRNAs change during cardiac hypertrophy. Furthermore, mis-expression of miRNAs, as well as loss-of-function experiments in mice, demonstrate that specific miRNAs can augment or attenuate the hypertrophic growth response and suggest the potential of these molecules as novel therapeutic targets. Here, we review the biology of miRNAs and their emerging roles in cardiac hypertrophy.

2 Biology of miRNAs

A little over ten years ago, the *lin-4* gene, which controls the timing of *C. elegans* larval development, was discovered to unexpectedly produce a 21-nucleotide-long noncoding RNA that suppressed *lin-14* protein expression without noticeably affecting *lin-14* mRNA levels. This small RNA was found to base pair to complementary sites in the 3′ untranslated region (UTR) of *lin-14* mRNA and negatively affect its translation (Lee et al. 1993; Wightman et al. 1993). Although this phenomenon was initially treated as a genetic oddity and virtually ignored for nearly a decade, we now recognize that thousands of these small RNAs, now called miRNAs, similar to lin-4 exist in the genomes of divergent species and post-transcriptionally regulate gene expression.

miRNAs are part of the RNA interference (RNAi) pathway, the general term for RNA-guided regulation of gene expression that is conserved in most eukaryotes (Meister and Tuschl 2004). Another class of non-coding RNAs, known as small interfering RNAs (siRNAs), also shares common downstream components of the RNAi pathway with miRNAs. Although mature miRNAs and siRNAs are structurally similar and both negatively regulate gene expression, their origins and upstream processing pathways differ significantly: siRNAs arise from foreign dsRNA, whereas miRNAs are genomically encoded and undergo more extensive post-transcriptional processing than siRNAs (Denli et al. 2004; Gregory et al. 2004). The RNAi pathway is thought to have first evolved using siRNAs as a form of innate immunity against viruses, and later endogenously encoded miRNAs were selected as beneficial post-transcriptional regulators of gene expression. The discovery of RNAi and miRNAs offers a new paradigm for understanding the control of gene expression during development and disease. Indeed, miRNAs are now recognized to regulate gene expression in a variety of fundamental biological processes including cell proliferation, differentiation, apoptosis, and tumorigenesis, and recently have been linked to cardiac hypertrophy and disease (Ambros and Chen 2007; Chien 2007; Mann 2007).

2.1 miRNA Biogenesis and Mechanism of miRNA Function

Mature miRNAs are approximately 18–24 nucleotides (nt) in length and their biogenesis begins with the transcription of primary-miRNA (pri-miRNA) (Bracht et al. 2004; Lee et al. 2004). RNA polymerase II transcribes pri-miRNAs as independent transcriptional units with lengths ranging from several hundred to several thousand nucleotides that may encode a single miRNA or sometimes two or more miRNAs (Bartel 2004). In addition to independent transcriptional units, some miRNAs originate from the introns of mRNA transcripts (Bartel 2004). The pri-miRNA enter the miRNA-processing pathway and undergo nuclear cleavage by the microprocessor complex in which RNase III endonucleases Drosha and DGCR8 produce an approximately 70-nt-long intermediate precursor-miRNA (pre-miRNA) whose hallmark is a stem-loop-like structure and a staggered cut left by microprocessor cleavage at the stem-loop base (Lee et al. 2003; Han et al. 2006). Exportin-5 recognizes the staggered cut and exports the pre-miRNA to the cytoplasm in a Ran-GTP-dependent manner (Yi et al. 2003; Lund et al. 2004). Once the pre-miRNA is cytoplasmic, Dicer, another RNAase III endonuclease, cleaves both stem arms of the pre-miRNA to generate the mature miRNA duplex (Grishok et al. 2001; Hutvagner et al. 2001; Ketting et al. 2001). A single stem of the resulting ~22-nt duplex is incorporated into the RNA-induced silencing complex (RISC), while the other stem arm is presumably degraded. Which stem-arm is incorporated into RISC is thought to be determined, at least in part, by thermal instability: the stem-arm with the weakest hydrogen bonding at its 5′ end is usually incorporated (Khvorova et al. 2003; Schwarz et al. 2003).

RISC is the ribonucleoprotein effector complex for miRNA-mediated gene expression regulation and consists of Argonaute protein family members and accessory factors such as R2D2, along with a miRNA and targeted mRNA (Hammond et al. 2000; Filipowicz 2005). Regulation of target gene expression by RISC is facilitated by miRNA complementary base pairing to target sequence(s) within the 3′ UTR of target mRNAs by a mechanism that is not yet fully understood (Doench and Sharp 2004; Pillai et al. 2004). Generally in animals, perfect or near-perfect complementary base pairing between RISC-bound miRNA and targeted mRNA results in immediate mRNA cleavage by Argonaute proteins, the catalytically-active components of RISC. However, the vast majority of miRNAs in animal are imperfectly complementary to their targeted mRNAs, which results in suppression of translation and subsequent mRNA degradation (Bartel 2004; Bagga et al. 2005).

Interestingly, miRNA-mediated translational repression and/or mRNA degradation has been connected to discrete cytoplasmic foci called processing bodies (P-bodies) (Liu et al. 2005a,b; Rossi 2005; Sen and Blau 2005; Chan and Slack 2006). P-bodies are sites of programmed mRNA degradation, where enzymes required for RNA turnover and RNA degradation intermediates accumulate. Argonaute proteins also localize to P-bodies, as well as miRNAs and their targeted mRNAs, indicating that miRNAs may mediate translation suppression and mRNA degradation by directing targeted mRNAs to P-bodies. Disruption of P-bodies in cells decreases the efficiency of RNAi, suggesting that they are an important component of the RNAi pathway (Jakymiw et al. 2005). Moreover, release of mRNAs targeted by miRNAs from P-bodies and subsequent re-expression of those mRNAs suggests that P-bodies may also function as mRNA storage centers (Bhattacharyya et al. 2006). Major advances have been made towards understanding the mechanisms underlying the RNAi phenomenon; nevertheless, many aspects of miRNA biogenesis, trafficking of RNAi machinery, RISC assembly, and the mechanisms underlying RISC function await clarification.

2.2 *Identification and Expression of miRNAs*

A variety of experimental approaches have been used to identify miRNAs and study their expression patterns. The cloning and sequencing of small RNAs from size fractioned RNA samples has uncovered many miRNAs that are tissue-specifically expressed (Lagos-Quintana et al. 2002, 2003). Complementing small RNA cloning approaches, bioinformatics screens that searched genomic databases for the characteristic stem-loop structures of precursor miRNAs have predicted the existence of hundreds of mammalian miRNAs (Lim et al. 2003; Bentwich et al. 2005). Other techniques, such as northern blotting, real-time RT-PCR, in situ hybridization, and repressible in vivo reporter transgenes have been adapted to verify such predictions and study the expression patterns of specific miRNAs (Mansfield et al. 2004; Shi and Chiang 2005; Wienholds et al. 2005). Recently, a comprehensive sequencing of over 250 small RNA libraries revealed additional new miRNAs and documented the expression patterns of most miRNAs (Landgraf et al. 2007).

To facilitate the analysis of global miRNA expression, microarray technology has been implemented with great success to quickly analyze the expression of hundreds of miRNA genes simultaneously (Thomson et al. 2004; Baskerville and Bartel 2005; Chen et al. 2006; van Rooij et al. 2006, Tatsuguchi et al. 2007). These types of studies have shown that miRNA expression, like that of protein-coding genes, is highly regulated according to the cell's developmental lineage and stage: whereas some miRNAs are ubiquitously expressed, others are expressed in a cell- and tissue-specific manner (Kloosterman and Plasterk 2006), implying that miRNAs may participate in a variety of biological processes.

2.3 *Prediction and Validation of miRNA Regulatory Targets*

Identifying the targets of specific miRNAs will be the key to our understanding the precise roles of miRNAs. Most animal miRNAs are imperfectly complementary to their target site, which thwarts using simple homology searches to identify animal miRNA target sites. To overcome this obstacle, several computational methods have been developed that incorporate sequence conservation and characteristics of known miRNA targets as criteria to predict new animal miRNA targets (Lewis et al. 2003; John et al. 2004; Kiriakidou et al. 2004; Rajewsky and Socci 2004; Krek et al. 2005; Zhao et al. 2005; Grimson et al. 2007). A major determinant for miRNA targeting is the perfect or near perfect complementary base pairing between the second and eighth nucleotides of the miRNA with its mRNA target site, known as the "seed" region. Other factors also deemed important for miRNA targeting include additional base pairing in the 3′ portion of the miRNA and the degree of local AU nucleotide content flanking the target site (Zhao et al. 2005; Grimson et al. 2007). The positive influence of increased AU content is attributed to a weaker secondary structure in the vicinity of the target site thus offering increased accessibility to RISC (Zhao et al. 2005; Grimson et al. 2007). Computational approaches taking these determinants into account, as well as sequence conservation of the target sites, have successfully predicted mammalian miRNA target sites, albeit the set of predictions produced for any particular miRNA almost certainly contains many false positives. Any prediction must be verified experimentally and, most importantly, placed into a relevant biological context before being considered a valid target. Given the vast number of known miRNAs and their potentially thousands of regulatory targets, it is hoped that a direct and facile method to identify miRNA target genes, possibly employing a proteomics-based strategy or from functional screening of cDNA libraries composed of 3′ UTRs of regulatory target genes, will become available.

3 miRNAs in Cardiovascular Development

The heart is the first organ to form and function during mammalian development (Srivastava and Olson 2000; Olson 2006). Cardiac progenitor cells are derived from the embryonic mesoderm, and they undergo cellular proliferation, differentiation and migration to eventually generate a mature, functional four-chambered heart. The molecular mechanisms that control cardiogenesis are not completely understood. Heart development and pathology are intimately linked to the regulation of complex genetic pathways, and much effort has been expended in attempt to understand these pathways. Most studies have focused on the role of transcription factors and regulatory enhancer sequences required for cardiac gene transcription. The regulation of cardiac gene expression has proved quite complex, with individual cardiac genes being controlled by multiple independent enhancers that direct very restricted expression patterns in the heart. Recent studies of miRNAs have reshaped our view of how cardiac gene expression is regulated by increasing this complexity even further by adding another layer of regulation at the post-transcriptional level.

3.1 Global Analysis of miRNA Function During Animal Development

A popular approach to determine the necessity of miRNAs in animal development has been to genetically ablate Dicer function, an endonuclease required for miRNA biogenesis. Dicer is encoded by a single locus in vertebrates and is required to fully process all vertebrate miRNAs to their mature, active form (Hutvagner et al. 2001; Ketting et al. 2001; Bernstein et al. 2003; Wienholds et al. 2003). In mice, abolishing Dicer function resulted in lethality by embryonic day 7.5 (Bernstein et al. 2003). The *Dicer* null mice did not express primitive streak marker *T(brachyury)*, indicating that development was likely arrested before the body plan was configured during gastrulation. Even though Dicer is required for RNAi (Hutvagner et al. 2001; Ketting et al. 2001), it is generally believed that the Dicer loss-of-function phenotype reflects the essential function of miRNAs since little RNAi activity is thought to be involved during embryogenesis. In zebrafish, zygotic *Dicer* mutants lived for almost 2 weeks before dying (Wienholds et al. 2003). Their prolonged survival was attributed to maternally contributed Dicer protein, prompting the creation of maternal-zygotic *Dicer* mutants, which revealed that loss of all miRNAs results in morphogenesis defects during gastrulation and early lethality. Somitogenesis, heart development, and brain formation all proved abnormal in the maternal-zygotic zebrafish *Dicer* mutants (Giraldez et al. 2005). These animal studies of global miRNA function by *Dicer* deletion indicate that miRNAs are important for early development. However, they were unable to address the roles of miRNAs during later development.

To circumvent the early embryonic lethality of the *Dicer* null mouse mutants and study the role of miRNAs in later development, Harfe et al. created a conditional *Dicer* knockout model to study the effect of tissue-specific *Dicer* deletion (Harfe et al. 2005). In this model, *loxP* sites were inserted around an exon encoding the majority of an RNAseIII domain required for miRNA processing. Upon mating the floxed *Dicer* mice to mice expressing Cre recombinase in specific tissues, the floxed exon is excised and miRNA processing inhibited in those tissues (Harfe et al. 2005; Andl et al. 2006; Harris et al. 2006; Yi et al. 2006). The Cre-inducible conditional *Dicer* model has revealed that Dicer function is essential for proper morphogenesis of the vertebrate limb, as well as the skin and lung (Harfe et al. 2005; Andl et al. 2006; Harris et al. 2006; Yi et al. 2006). In addition, Dicer was recently shown to play a critical role in female germline development (Murchison et al. 2007; Tang et al. 2007).

Recently, Zhao and colleagues applied the conditional *Dicer* mouse approach to examine the role of miRNAs in the developing heart (Zhao et al. 2007). Using mice expressing Cre recombinase controlled by the *Nkx2.5* promoter region that directs expression in cardiac progenitors as early as embryonic day (E) 8.5, they report that deletion of *Dicer* specifically in the heart results in death by E 12.5. The *Dicer* mutant mice suffered from cardiac failure and exhibited a range of developmental defects during morphogenesis, including pericardial edema and poorly developed ventricular myocardium. The early lethality in the *Dicer* mutant indicates that miRNAs are required for proper heart development. However, deletion of *Dicer* occurred after gastrulation, leaving questions about the importance of miRNAs in cardiac specification and patterning. Meanwhile, because cardiac-specific *Dicer* mutant mice die during embryogenesis, it is not clear what are the roles of miRNAs in the postnatal and adult hearts.

Collectively, the *Dicer* deletion studies demonstrate that miRNAs play an essential role in animal development. However, since *Dicer* deletion inhibits the processing of all miRNAs, those studies have not provided insight into the precise functions of specific miRNAs. The tissue-specific expression patterns of certain miRNAs, including miRNAs expressed solely in muscle tissues, allude to specific biological roles in those tissues.

3.2 *Regulation of Muscle miRNA Expression*

Several microRNA genes are specifically expressed or highly enriched in skeletal and/or cardiac muscle, so-called "muscle miRNAs." These include miR-1, miR-133, miR-206, and miR-208. Thus far, the expression of muscle miRNAs appears largely regulated by well-established muscle transcriptional networks involving SRF, MyoD, Twist, MEF2, and myocardin (Sokol and Ambros 2005; Zhao et al. 2005; Chen et al. 2006; Rao et al. 2006). For example, miR-1 and miR-133 are expressed solely in skeletal and cardiac muscle tissues by two genes that encode polycistronic transcripts for miR-1 and miR-133. That is to say, miR-1-1 and miR-133a-2 are clustered on mouse chromosome 2, while miR-1-2 and miR-133a-1 are clustered on chromosome 18 (Chen et al. 2006). Promoter analyses demonstrate that both mouse chromosome 2 and chromosome 18 clusters contain upstream enhancers with SRF binding sites, and that myocardin activity increases the expression of those promoters (Zhao et al. 2005; Chen et al. 2006). miR-1 was highly conserved during evolution and, in addition to mouse and human, it is found in the genomes of organisms as diverse as worm, fly, zebrafish, and chicken. The pathways controlling miR-1 expression also appear highly conserved: *Drosophila* miR-1 expression in the presumptive and early mesoderm occurs downstream of Twist and MEF2, two transcription factors that are major regulators of mammalian muscle development (Sokol and Ambros 2005; Zhao et al. 2005).

More than 127 human miRNAs have been identified within the introns of protein-coding genes, and findings support the idea that these intronic miRNAs are generally co-expressed with their host genes (Rodriguez et al. 2004; Baskerville and Bartel 2005; Kim and Kim 2007; van Rooij et al. 2007). Whereas several muscle miRNAs are expressed in both skeletal and cardiac muscle tissues, miR-208 is a heart-specific miRNA that is encoded by an intron of the cardiac muscle *α-myosin heavy chain* (*α-MHC*) gene, whose transcription is largely controlled by thyroid hormone signaling (Lompre et al. 1984; Lagos-Quintana et al. 2003; van Rooij et al. 2007). The expression of miR-208 mirrors the expression pattern of its host gene *α-MHC* during development, suggesting that miR-208 and α-MHC expression are regulated by a common element (van Rooij et al. 2007).

These studies suggest the tissue-specific expression of muscle miRNAs is largely regulated at the transcriptional level. They also suggest that tight temporal and spatial regulation of miRNA expression is important for their function. Interestingly, regulated pre-miRNA processing has been proposed as an alternative mechanism to post-transcriptional control miRNA function (Obernosterer et al. 2006; Thomson et al. 2006). Future studies on both the transcriptional and post-transcriptional mechanisms controlling miRNA expression are needed to provide greater insight into the role of miRNA-mediated regulation in cardiac genetic networks important for human disease. Providing insight into the role of miRNAs during heart development, genetic studies of individual miRNAs expressed specifically in muscle tissues have been undertaken.

3.3 Genetic Studies of Specific miRNAs Reveal Roles in Cardiac Development and Function

Conditional deletion of *Dicer* in the heart revealed a requirement for miRNAs during heart development by downregulating the expression of potentially hundreds of miRNAs. To begin assigning biological roles to individual miRNAs, several groups have undertook gene deletion studies of muscle-specific miRNAs and identified distinct roles for miRNAs in cardiac development and function (Fig. 1)

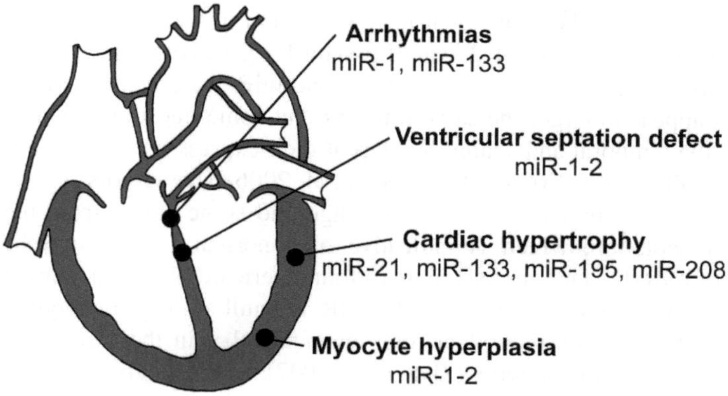

Fig. 1 Roles of miRNAs in heart development and function. The several miRNAs shown are associated with various cardiac abnormalities affecting heart development and function. Arrhythmias relate to electrophysiological defects that result in an irregular heartbeat. Defects in ventricular septation are characterized by an abnormal opening between the cardiac ventricles that allows blood to pass directly from the left to the right ventricle, thereby reducing the efficiency of cardiac output. Cardiac hypertrophy is the primary response of the heart to increased cardiac load and involves increased size of ventricular myocytes. In contrast, the number of myocytes increases during hyperplasia

In *Drosophila,* miR-1 is encoded by a single gene located on chromosome 2. Gene deletion of *Drosophila* miR-1 revealed miR-1 as an essential gene required for viability (Kwon et al. 2005; Sokol and Ambros 2005). Without miR-1, homozygous mutant larvae exhibited decreased locomotion that ultimately progressed to death accompanied by severe gross disruption of the larval musculature (Sokol and Ambros 2005). A subset of severely affected miR-1 null embryos exhibited an enlarged pool of cardiac progenitors, suggesting that miR-1 may modulate differentiation of heart (Kwon et al. 2005). Re-introduction of miR-1 into developing muscle partially rescued the mutant phenotype, strongly supporting a muscle-specific role for miR-1 (Kwon et al. 2005; Sokol and Ambros 2005).

Consistent with the role for miR-1 in muscle differentiation, over-expression of miR-1 in the developing mouse heart resulted in reduced ventricular myocyte expansion and decreased the number of proliferating myocytes (Zhao et al. 2005). This phenotype was explained in part by the presence of a miR-1 target site in the 3′ UTR of the Hand2 cardiac transcription factor (Zhao et al. 2005), whose genetic ablation in the mouse produced a similar failure in ventricular myocyte expansion (Srivastava et al. 1997). In agreement with this notion, over-expression of miR-1 in the mouse heart decreased Hand2 protein levels (Zhao et al. 2005). Similarly, introduction of miR-1 into *Xenopus* embryos interfered with heart development (Chen et al. 2006).

Further extending the analysis of miR-1 function in vertebrates, Zhao and colleagues targeted mouse miR-1-2, one of two miR-1 genes expressed in skeletal and cardiac muscle (Zhao et al. 2007). miR-1-1 and miR-1-2 are encoded by separate genes and appear to target the same mRNAs, but some questions seem to remain regarding the temporal and spatial overlap of their expression patterns (Wienholds et al. 2005; Zhao et al. 2005, 2007; Chen et al. 2006).The authors report that half of miR-1-2 null animals die by weaning age and some suffer from incomplete ventricular septation, which is indicative of abnormal cardiac morphogenesis. Analysis of miR-1-2 null animals in utero found pericardial edema, consistent with embryonic myocardial dysfunction. The miR-1-2 null animal phenotype indicates that miR-1-2 plays non-redundant roles with miR-1-1 in the heart despite their overlapping expression patterns (Zhao et al. 2007). Interestingly, miR-1-2 deletion did not appear to affect skeletal muscle development by gross morphological analysis. It is intriguing to speculate that the different requirement for miR-1 in cardiac versus skeletal muscle development reflects distinct target genes affected in those two muscle tissues.

miR-208 is expressed specifically in the heart and was recently deleted from the mouse genome by van Rooij et al. (2007). miR-208 null animals were viable and appeared normal without any apparent gross developmental defects. However, the miR-208 null animals exhibited a slight reduction in contractility at 2 months of age and a continued reduction in cardiac function in later life. Further analyses revealed a requirement for miR-208 in the cardiac hypertrophic growth response. Although miR-208 does not appear to be necessary for cardiogenesis, the physiological defects exhibited in the absence of miR-208 during adulthood indicate that miR-208 plays distinct roles in the adult heart. Collectively, compelling genetic evidences have established that miRNAs are indeed an emerging class of molecules important for cardiac muscle development (Table 1).

4 miRNAs in the Hypertrophic Growth Response

Cardiac myocytes proliferate rapidly during embryogenesis, but adult cardiac myocytes lose their proliferative capacity, and they respond to mechanical and pathological stimuli by hypertrophic growth, defined by an increase in myocyte size and/or myofibrillar volume without a change in myocyte number (Frey and Olson 2003). Cardiac hypertrophy is accompanied by the activation of a set of fetal cardiac genes that are normally expressed in the heart only before birth. The reactivation of cardiac fetal genes in post-natal cardiomyocytes in response to hypertrophic signals suggests that the molecular events that control cardiac gene expression during development may be redeployed to regulate hypertrophic cardiac growth or heart regeneration (Heineke and Molkentin 2006). Recent studies indicate that a change in the expression of miRNAs also accompanies cardiac hypertrophy and also indicate that miRNAs are novel regulators of the hypertrophic growth response.

Table 1 microRNAs implicated in the cardiac hypertrophic growth response

microRNA	Expression pattern	Expression change during hypertrophy	Validated targets	References
miR-1	Heart, skeletal muscle	Downregulated	Cdk9, Delta, Fibronectin, GJA1, Hand2, Irx5, KCNJ2, HDAC4, KCNE1, RasGAP, Rheb	Kwon et al. (2005), Zhao et al. (2005), Chen et al. (2006), Rao et al. (2006), Luo et al. (2007), Yang et al. (2007), Zhao et al. (2007)
miR-21	Heart, spleen, small intestine, colon	Upregulated	PTEN, TPM1	van Rooij et al. (2006), Cheng et al. (2007), Meng et al. (2007), Tatsuguchi et al. (2007), Zhu et al. (2007)
miR-133	Heart, skeletal muscle	Downregulated	Cdc42, ERG, KCNQ1, WHSC2, SRF, RhoA	Chen et al. (2006), Rao et al. (2006), Care et al. (2007), Luo et al. (2007), Xiao et al. (2007)
miR-195	Heart, lung, kidney, skin	Upregulated	None reported.	Lagos-Quintana et al. (2003), van Rooij et al. (2006)
miR-208	Heart	Not reported	Thrap1	van Rooij et al. (2007)

Cdc42 Cell division cycle 42; *Cdk9* cyclin-dependent kinase 9; *ERG* ether-a-go-go postassium channel; *GJA1* gap junction protein alpha 1; *Hand2* heart and neural crest derivatives expressed 2; *HDAC4* histone deacetylase 4; *Irx5* iroquois homeobox protein; *KCNE1* potassium voltage-gated channel; Isk-related family, member 1; *KCNJ2* potassium inwardly-rectifying channel, subfamily J, member 2; *KCNQ1* potassium voltage-gated channel, KQT-like subfamily, member 1; *PTEN* phosphatase and tensin homolog; *RasGAP* Ras GTPase-activating protein; *Rheb* Ras homolog enriched in brain; *RhoA* Ras homolog A; *SRF* serum response factor; *Thrap1* thyroid hormone receptor associated protein 1; *TPM1* tropomyosin 1; *WHSC2* Wolf-Hirschhorn syndrome candidate 2

4.1 miRNA Expression is Dynamically Regulated in Response to Cardiac Hypertrophy

Using miRNA microarrays, the global miRNA expression profile is regulated during cardiac hypertrophy in both in vitro and in vivo model systems, suggesting that miRNAs are involved in this disease process (van Rooij et al. 2006; Cheng et al. 2007; Tatsuguchi et al. 2007). Olson and colleagues are among the first to report the change of miRNA expression in thoracic aortic-banded hearts (TAB) as well as the calcineurin over-expression transgenic mice, two animal models of pathological cardiac hypertrophy (van Rooij et al. 2006). In their report, the expression of only a relatively small fraction of miRNAs were changed in response to cardiac hypertrophy (van Rooij et al. 2006), which is consistent with similar reports from other groups (Ji et al. 2007; Tatsuguchi et al. 2007). In contrast, another report documented differential expression of over 100 miRNAs during cardiac hypertrophy (Cheng et al. 2007). The cause of those differences is not known. Additional experiments, in particular those using alternative approaches, such as northern blotting and real-time RT-PCR, seem necessary to verify some of the prior results. Regardless of those differences, an emerging picture has convincingly demonstrated that miRNAs are a new class of previously unrecognized regulators for cardiac hypertrophy. Furthermore, dysregulated miRNA expression has been shown in human patients with failing hearts (van Rooij et al. 2006; Yang et al. 2007). One of the surprising observations from those miRNA profiling studies is that there were more miRNAs found upregulated than downregulated in response to cardiac hypertrophy. It is not clear what this means, although one intriguing explanation is that the translation/expression of many proteins could be inhibited during cardiac hypertrophy, given that miRNAs are repressors of gene expression. Functional analyses using both gain- and loss-of-function approaches have begun to establish a correlation between miRNAs and cardiac hypertrophy by demonstrating that stress-regulated miRNAs can both positively and negatively influence the cardiac hypertrophic growth response (Table 1) (van Rooij et al. 2006, 2007; Care et al. 2007; Cheng et al. 2007; Sayed et al. 2007; Tatsuguchi et al. 2007).

4.2 Genetic Studies Reveal that miRNAs Can Modulate Pathological Hypertrophy

Muscle-specifically expressed miR-1 and miR-133 play a critical role in skeletal muscle proliferation and differentiation (Chen et al. 2006). miR-1 was further shown as essential for heart development (Zhao et al. 2007). Interestingly, expression of miR-1 and miR-133 are both downregulated during cardiac hypertrophy. Downregulation of these two miRNAs is proposed to be necessary for the expression of targeted growth-related genes and induction of hypertrophic growth (Care et al. 2007; Sayed et al. 2007). In support, ectopic expression of miR-1 or miR-133 inhibited target gene expression and the hypertrophic growth response in a tissue-culture model of cardiac hypertrophy (Care et al. 2007; Sayed et al. 2007). Conversely, blocking endogenous miR-133 function in isolated cardiomyocytes augmented agonist-induced hypertrophy (Care et al. 2007). Furthermore, prolonged inhibition of miR-133 in vivo using chemically-modified oligonucleotides antisense to miR-133, delivered by an osmotic minipump implanted into the heart, caused a marked hypertrophic response (Care et al. 2007). It should be pointed out that while the expression of miR-1 appears dysregulated in cardiac hypertrophy (Sayed et al. 2007), there is not yet direct genetic evidence supporting a role for this miRNA in the regulation of hypertrophy. Instead, miR-1 was found to have arrhythmogenic potential when overexpressed in adult rat hearts (Yang et al. 2007), suggesting that miR-1 may play an essential role in cardiac electrophysiology, in addition to its role in heart development.

Whereas miR-1 and -133 are downregulated during cardiac hypertrophy, miR-195 is upregulated and was found sufficient to induce hypertrophic growth in cultured cardiomyocytes as well as in transgenic mice (van Rooij et al. 2006). In contrast, transgenic mice over-expressing miR-214, a miRNA also upregulated during hypertrophy, caused no detectable phenotypic effect in the heart (van Rooij et al. 2006). Those studies indicate that some miRNAs, but not others, are sufficient to induce cardiac hypertrophy. It will be interesting to investigate whether they are necessary for the hypertrophic response using a loss-of-function approach. Several growth-related genes have been identified as specific mRNA targets for miR-1, miR-133, and miR-195 (Care et al. 2007; Meng et al. 2007; Sayed et al. 2007; Zhu et al. 2007), but how those miRNAs integrate into relevant genetic pathways to modulate the hypertrophic response is unclear.

Although genetic ablation did not identify a critical role for miR-208 in the developing mouse, a striking postnatal role for miR-208 was revealed (van Rooij et al. 2007). Loss of miR-208 protects mice against cardiac hypertrophy and upregulation of β-MHC induced by hypothyroidism, activated calcineuron signaling and cardiac pressure-overload-induced stress (van Rooij et al. 2007). These results suggest that the genetic pathways coordinating cardiac hypertrophy share a common component regulated by miR-208. One candidate proposed is Thyroid hormone receptor associated protein 1 (Thrap1), a co-factor of the thyroid hormone nuclear receptor, which can positively and negatively influence transcription. Expression of Thrap1 mRNA is targeted by miR-208, and Thrap1 protein levels are elevated in miR-208 mutant hearts, suggesting that miR-208 may function, at least in part, by regulating the expression of a thyroid hormone signaling pathway component (van Rooij et al. 2007).

miR-21, a miRNA implicated in tumor-related cell growth and apoptosis (Chan et al. 2005; Cheng et al. 2005; Si et al. 2006), is upregulated in response to agonist-induced cardiac hypertrophy in cell culture experiments and in pressure-overload-induced hypertrophy in vivo (van Rooij et al. 2006; Cheng et al. 2007; Sayed et al. 2007; Tatsuguchi et al. 2007). Inhibition of miR-21 by transfection of locked nucleic acid (LNA)-modified antisense oligionucleotides induced hypertrophic growth in rat neonatal cardiomyocytes, as assessed by cardiomyocyte cell size and expression of hypertrophic markers such as ANF, α-actinin, and skeletal muscle α-actin (Tatsuguchi et al. 2007). A role for miR-21 in hypertrophy was further supported by a gain-of-function approach where ectopic miR-21 expression slightly decreased cardiomyocyte size and reduced expression of hypertrophic marker genes (Tatsuguchi et al. 2007). In contrast, the inhibition of miR-21 by transfection of 2′*O*-Methyl antisense oligonucleotides attenuated global protein synthesis and cell size in an agonist-induced model of hypertrophy in isolated cardiomyocytes (Cheng et al. 2007). Interestingly, other reports on miR-21 function also appear contradictory: while one study documented that miR-21 inhibition provoked cell growth in HeLa cells (Cheng et al. 2005), others showed that miR-21 inhibition led to activation of apoptosis and decreased cell proliferation (Chan et al. 2005; Si et al. 2006). Clearly, the identification of the miR-21 regulatory targets and further analysis of the molecular pathways modulated by miR-21 in different biological systems are needed to better understand the biological function of this miRNA.

Collectively, these studies support miR-1, miR-21, miR-133, miR-195, and miR-208 as important and newly identified regulators of cardiac hypertrophy. In addition, identification of the hypertrophic miRNA expression signature has yielded many hitherto unrecognized candidate genes involved in cardiac hypertrophy, and those candidates are awaiting closer examination. Given the complexity of the cardiac remodeling occurring during hypertrophy, the identification of specific targets of miRNAs involved in the hypertrophic response will provide more insight into the molecular mechanisms underlying this disease process.

5 Conclusions and Future Directions

It is becoming more and more evident that miRNAs are important regulators of normal and pathological processes, including cardiac hypertrophy. These findings that miRNAs participate in heart formation and function offers a new paradigm for the post-transcriptional regulation of gene expression in cardiovascular biology. However, we are still left to answer many basic questions regarding miRNAs: how many miRNAs exist, where are they expressed, and which genes do they target? Are many miRNAs important for cardiac function or just a handful of them? And how are those miRNAs integrated into the complex genetic networks that regulate specific processes in the heart? Understanding the role of miRNAs in the heart will provide more insight into the molecular basis of cardiovascular disease, and it is very intriguing to speculate that miRNAs might one day be exploited as therapeutic targets.

References

Ambros V, Chen X (2007) The regulation of genes and genomes by small RNAs. Development 134:1635–1641

Andl T, Murchison EP, Liu F et al (2006) The miRNA-processing enzyme dicer is essential for the morphogenesis and maintenance of hair follicles. Curr Biol 16:1041–1049

Bagga S, Bracht J, Hunter S et al (2005) Regulation by let-7 and lin-4 miRNAs results in target mRNA degradation. Cell 122:553–563

Bartel DP (2004) MicroRNAs: genomics, biogenesis, mechanism, and function. Cell 116:281–297

Baskerville S, Bartel DP (2005) Microarray profiling of microRNAs reveals frequent coexpression with neighboring miRNAs and host genes. RNA 11:241–247

Bentwich I, Avniel A, Karov Y et al (2005) Identification of hundreds of conserved and nonconserved human microRNAs. Nat Genet 37:766–770

Bernstein E, Kim SY, Carmell MA et al (2003) Dicer is essential for mouse development. Nat Genet 35:215–217

Bhattacharyya SN, Habermacher R, Martine U et al (2006) Stress-induced reversal of microRNA repression and mRNA P-body localization in human cells. Cold Spring Harb Symp Quant Biol 71:513–521

Bracht J, Hunter S, Eachus R et al (2004) Trans-splicing and polyadenylation of let-7 microRNA primary transcripts. RNA 10:1586–1594

Care A, Catalucci D, Felicetti F et al (2007) MicroRNA-133 controls cardiac hypertrophy. Nat Med 13:613–618

Chan SP, Slack FJ (2006) microRNA-mediated silencing inside P-bodies. RNA Biol 3:97–100

Chan JA, Krichevsky AM, Kosik KS (2005) MicroRNA-21 is an antiapoptotic factor in human glioblastoma cells. Cancer Res 65:6029–6033

Chen JF, Mandel EM, Thomson JM, Wu Q, Callis TE, Hammond SM, Conlon FL, Wang DZ (2006) The role of microRNA-1 and microRNA-133 in skeletal muscle proliferation and differentiation. Nat Genet 38:228–233

Cheng AM, Byrom MW, Shelton J et al (2005) Antisense inhibition of human miRNAs and indications for an involvement of miRNA in cell growth and apoptosis. Nucleic Acids Res 33:1290–1297

Cheng Y, Ji R, Yue J et al (2007) MicroRNAs are aberrantly expressed in hypertrophic heart: do they play a role in cardiac hypertrophy? Am J Pathol 170:1831–1840

Chien KR (2007) Molecular medicine: microRNAs and the tell-tale heart. Nature 447:389–390

Denli AM, Tops BB, Plasterk RH et al (2004) Processing of primary microRNAs by the Microprocessor complex. Nature 432:231–235

Doench JG, Sharp PA (2004) Specificity of microRNA target selection in translational repression. Genes Dev 18:504–511

Filipowicz W (2005) RNAi: the nuts and bolts of the RISC machine. Cell 122:17–20

Frey N, Olson EN (2003) Cardiac hypertrophy: the good, the bad, and the ugly. Annu Rev Physiol 65:45–79

Giraldez AJ, Cinalli RM, Glasner ME et al (2005) MicroRNAs regulate brain morphogenesis in zebrafish. Science 308:833–838

Gregory RI, Yan KP, Amuthan G et al (2004) The Microprocessor complex mediates the genesis of microRNAs. Nature 432:235–240

Griffiths-Jones S (2004) The microRNA Registry. Nucleic Acids Res 32:D109–D111

Grimson A, Farh KK, Johnston WK et al (2007) MicroRNA targeting specificity in mammals: determinants beyond seed pairing. Mol Cell 27:91–105

Grishok A, Pasquinelli AE, Conte D et al (2001) Genes and mechanisms related to RNA interference regulate expression of the small temporal RNAs that control C. elegans developmental timing. Cell 106:23–34

Hammond SM, Bernstein E, Beach D et al (2000) An RNA-directed nuclease mediates post-transcriptional gene silencing in Drosophila cells. Nature 404:293–296

Han J, Lee Y, Yeom KH et al (2006) Molecular basis for the recognition of primary microRNAs by the Drosha-DGCR8 complex. Cell 125:887–901

Harfe BD, McManus MT, Mansfield JH et al (2005) The RNaseIII enzyme Dicer is required for morphogenesis but not patterning of the vertebrate limb. Proc Natl Acad Sci USA 102:10898–10903

Harris KS, Zhang Z, McManus MT et al (2006) Dicer function is essential for lung epithelium morphogenesis. Proc Natl Acad Sci USA 103:2208–2213

Heineke J, Molkentin JD (2006) Regulation of cardiac hypertrophy by intracellular signalling pathways. Nat Rev Mol Cell Biol 7:589–600

Hunter JJ, Chien KR (1999) Signaling pathways for cardiac hypertrophy and failure. N Engl J Med 341:1276–1283

Hutvagner G, McLachlan J, Pasquinelli AE et al (2001) A cellular function for the RNA-interference enzyme Dicer in the maturation of the let-7 small temporal RNA. Science 293:834–838

Jakymiw A, Lian S, Eystathioy T et al (2005) Disruption of GW bodies impairs mammalian RNA interference. Nat Cell Biol 7:1267–1274

Ji R, Cheng Y, Yue J, Yang J et al (2007) MicroRNA expression signature and antisense-mediated depletion reveal an essential role of MicroRNA in vascular neointimal lesion formation. Circ Res 100:1579–1588

John B, Enright AJ, Aravin A et al (2004) Human MicroRNA targets. PLoS Biol 2:e363

Ketting RF, Fischer SE, Bernstein E et al (2001) Dicer functions in RNA interference and in synthesis of small RNA involved in developmental timing in C. elegans. Genes Dev 15:2654–2659

Khvorova A, Reynolds A, Jayasena SD (2003) Functional siRNAs and miRNAs exhibit strand bias. Cell 115:209–216

Kim YK, Kim VN (2007) Processing of intronic microRNAs. Embo J 26:775–783

Kiriakidou M, Nelson PT, Kouranov A et al (2004) A combined computational-experimental approach predicts human microRNA targets. Genes Dev 18:1165–1178

Kloosterman WP, Plasterk RH (2006) The diverse functions of microRNAs in animal development and disease. Dev Cell 11:441–450

Krek A, Grun D, Poy MN et al (2005) Combinatorial microRNA target predictions. Nat Genet 37:495–500

Kwon C, Han Z, Olson EN et al (2005) MicroRNA1 influences cardiac differentiation in Drosophila and regulates Notch signaling. Proc Natl Acad Sci USA 102:18986–18991

Lagos-Quintana M, Rauhut R, Yalcin A et al (2002) Identification of tissue-specific microRNAs from mouse. Curr Biol 12:735–739

Lagos-Quintana M, Rauhut R, Meyer J et al (2003) New microRNAs from mouse and human. RNA 9:175–179

Landgraf P, Rusu M, Sheridan R et al (2007) A mammalian microRNA expression atlas based on small RNA library sequencing. Cell 129:1401–1414

Lee RC, Feinbaum RL, Ambros V (1993) The C. elegans heterochronic gene lin-4 encodes small RNAs with antisense complementarity to lin-14. Cell 75:843–854

Lee Y, Ahn C, Han J et al (2003) The nuclear RNase III Drosha initiates microRNA processing. Nature 425:415–419

Lee Y, Kim M, Han J et al (2004) MicroRNA genes are transcribed by RNA polymerase II. EMBO J 23:4051–4060

Lewis BP, Shih IH, Jones-Rhoades MW et al (2003) Prediction of mammalian microRNA targets. Cell 115:787–798

Lim LP, Glasner ME, Yekta S et al (2003) Vertebrate microRNA genes. Science 299:1540

Liu J, Rivas FV, Wohlschlegel J et al (2005a) A role for the P-body component GW182 in microRNA function. Nat Cell Biol 7:1261–1266

Liu J, Valencia-Sanchez MA, Hannon GJ et al (2005b) MicroRNA-dependent localization of targeted mRNAs to mammalian P-bodies. Nat Cell Biol 7:719–723

Lompre AM, Nadal-Ginard B, Mahdavi V (1984) Expression of the cardiac ventricular alpha- and beta-myosin heavy chain genes is developmentally and hormonally regulated. J Biol Chem 259:6437–6446

Lund E, Guttinger S, Calado A et al (2004) Nuclear export of microRNA precursors. Science 303:95–98

Luo X, Xiao J, Lin H et al (2007) Transcriptional activation by stimulating protein 1 and post-transcriptional repression by muscle-specific microRNAs of I(Ks)-encoding genes and potential implications in regional heterogeneity of their expressions. J Cell Physiol 212:358–367

Mann DL (2007) MicroRNAs and the failing heart. N Engl J Med 356:2644–2645

Mansfield JH, Harfe BD, Nissen R et al (2004) MicroRNA-responsive 'sensor' transgenes uncover Hox-like and other developmentally regulated patterns of vertebrate microRNA expression. Nat Genet 36:1079–1083

Meister G, Tuschl T (2004) Mechanisms of gene silencing by double-stranded RNA. Nature 431:343–349

Meng F, Henson R, Wehbe-Janek H et al (2007) MicroRNA-21 Regulates expression of the PTEN tumor suppressor gene in human hepatocellular cancer. Gastroenterology 133:647–658
Murchison EP, Stein P, Xuan Z et al (2007) Critical roles for Dicer in the female germline. Genes Dev 21:682–693
Obernosterer G, Leuschner PJ, Alenius M et al (2006) Post-transcriptional regulation of microRNA expression. RNA 12:1161–1167
Olson EN (2006) Gene regulatory networks in the evolution and development of the heart. Science 313:1922–1927
Pillai RS, Artus CG, Filipowicz W (2004) Tethering of human Ago proteins to mRNA mimics the miRNA-mediated repression of protein synthesis. RNA 10:1518–1525
Rajewsky N, Socci ND (2004) Computational identification of microRNA targets. Dev Biol 267:529–535
Rao PK, Kumar RM, Farkhondeh M et al (2006) Myogenic factors that regulate expression of muscle-specific microRNAs. Proc Natl Acad Sci USA 103:8721–8726
Rodriguez A, Griffiths-Jones S, Ashurst JL et al (2004) Identification of mammalian microRNA host genes and transcription units. Genome Res 14:1902–1910
Rossi JJ (2005) RNAi and the P-body connection. Nat Cell Biol 7:643–644
Sayed D, Hong C, Chen IY et al (2007) MicroRNAs play an essential role in the development of cardiac hypertrophy. Circ Res 100:416–424
Schwarz DS, Hutvagner G, Du T et al (2003) Asymmetry in the assembly of the RNAi enzyme complex. Cell 115:199–208
Sen GL, Blau HM (2005) Argonaute 2/RISC resides in sites of mammalian mRNA decay known as cytoplasmic bodies. Nat Cell Biol 7:633–636
Shi R, Chiang VL (2005) Facile means for quantifying microRNA expression by real-time PCR. Biotechniques 39:519–525
Si ML, Zhu S, Wu H et al (2006) miR-21-mediated tumor growth. Oncogene 26:2799–2803
Sokol NS, Ambros V (2005) Mesodermally expressed Drosophila microRNA-1 is regulated by Twist and is required in muscles during larval growth. Genes Dev 19:2343–2354
Srivastava D, Olson EN (2000) A genetic blueprint for cardiac development. Nature 407:221–226
Srivastava D, Thomas T, Lin Q et al (1997) Regulation of cardiac mesodermal and neural crest development by the bHLH transcription factor, dHAND. Nat Genet 16:154–160
Tang F, Kaneda M, O'Carroll D et al (2007) Maternal microRNAs are essential for mouse zygotic development. Genes Dev 21:644–648
Tatsuguchi M, Seok HY, Callis TE et al (2007) Expression of microRNAs is dynamically regulated during cardiomyocyte hypertrophy. J Mol Cell Cardiol 42:1137–1141
Thomson JM, Parker J, Perou CM et al (2004) A custom microarray platform for analysis of microRNA gene expression. Nat Methods 1:47–53
Thomson JM, Newman M, Parker JS et al (2006) Extensive post-transcriptional regulation of microRNAs and its implications for cancer. Genes Dev 20:2202–2207
van Rooij E, Sutherland LB, Liu N et al (2006) A signature pattern of stress-responsive microRNAs that can evoke cardiac hypertrophy and heart failure. Proc Natl Acad Sci USA 103:18255–18260
van Rooij E, Sutherland LB, Qi X et al (2007) Control of stress-dependent cardiac growth and gene expression by a microRNA. Science 316:575–579
Wienholds E, Koudijs MJ, van Eeden FJ et al (2003) The microRNA-producing enzyme Dicer1 is essential for zebrafish development. Nat Genet 35:217–218
Wienholds E, Kloosterman WP, Miska E et al (2005) MicroRNA expression in zebrafish embryonic development. Science 309:310–311
Wightman B, Ha I, Ruvkun G (1993) Posttranscriptional regulation of the heterochronic gene lin-14 by lin-4 mediates temporal pattern formation in C. elegans. Cell 75:855–862
Xiao J, Luo X, Lin H et al (2007) MicroRNA miR-133 represses HERG K+ channel expression contributing to QT prolongation in diabetic hearts. J Biol Chem 282:12363–12367

Yang B, Lin H, Xiao J et al (2007) The muscle-specific microRNA miR-1 regulates cardiac arrhythmogenic potential by targeting GJA1 and KCNJ2. Nat Med 13:486–491

Yi R, Qin Y, Macara IG et al (2003) Exportin-5 mediates the nuclear export of pre-microRNAs and short hairpin RNAs. Genes Dev 17:3011–3016

Yi R, O'Carroll D, Pasolli HA et al (2006) Morphogenesis in skin is governed by discrete sets of differentially expressed microRNAs. Nat Genet 38:356–362

Zhao Y, Samal E, Srivastava D (2005) Serum response factor regulates a muscle-specific microRNA that targets Hand2 during cardiogenesis. Nature 436:214–220

Zhao Y, Ransom JF, Li A et al (2007) Dysregulation of cardiogenesis, cardiac conduction, and cell cycle in mice lacking miRNA-1-2. Cell 129:303–317

Zhu S, Si ML, Wu H et al (2007) MicroRNA-21 targets the tumor suppressor gene tropomyosin 1 (TPM1). J Biol Chem 282:14328–14336

MicroRNAs and the Control of Heart Pathophysiology

D. Catalucci(✉), M.V.G. Latronico, and G. Condorelli

Contents

Abstract MicroRNAs (miRNAs) are small, endogenous, non-coding RNAs that control gene expression at a posttranscriptional level. They are one of the most abundant class of gene regulatory molecules in multicellular organisms, and are being implicated in an increasing number of biological processes. Many miRNAs are expressed in a tissue-specific manner during development and take part in the regulation of cell-lineage decisions and morphogenesis. Moreover, much work is now highlighting their importance for malformations and disease. This chapter presents knowledge on miRNA biology, and then describes the recent findings regarding aspects of cardiovascular pathophysiology.

1 miRNA: Biogenesis and Mechanisms of Action

MicroRNAs (miRNAs) are 21–26-nucleotide (nt)-long, single-stranded RNAs. They derive from characteristic hairpin-shaped precursors with a structure that consists of two portions: a terminal loop made up of approximately 80 nts and a double-stranded (ds) RNA stem of varying length. This stem portion contains unpaired sequences, forming mismatches and/or bulges along the stem, which disrupt the normal helix structure of the molecule. In mammals, the majority of

D. Catalucci
Division of Cardiology, Department of Medicine, University of California San Diego, 9500 Gilman Drive, BSB 5022, LaJolla, San Diego, Ca 92093-0613, USA and Scientific and Technology Pole, IRCCS MultiMedica, Via G. Fantoli 16/15, 20138 Milan, Italy
e-mail: dcatalucci@ucsd.edu

V.A. Erdmann et al. (eds.) *RNA Technologies in Cardiovascular Medicine and Research*, doi: 10.1007/978-3-540-78709-9,

miRNAs are encoded by sequences of DNA located in introns of either protein-coding or non-coding host genes (Rodriguez et al. 2004). A smaller percentage of miRNA genes are found in exons of protein-coding genes. Many, however, still have an uncertain origin. Apart from singularly expressed miRNAs, a significant number are assembled in clusters in which two or three miRNAs are generated from a common parent mRNA. To complicate matters further, miRNA genes can be found on either the sense or anti-sense DNA strand. Some miRNAs also seem to be independently expressed in that they have their own independent promoters and enhancers (Zhao et al. 2005b).

RNA pol II transcribes a primary transcript (primary miR precursor or pri-miRNA) from the transcriptional units of miR genes (Fig. 1a) (Lee et al. 2004). These genes differ from those of protein-coding ones in that they do not possess canonical TATA boxes and are intronless. Pri-miRNAs are several hundreds or thousands of nts long and have the distinctive stem-loop configuration described above, and, in addition, the 5′ end cap structure and poly-adenylated 3′ tail sequence typical of RNA pol II transcripts. In the nucleus, pri-miRNA is cropped into a shorter 60–70-nt-long transcript, termed pre-miRNA, by the enzyme, Drosha (Lee et al. 2002, 2003). Pre-miRNA is then exported to the cytoplasm by the export

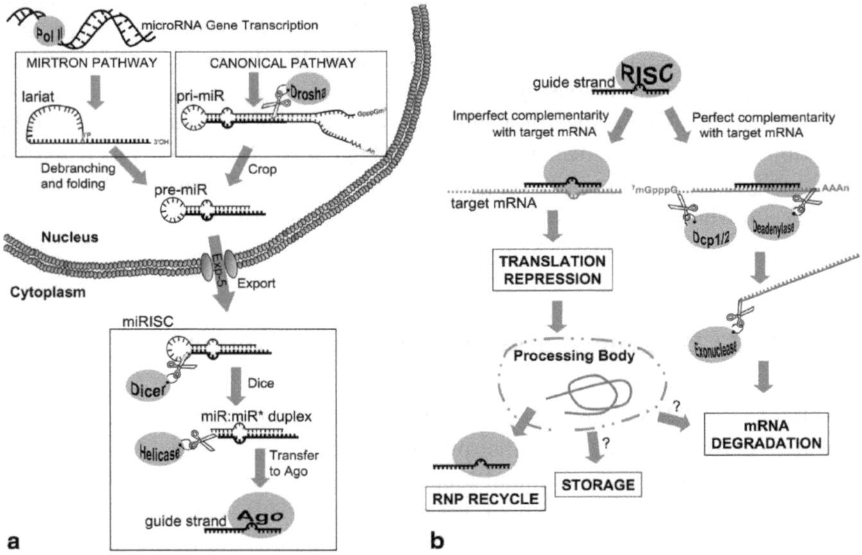

Fig. 1 Biogenesis and mechanisms of action of microRNAs. (**a**) Canonical (*upper right box*) and mirtron (*upper left box*) transcription converge at the pre-miRNA stage. After nuclear export, the microRNA is incorporated and processed in the miRISC (*lower box*). (**b**) When binding of a microRNA with the target mRNA is imperfect, translation seems to be repressed and the mRNA, stripped of ribosomes, is transported into processing bodies. When binding is perfectly complementary, the mRNA is degraded by an exonuclease after deadenylation (deadenylase) and decapping (Dcp1/2). Repressed mRNA can either be stored or degraded, and RISC components recycled

receptor, exportin-5, in a Ran-GTP-dependent process (Kim 2004). In the cytoplasm, the miRNA precursor undergoes a second cleavage step by Dicer, another RNase III of ~200 kDa, which forms a complex with a dsRNA binding domain-containing cofactor, TRBP (HIV-1 transactivation response element RNA-binding protein), to form part of the miRISC (see below). Dicer chops off the terminal loop of the hairpin leaving an 18–22-nt imperfectly paired duplex comprised of a mature miRNA guide strand and a similarly sized passenger strand (or miRNA* fragment) derived from opposite arms of the pre-miRNA. In addition to the above described pathway of miRNA biogenesis, an alternative pathway has been uncovered by Ruby et al. that does not necessitate Drosha processing (Okamura et al. 2007; Ruby et al. 2007). These authors have documented the presence in *Drosophila melanogaster* and *C. elegans* of pre-miRNA-sized genes located within introns that they have termed mirtrons. Mirtrons give rise to pre-miRNAs after splicing of the intron from the host gene, passing through a lariat intermediate (Fig. 1a, upper left box). Lariats are formed when the 5′ end of spliced intronic RNA forms a bond with an adenosine contained within the intron sequence to form a looped structure. After debranching of the lariat and folding, a pre-miRNA is formed which converges with the canonical miRNA biogenesis pathway. The mirtron pathway has been speculated to represent the ancestral pathway of miRNA biogenesis from which the canonical biosynthetic pathway evolved after the emergence of Drosha. There are few pre-miRNA-sized introns within higher animal species, and no mirtrons have been found in mammals to date. However, the presence of mirtrons even in mammals has not been totally excluded.

In order for miRNAs to function, they must first become associated with proteins to form a ribonucleoprotein particle (RNP) called the miRNA-induced silencing complex (miRISC). At least three proteins are needed to create a properly functioning miRISC: Dicer, TRBP and a member of the Argonaute (Ago) family (Gregory et al. 2005). Ago is responsible for the enzymatic 'slicer' activity of the miRISC, while the loaded miRNA is responsible for target recognition. Consequently, after cleavage of the pre-mRNA by Dicer, the miRNA:miRNA* duplex is unwound by a helicase, the passenger strand pealed away and degraded, and the guide strand associated with the miRISC. (Fig. 1a, bottom box) (for review, see Peters and Meister 2007).

As pointed out above, the effects of miRNAs are produced in the cytoplasm of cells through their base pairing with complementary sequences present at the 3′ UTR of target mRNAs. Binding specificity is dictated by only 6–7 nts that contribute disproportionately to this interaction. This sequence, located at the 5′ end of the miRNA molecule, is called the seed and nucleates binding, while the rest of the molecule subsequently 'zips-up', usually with only partial complementarity, to produce characteristic mismatches and bulges especially in the central region and, to a lesser extent, at the 3′ end of the miRNA (Pillai et al. 2005). The degree of complementarity that the segment of the miRNA external to the seed sequence has with the binding site present at the 3′ UTR of the target mRNA is important for the type of posttranslational control the miRNA will produce (Fig. 1b). When binding of this segment is partial, miRNAs are responsible for reduced translation of targeted mRNAs due to reduced efficiency of translation rather than a decrease in mRNA abundance. However, the exact mechanism of action is still not totally clear. In fact, inhibition of initiation (Humphreys et al. 2005; Olsen and Ambros 1999) and elongation (Lee et al. 1993; Maroney et al. 2006; Wightman et al. 1993) have been documented; moreover, degradation of nascent protein has also been suggested to occur (Nottrott et al. 2006; Tang 2005). When miRNAs bind with precise complementarity to target mRNAs, they behave similarly to siRNAs and signal for mRNA degradation (Behm-Ansmant et al. 2006; Chendrimada et al. 2007; Giraldez et al. 2006; Schmitter et al. 2006; Wu et al. 2006; Zhao et al. 2005a). However, miRNA-mediated mRNA degradation occurs not via an siRNA-like mechanism of endonucleolytic cleavage and then degradation of the two produced fragments, but rather through deadenylation followed by decapping and subsequent degradation by an exonuclease.

What factors contribute to target recognition? First, Ago with its ability to bind both ds- and ssRNA makes this protein a suitable candidate for presenting the seed and stabilising pairing (Lingel et al. 2003). The presentation of the nucleotides of the seed prearranged in the geometry of an A-form helix enhances the affinity with the matched mRNA segment. Secondly, it seems that target specificity is determined not only by sequence matching, but also at least by target accessibility. The study of validated targets has evidenced that miRNAs preferentially target UTR sites that do not have a complex secondary structure and that are located in accessible regions of the RNA based on favourable thermodynamics (Zhao et al. 2005b). In fact, it has been shown that areas rich in AUs are found in the vicinity of genuine target sites and that these areas are associated with destabilised mRNA secondary structure (Grimson et al. 2007). Inaccessible sites might become accessible under certain conditions that promote unfolding of a secondary structure, and a hypothesis has been brought forward by Zhao et al. (2005) whereby the binding of an miRNA produces a 'melting effect' that alters the secondary structure of neighbouring sites, allowing subsequent binding of other miRNAs. Moreover, sites found at the extremities of the 3′ UTR rather than at its centre, and at least 15 nts away from the stop codon, better represent effective binding sites in that these characteristics allow for improved proximation to needed complexes and lessened interference by translating ribosomes, respectively (Grimson et al. 2007). Therefore, sites with high sequence complementarity are not necessarily real targets, as is also sustained by the fact that target prediction algorithms tend to over-predict the number of targets that a given miRNA has.

Interestingly, miRISC components localise to structures called processing bodies (P-bodies) (Liu et al. 2005). These are cytoplasmic foci containing enzymes important in the normal pathway of mRNA degradation. Within P-bodies, translationally repressed mRNA is either sequestered in storage structures or processed for degradation, but their exact function and mode of action remain unknown.

The distinct tissue-specific distribution of many miRNAs has led to the idea that these small RNAs must be involved in tissue differentiation and development. In fact, the early reports on miRNAs documented an important switch effect during development of nematodes (Lee et al. 1993; Wightman et al. 1993): when a threshold was reached in the expression of a miRNA (i.e., *lin-4*), the inhibition of a target (i.e., *lin-14*) allowed the continued development of the worm. The importance of miRNAs was then determined also for the normal development of mammals. Mouse and human embryonic stem cells, for example, were found to specifically express a set of miRNAs that, upon differentiation, were downregulated (Houbaviy et al. 2003; Suh et al. 2004). It has also been proposed that tissue-specific miRNAs might act in a fail-safe mechanism, conferring robustness to tissue-specific gene expression by blocking the expression of sets of mRNAs in tissues in which their presence would be detrimental. In fact, analysis of *D. melanogaster* revealed that mRNAs with miR-1 target sites are expressed largely where this miRNA is not expressed (i.e., in non-muscle tissues; Stark et al. 2005), and, in mice, mRNAs were shown to be co-expressed with miRNAs that have evolved to be devoid of the target sequences for these miRNAs (Farh et al. 2005). In addition, deletion of Dicer was shown to produce defects of growth but not of basic tissue-specific differentiation, and this pointed to a mechanism of fine-tuning of protein expression (Harfe et al. 2005). Importantly, bioinformatic analyses have predicted that each miRNA may regulate hundreds of targets, and it is thought that up to one-third of a complete genome may be regulated by miRNAs. Many reports have indicated that miRNAs are implicated in a variety of basic biological processes apart from development, such as metabolism (Poy et al. 2004; Xu et al. 2003), proliferation (Poy et al. 2004; Xu et al. 2004), stress (Obernosterer et al. 2006), apoptosis (Obernosterer et al. 2006; Okamura et al. 2007), neural development (McFadden et al. 2005), haematopoiesis (Chen et al. 2004), and oncogenesis (Brennecke et al. 2005; Pillai et al. 2005; Poy et al. 2004; Prasanth et al. 2005). In most of these studies, however, the actual role of the miRNA is fairly unclear.

Apart from the mechanisms of action and functions described above, recent reports have started to document a number of novel mechanisms that alter miRNA function (Fig. 2). As we have seen, the canonical mechanism of action of miRNAs relies on post-translational repression within the cytoplasmic compartment of the cell, either via reduced translation efficiency or degradation of the targeted mRNA. However, the view that this is the sole mechanism through which miRNAs act has been challenged in a report by Hwang et al. (2007). Even if relatively small in size, miRNAs can still contain specific sequences at the 3′ end that are responsible for controlling their posttranscriptional behaviour. For example, miR-29b contains a distinctive 3′ hexanucleotide terminal motif responsible for relocating this miRNA back into the nucleus during the cycling phase of cells. The exact function of nuclear re-located miRNAs, however, is not understood as yet, but has been speculated to involve either transcriptional control or regulation of splicing. Moreover, differential precursor processing has been documented for some miRNAs, such as miR-138. The pre-miRNA can be ubiquitously expressed, but the mature miRNA is characteristic of selective maturation occurring in certain cell types or at a particular developmental stage (Obernosterer et al. 2006). It seems that an unknown inhibitor binds to the pre-miRNA, preventing further processing by Dicer. The accumulation of the pre-miRNA within the cytoplasm might, then, represent not only an addition level of control of miRNA expression (thought to occur primarily at the transcriptional level) but may also suggest the presence of a novel, and as yet

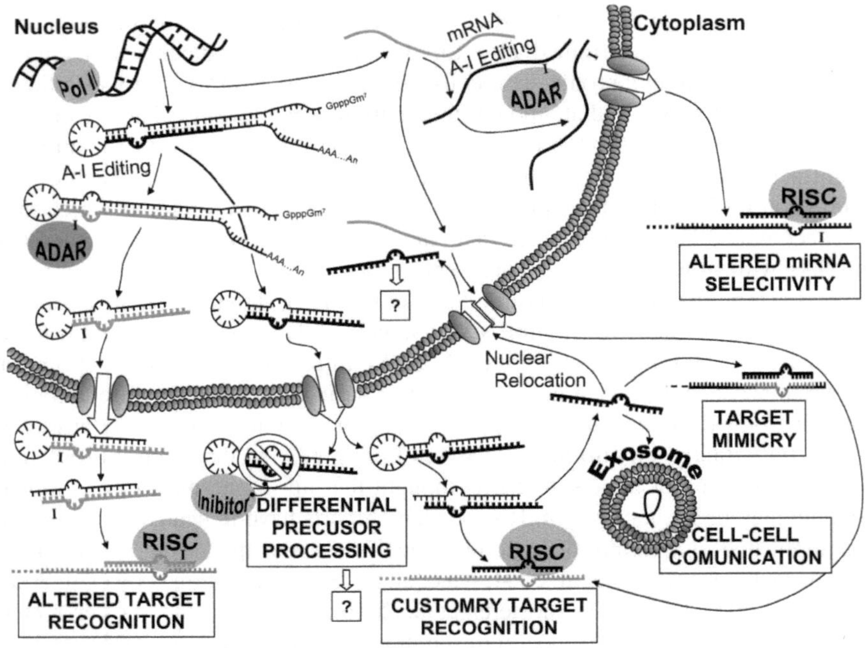

Fig. 2 Some interesting new additions to microRNA biology. Adenosine deaminases that act on RNA (ADAR) can edit both mRNA and miRNA by transforming an adenosine (*A*) into an inosine (*I*). This can alter the selectivity of an mRNA for a miRNA or the recognition of a microRNA for it canonical target. At least in plants, decoy mRNAs mimic targets but do not permit degradation, thus sequestering microRNAs and permitting synthesis of protein that would otherwise not take place. MicroRNA function may also be regulated by inhibitors that impede processing. Moreover, microRNAs may be secreted by in exosomes and, thus, affect other cell types

unknown, function for miRNA precursors. miRNAs have also been recently implicated in an exosome-mediated cell–cell communication mechanism (Valadi et al. 2007). In fact, exosomes have been identified containing RNAs, including miRNA. This finding implies the extraordinary possibility that cells can modify gene expression of neighbouring cells in a paracrine fashion, and, moreover, if released into the circulation, of cells at distant sites, too.

Editing by adenosine deaminases that act on RNA (ADARs) has also been linked to miRNA biology. ADARs convert adenosine to inosine at crucial points of dsRNA. As inosine is translated as if it were a guanosine, a different amino acid is incorporated into the nascent protein at the site of editing that may radically alter the protein's function. It has been documented that ADARs edit miRNAs and profoundly alter their action, and that ADAR-mediated editing is not only selective for certain miRNAs but is also selective for certain adenosines within the edited miRNA. It has been shown that pri-miR-22, a widely expressed miRNA, is edited by the ADAR isoform, ADAR1p150, at points that may affect strand choice (Luciano et al. 2004), and that a number of pri-miRNAs, and especially pri-miR-142, are edited by the nuclear located ADARs, ADAR1p110 and ADAR2 (Yang et al. 2006). Pri-miRNAs edited at some sites are transported to the cytoplasm, by an unknown mechanism, where they are rapidly degraded by Tudor-SN, a component of the RISC complex. On the other hand, pri-miRNAs edited at other sites continue the normal maturation process. In this case, editing might be responsible for the generation of new miRNA isoforms that recognise a different set of mRNAs to the unedited version, such as shown to occur for miR-376a-5p in a tissue-specific manner (Kawahara et al. 2007). Moreover, ADAR hyper-editing of mRNA 3′ UTRs is involved in a mechanism that sequesters mRNA within the nucleus, delaying further processing (Prasanth et al. 2005). It has been therefore suggested that more subtle editing of 3′ UTRs might be a mechanism whereby mRNAs are recoded for differential miRNA processing (Das and Carmichael 2007). Another fascinating recent addition to the biology of miRNAs is illustrated by miR-399, a miRNA contained in plants (Franco-Zorrilla et al. 2007). MiR-399 targets PHO2 mRNA, which encodes for a protein that negatively affects the phosphate content of shoots, and signals for its cleavage. IPS1, a non-protein coding RNA, also has a miR-399 recognition site, but pairing is interrupted by a mismatched loop at the cleavage site, thus precluding degradation of this RNA. It has been shown that upregulation of IPS1 is responsible for sequestration of miR-399, which inhibits miR-399-mediated cleavage of PHO2. This regulatory mechanism, consisting of a noncleavable RNA that forms a non-productive interaction with a complementary miRNA, has been called target mimicry. Target mimicry has so far only been demonstrated in plants but has not been excluded to occur in other organisms.

Bioinformatic approaches are being continuously developed and improved to predict both new miRNAs and their targets. A number of algorithms based on various methods are available to search for putative miRNAs. MiRscan, PalGrade and Triplet-SVM are three algorithms that can be applied to humans (for reviews, see Bentwich 2005; Brown and Sanseau 2005; Chaudhuri and Chatterjee 2007). A predicted 1,000 miRNAs may exist in the human genome, and it has been estimated that at least 1% of genes contain miRNAs (Berezikov et al. 2005; Griffiths-Jones 2004) making them one of the most abundant class of regulators (Stark et al. 2005). Currently, around 560 miRNA sequences have been identified and are catalogued in a searchable Web-based data register (Griffiths-Jones 2004; Griffiths-Jones et al. 2006).

Many algorithms also exist for target prediction. For this too the different algorithms are based on different selection criteria. Examples include Diana-MicroT, PicTar, miRanda and TargetScan.

2 miRNAs and Muscle Development

Various miRNAs have been identified that are expressed in a tissue-specific manner and, thus, may regulate tissue-specific functions (Wienholds et al. 2005). In muscle, for example, miR-1 and miR-133 have been found to be preferentially expressed

(Baskerville and Bartel 2005, Sempere et al. 2004). To date, only miR-208 has been found to be purely cardiac-specific.

The miR-133 family (comprised of miR-133a-1, miR-133a-2 and miR-133b) is expressed from bicistronic units together with miR-1, which are among the most highly conserved miRNAs (Mansfield et al. 2004, Sokol and Ambros 2005, Wienholds et al. 2005, Zhao et al. 2005b). An ancient genomic duplication is thought to have resulted in two distinct loci for the miR-1/miR-133 cluster in vertebrates (Chen et al. 2006, Rao et al. 2006). The miR-1 family is comprised of miR-1-1 and miR-1-2 together with miR-206, which is not expressed in the heart. In flies, transcription of miR-1 is activated in a broad pan-mesodermal domain prior to gastrulation, whereas the two mouse miR-1 genes are detected later, at the beginning of muscle differentiation, and then become progressively more expressed. The differences in the spatiotemporal occurrence of miR-1 reported between the species studied to date seem to imply that miR-1 has evolved as a mesodermal/muscle-specific miRNA early in animal evolution, but then has been integrated into a hierarchy of muscle transcription networks. Therefore, the role of a given miRNA may be slightly different from one species to the next (Brennecke et al. 2005).

During the last year or so, elegant experimental studies have started to unravel the important role of miRNAs in muscle biology. Mammalian cardiac miR-1 and miR-133 are controlled by serum response factor (SRF). SRF recruits a coactivator, myocardin, to muscle-specific genes that control differentiation (Zhao et al. 2005b). In addition, miR-133 has been reported to repress SRF, suggesting the presence of a possible regulatory loop (Chen et al. 2006). The regulation of cardiac miR-1 is different to that occurring in skeletal muscle in which expression requires myogenic transcription factors such as myogenic differentiation 1 (MyoD), myocyte enhancer factor 2 (Mef2), and myogenin. In addition, the presence of putative transcription factor binding sites in between miR-1-1 and miR-133a-2 suggests the possibility that the individual miRNAs contained in a polycistronic unit may be independently regulated (Rao et al. 2006).

During early embryonic development, skeletal and heart muscle expression of miR-1 and miR-133 is limited (Carè et al. 2007, Chen et al. 2006, Wienholds et al. 2005). Expression then increases during the later embryonic stages and during the neonatal period, and becomes maximal in adulthood. In mice, analysis of spatiotemporal patterning for miR-1 revealed that miR-1-1 is found initially strongly expressed in the less proliferative inner curvature of the heart loop and in atria, but becomes ubiquitously expressed in the heart as development progresses; on the other hand, miR-1-2 can be found mainly in the ventricles (Zhao et al. 2005b). The temporal pattern of expression is recapitulated in C_2C_{12} cells and human myoblasts in vitro in that undifferentiated myoblasts do not express muscle-specific miRNAs, while expression increases with differentiation into myotubes (Chen et al. 2006, Rao et al. 2006).

Analysis of *D. melanogaster* mutants revealed that miR-1 is not required for specification or patterning of muscle, but rather for post-mitotic muscle growth of larvae (Sokol and Ambros 2005). Moreover, in developing *Xenopus laevi*, the introduction of miR-1 produced an altered phenotype characterised by less cell proliferation and no development of cardiac tissue (Chen et al. 2006). miR-1 inhibits cardiomyocyte progenitor proliferation also in mice (Zhao et al. 2005b). This was found to occur via inhibiting translation of Hand2, a transcription factor regulating ventricular cardiomyocyte expansion (McFadden et al. 2005). The overexpression of miR-1 resulted in thin-walled ventricles, due to premature differentiation and early withdrawal of cardiomyocytes from the cell cycle. Oppositely, adult miR-1-2 knockout mice presented with thickened chamber walls due to hyperplasia that had continued into adult life, and many of the embryos from these mice often suffered from septal defects, further demonstrating the fundamental role of miR-1 for heart development (Zhao et al. 2007). The inhibitory effect of miR-1 for proliferation in the heart is coherent with its function in skeletal muscle since overexpression of miR-1 in myoblasts decreases proliferation and promotes skeletal muscle differentiation by targeting of histone deacetylase 4 (HDAC4) (Chen et al. 2006). Moreover, the expression of predicted miR-1 mRNA targets were reduced after differentiation into myotubes, when this miRNA is functional, indicating that miRNAs may destabilise pre-existing mRNAs, permitting a more robust transition towards myogenic differentiation (Farh et al. 2005).

In contrast to miR-1, miR-133 is capable of promoting proliferation by targeting SRF in skeletal muscle (Chen et al. 2006). In fact, miR-133 misexpression increased proliferation and altered cardiac tissue formation with disorganised looping and chamber formation (Chen et al. 2006).

All these results demonstrate the critical importance not only of proper timing but also of correct dosing of miRNAs for heart development. Moreover, the opposing effects of miR-1 and miR-133 on proliferation and differentiation and the finding that both these miRNAs are induced with development, demonstrates the complexity of miRNA developmental control.

3 miRNAs and Heart Pathophysiology

MiRs are estimated to regulate ~30% of the human genome, and it is not surprising that perturbed patterns of expression occur in disease states. In skeletal muscle, for example, alterations of phenotype occur with the dysregulation of miRNA expression and, incidentally, also with mutations of the targeted sequence of mRNAs. In fact, in mouse hindlimb muscles with functional overload-induced hypertrophy, miR-206 was found up-regulated and miR-1 and miR-133a down-regulated (McCarthy and Esser 2007). Moreover, a gain-of-function polymorphism leading to inappropriate targeting of the myostatin gene has been ascribed as being responsible for the increased muscle mass of Texel sheep. In these animals, a mutation in the 3′ UTR of myostatin mRNA creates a binding site for miR-1/miR-206, which is otherwise absent, producing a hypertrophic phenotype similar to the myostain-null mouse (Clop et al. 2006).

With regard to the heart, alteration of miRNA expression in disease states is just starting to be documented. Stress of the cardiovascular system can lead to a state distinguished by an increase in the size of the heart, called hypertrophy. Hypertrophy has a number of basic characteristics: an increase in cardiomyocyte size; alterations in contractile function of the heart; and re-expression of foetal-type genes such as β-myosin heavy chain (β-MHC), atrial natriuretic peptide (ANF), and skeletal alpha-actin (SkAc) (for a review, see Catalucci et al. (2008)). Functionally, hypertrophy is initially a compensatory process, but with chronic exposure to stress signals, the heart eventually undergoes maladaptive changes that lead to heart failure and even death.

Dysregulation of miRNA expression during hypertrophic growth of the heart subsequent to pathological stress was reported first by van Rooij et al. (2006). For their studies, these authors used two experimental models of hypertrophy, transverse aortic constriction (TAC), a widely used in vivo model of hypertrophy that induces pressure overload of the left ventricle of animals, and calcineurin-transgenic mice. Both of these models were found to have in common a number of up- or down-regulated miRNAs. Intriguingly, the overexpression of individual miRNAs that were up-regulated with hypertrophy was capable of inducing an increase in the size of cardiomyocytes in vitro. Moreover, pathological cardiac remodelling and heart failure were recapitulated in a transgenic mouse overexpressing miR-195, one of the stress-induced miRNAs. Importantly, theses authors also analyzed miRNA expression in idiopathic end-stage failing human hearts by Northern blotting, and documented increased expression of miR-24, miR-125b, miR-195, miR-199a and miR-214. Since this initial publication, a surge of reports have corroborated the finding of a complex array of dysregulated miRNAs in stress-related hypertrophy of the postnatal heart, even if the pattern of expression found has been slightly different from report to report (Cheng et al. 2007, Sayed et al. 2007, Tatsuguchi et al. 2007). In addition, the miRNA expression pattern occurring with end-stage human failing hearts has been ascertained to be highly congruent to that seen in the foetus (Thum et al. 2007). This finding is consistent with the known reactivation of the foetal gene programme occurring during pathological forms of hypertrophy. Thus, the reexpression of a set of foetal miRNAs seems to be at the basis of the modification of the cardiac transcriptosome occurring with hypertrophy and failure.

Central to our understanding of the role of miRNAs, is correlation with specific targets. To date, only a few miRNAs have been studied in this respect. MiR-1 down-regulation has been documented as a feature of hypertrophy in TAC-operated mice and in hearts of cardiomyopathic patients (Carè et al. 2007). Its down-regulation was reported as being a relative early event, in that reduced miR-1 was detectable after only 1 day of TAC in mice (Sayed et al. 2007). Four in silico-predicted targets of miR-1 have been found increased after TAC: Ras GTPase–activating protein (RasGAP), cyclin-dependent kinase 9 (Cdk9), Ras homolog enriched in brain (Rheb) and fibronectin. The roles that these proteins play in hypertrophy are not clear but may involve growth-factor signalling, the regulation of translation and extracellular matrix reorganisation.

MiR-133 has also been found decreased in the ventricular tissue of mice subjected to TAC (Carè et al. 2007, Cheng et al. 2007, van Rooij et al. 2006) and in cultured cardiomyocytes treated with phenylephrine (PHE), a hypertrophic stimulus (Carè et al. 2007). Interestingly, miR-133 expression was also decreased in atria and ventricles from Akt-overexpressing transgenic mice, a physiological hypertrophic model (Carè et al. 2007). Since Akt is part of the insulin growth factor-1 (IGF-1) pathway activated during hypertrophy induced by physiological stimuli, hearts of exercise-trained rats were examined and found to have reduced levels of miR-133, along with miR-1 (Carè et al. 2007). Thus, these two miRNAs seem to be implicated in the increase of cardiomyocyte size irrespectively of whether the underlying cause is pathological or physiological in nature. Importantly, a significant reduction of miR-133 was found in myectomies from hearts of cardiomyopathic patients (Carè et al. 2007). To determine the functional significance of reduced miR-133 expression during hypertrophy, gain-of-function and loss-of-function studies were performed. Transduction of cardiomyocytes with a miR-133a-2-expressing viral vector blunted the hypertrophic response of cardiomyocytes to PHE treatment. Moreover, in vivo administration of antagomir-133 (an antisense RNA oligonucleotide capable of knocking down miR-133 expression) induced hypertrophy. Of the possible targets predicted, expression of Cdc42 (implicated in cytoskeletal modifications during hypertrophy), Rho-A (a GTP-GDP binding molecule, also critical for hypertrophy) and NELF-A/WHSC2 (a nuclear factor involved in heart genesis) were found to correlate with hypertrophy. Rho-A and Cdc42 have already been associated with cell growth, cytoskeletal and myofibrillar rearrangements, and the regulation of contractility in cardiomyocytes (Brown et al. 2006, Ke et al. 2004). Wolf-Hirschhorn Syndrome Complex 2 (WHSC2), on the other hand, has not been particularly studied in relation to hypertrophy. A few studies have limited its identification as a repressor of transcription, probably operating at the RNA elongation step (Mariotti et al. 2000). In the Wolf-Hirschhorn Syndrome, a congenital condition characterised by mental retardation, a number of abnormalities including cardiovascular ones are common. Interestingly, transduction of cardiomyocytes in vitro and in vivo with an adenoviral vector containing a *Whsc2* transgene resulted in decreased protein synthesis, but induced the foetal gene programme, a characteristic of hypertrophic response (Carè et al. 2007). Interestingly, up-regulation of Rho-A was also noted, corroborating the idea that WHSC2 could play a selective role in hypertrophy.

Another miRNA with a validated target is miR-208. This miRNA is encoded by intron 29 of the α-MHC gene. In human samples of idiopathic cardiomyopathy, a correlation was found with the expression of pre-miR-208 (van Rooij et al. 2007). In mice, knockout of miR-208 produced viable animals with no obvious cardiac phenotype apart from the upregulation of fast skeletal muscle contractile proteins and stress proteins, but their mRNAs were not found to be miR-208 targets. Interestingly, miR-208 knockout did produce an altered phenotype after TAC, in that operated mice presented with blunted hypertrophic and fibrotic responses. Moreover, upregulation of stress markers (such as ANF and brain natriuretic peptide, BNP) were increased in the hearts of these mice, as predicted, but the increase in β-MHC was absent. In contrast, α-MHC expression was increased, rather than reduced. This miRNA, then, seems to control regulation of β-MHC during conditions of stress, but not during normal development. Of the predicted miR-208 targets, thyroid hormone receptor (TR) associated protein 1 (THRAP1), the TR co-regulator, was found to be increased in miR-208-knockout mice. Therefore stress stimuli, responsible for the reduction of α-MHC transcription, consequentially also reduce the level of the miR-208 transcript, which, in turn, relieves transcriptional repression on its target mRNA, *thrap*. The resulting increase in THRAP1 protein influences the TR-regulated expression of α- and β-MHCs, which are inversely affected through a positive and negative TRE, respectively.

Membrane excitability is another fundamental characteristic of the cardiomyocyte that can be altered with disease. Reports have documented that also this aspect of cardiac pathophysiology is regulated by miRNAs. In fact, an important role for miRNAs in the physiological distribution of K^+ channels has been reported (Xiao et al. 2007a). KCNQ1 and KCNE1 are two subunits that assemble in the heart to form the slow delayed rectifier K^+ current (I_{Ks}). An important spatial patterning exists for this channel in the normal heart, such as apex to base, and epi/endocardium to midmyocardium gradients. Interestingly, miR-1 and miR-133 expression is also spatially heterogeneous and, intriguingly, specular in many aspects to that of I_{Ks}. Thus, in many of the areas were I_{Ks} is more densely expressed, miR-1 and miR-133 are less abundant. However, even if computational predictions did not evidence that KCNQ1 or KCNE1 were targets of either miR-1 or miR-133, a careful analysis of the 3′ UTRs revealed putative binding sites for miR-1 on KCNE1, and for miR-133 on KCNQ1. Thus, spatial differences in the expression of miRNAs can exist within an adult organ that may be responsible for modulating, at least in part, the expression pattern of target proteins.

A relationship between altered cardiac electrical mechanisms and dysregulation of miRNAs has been shown also in disease states. Down-regulation of *ERG* (ether-a-go-go related gene), which encodes for the rapid delayed rectifier K^+ current, I_{Kr}, has been directly linked to miR-133 up-regulation in the hearts of a rabbit model of diabetes and in ventricular samples from diabetic patients (Xiao et al. 2007a). Furthermore, up-regulation of miR-1 has been reported in individuals with coronary artery disease, the leading cause of death in industrialised countries, many of which are attributable to arrhythmias (Yang et al. 2007). Ablation of miR-1 in vivo, using an antisense inhibitor, was found to be sufficient to relieve arrhythmogenesis in infarcted rat hearts. Validated miR-1 targets were demonstrated to be *KCNJ2* (which encodes the K^+ channel subunit, Kir2.1, responsible for I_{K1}) and *GJA1* (which encodes for connexin 43, involved in intercellular conductance). These proteins are down-regulated in mice after myocardial infarction and in samples from coronary artery diseased patients. Abnormal propagation of cardiac electrical activity was a feature reported also for surviving miR-1-2 knockout mice. These mice had apparently normal anatomy and function but suffered from a slowed heart rate (Zhao et al. 2007). In that report, *Irx5* (Iroquois family of homeodomain-containing transcription factor), which regulates cardiac repolarisation by repressing transcription of a key potassium channel, Kcnd2, was reported as a validated target of miR-1.

4 Conclusion and Future Prospects

Since the discovery of the first miRNA, an exponential advancement has been made in the understanding of mechanisms and roles of these small RNAs. High throughput miRNA expression profiling, computational target prediction, in vitro target validation, as well as elegant in vivo experimental approaches, have contributed remarkably to shedding light on the importance of these molecules as relevant players in the regulation of gene expression and, particularly, as potential clinical tools in a variety of diseases. Importantly, it has been experimentally shown that knocking-down inappropriately overexpressed miRNAs with antisense oligonucleotides, modified to improve their delivery and stability, such as *anti-miRNA oligonucleotides* (*AMOs*) (Weiler et al. 2006) and *antagomirs* (Krutzfeldt et al. 2005), is conceptually feasible, and their use in a number of studies on the heart has already been reported. Moreover, miRNA can themselves be used as drugs if an increased expression is required, and these can be relatively easily synthesised and tailored to affect single target mRNAs, thus minimising unwanted effects (Xiao et al. 2007b).

Even if no clinical trials have been reported at the moment, several miRNA-based therapeutical treatments have recently been envisioned, mostly though in relation to cancer biology. However, advancement in the knowledge of cancer biology and treatment may filter through to the cardiovascular setting where, as above described in this chapter, important findings have already been made.

References

Baskerville S, Bartel DP (2005) Microarray profiling of microRNAs reveals frequent coexpression with neighboring miRNAs and host genes. RNA 11:241–247

Behm-Ansmant I, Rehwinkel J, Doerks T et al (2006) mRNA degradation by miRNAs and GW182 requires both CCR4:NOT deadenylase and DCP1:DCP2 decapping complexes. Genes Dev 20:1885–1898

Bentwich I (2005) Prediction and validation of microRNAs and their targets. FEBS Lett 579:5904–5910

Berezikov E, Guryev V, van de Belt J et al (2005) Phylogenetic shadowing and computational identification of human microRNA genes. Cell 120:21–24

Brennecke J, Stark A, Cohen SM (2005) Not miR-ly muscular: microRNAs and muscle development. Genes Dev 19:2261–2264

Brown JR, Sanseau P (2005) A computational view of microRNAs and their targets. Drug Discov Today 10:595–601

Brown JH, Del Re DP, Sussman MA (2006) The Rac and Rho hall of fame: a decade of hypertrophic signaling hits. Circ Res 98:730–742

Carè A, Catalucci D, Felicetti F et al (2007) MicroRNA-133 controls cardiac hypertrophy. Nat Med 13:613–618

Catalucci D, Latronico M, Condorelli G (2008) Physiological myocardial hypertrophy: how and why? Front Biosci 13:312–324

Chaudhuri K, Chatterjee R (2007) MicroRNA detection and target prediction: integration of computational and experimental approaches. DNA Cell Biol 26:321–337

Chen CZ, Li L, Lodish HF et al (2004) MicroRNAs modulate hematopoietic lineage differentiation. Science 303:83–86
Chen JF, Mandel EM, Thomson JM et al (2006) The role of microRNA-1 and microRNA-133 in skeletal muscle proliferation and differentiation. Nat Genet 38:228–233
Chendrimada TP, Finn KJ, Ji X et al (2007) MicroRNA silencing through RISC recruitment of eIF6. Nature 447:823–828
Cheng Y, Ji R, Yue J et al (2007) MicroRNAs are aberrantly expressed in hypertrophic heart. Do they play a role in cardiac hypertrophy? Am J Pathol 170(6):1831–1840
Clop A, Marcq F, Takeda H et al (2006) A mutation creating a potential illegitimate microRNA target site in the myostatin gene affects muscularity in sheep. Nat Genet 38:813–818
Das AK, Carmichael GG (2007) ADAR editing wobbles the microRNA world. Chem Biol 2:217–220
Farh KK, Grimson A, Jan C et al (2005) The widespread impact of mammalian MicroRNAs on mRNA repression and evolution. Science 310:1817–1821
Franco-Zorrilla JM, Valli A, Todesco M et al (2007) Target mimicry provides a new mechanism for regulation of microRNA activity. Nat Genet 39:1033–1037
Giraldez AJ, Mishima Y, Rihel J et al (2006) Zebrafish MiR-430 promotes deadenylation and clearance of maternal mRNAs. Science 312:75–79
Gregory RI, Chendrimada TP, Cooch N et al (2005) Human RISC couples microRNA biogenesis and posttranscriptional gene silencing. Cell 123:631–640
Griffiths-Jones S (2004) The microRNA Registry. Nucleic Acids Res 32:D109–D111
Griffiths-Jones S, Grocock RJ, van Dongen S et al (2006) miRBase: microRNA sequences, targets and gene nomenclature. Nucleic Acids Res 34:D140–D144
Grimson A, Farh KK, Johnston WK et al (2007) MicroRNA targeting specificity in mammals: determinants beyond seed pairing. Mol Cell 27:91–105
Harfe BD, McManus MT, Mansfield JH et al (2005) The RNaseIII enzyme Dicer is required for morphogenesis but not patterning of the vertebrate limb. Proc Natl Acad Sci USA 102:10898–10903
Houbaviy HB, Murray MF, Sharp PA (2003) Embryonic stem cell-specific MicroRNAs. Dev Cell 5:351–358
Humphreys DT, Westman BJ, Martin DI et al (2005) MicroRNAs control translation initiation by inhibiting eukaryotic initiation factor 4E/cap and poly(A) tail function. Proc Natl Acad Sci USA 102:16961–16966
Hwang HW, Wentzel EA, Mendell JT (2007) A hexanucleotide element directs microRNA nuclear import. Science 315:97–100.
Kawahara Y, Zinshteyn B, Sethupathy P et al (2007) Redirection of silencing targets by adenosine-to-inosine editing of miRNAs. Science 315:1137–1140
Ke Y, Wang L, Pyle WG et al (2004) Intracellular localization and functional effects of P21-activated kinase-1 (Pak1) in cardiac myocytes. Circ Res 94:194–200
Kim VN (2004) MicroRNA precursors in motion: exportin-5 mediates their nuclear export. Trends Cell Biol 14:156–159
Krutzfeldt J, Rajewsky N, Braich R et al (2005) Silencing of microRNAs in vivo with 'antagomirs'. Nature 438:685–689
Lee RC, Feinbaum RL, Ambros V (1993) The *C. elegans* heterochronic gene lin-4 encodes small RNAs with antisense complementarity to lin-14. Cell 75:843–854
Lee Y, Jeon K, Lee JT et al (2002) MicroRNA maturation: stepwise processing and subcellular localization. EMBO J 21:4663–4670
Lee Y, Ahn C, Han J et al (2003) The nuclear RNase III Drosha initiates microRNA processing. Nature 425:415–419
Lee Y, Kim M, Han J et al (2004) MicroRNA genes are transcribed by RNA polymerase II. EMBO J 23:4051–4060
Lingel A, Simon B, Izaurralde E et al (2003) Structure and nucleic-acid binding of the Drosophila Argonaute 2 PAZ domain. Nature 426:465–469
Liu J, Valencia-Sanchez MA, Hannon GJ et al (2005) MicroRNA-dependent localization of targeted mRNAs to mammalian P-bodies. Nat Cell Biol 7:719–723

Luciano DJ, Mirsky H, Vendetti NJ et al (2004) RNA editing of a miRNA precursor. RNA 10:1174–1177
Mansfield JH, Harfe BD, Nissen R et al (2004) MicroRNA-responsive 'sensor' transgenes uncover Hox-like and other developmentally regulated patterns of vertebrate microRNA expression. Nat Genet 36:1079–1083
Mariotti M, Manganini M, Maier JA (2000) Modulation of WHSC2 expression in human endothelial cells. FEBS Lett 487:166–170
Maroney PA, Yu Y, Fisher J et al (2006) Evidence that microRNAs are associated with translating messenger RNAs in human cells. Nat Struct Mol Biol 13:1102–1107
McCarthy JJ, Esser KA (2007) MicroRNA-1 and microRNA-133a expression are decreased during skeletal muscle hypertrophy. J Appl Physiol 102:306–313
McFadden DG, Barbosa AC, Richardson JA et al (2005) The Hand1 and Hand2 transcription factors regulate expansion of the embryonic cardiac ventricles in a gene dosage-dependent manner. Development 132:189–201
Nottrott S, Simard MJ, Richter JD (2006) Human let-7a miRNA blocks protein production on actively translating polyribosomes. Nat Struct Mol Biol 13:1108–1114
Obernosterer G, Leuschner PJ, Alenius M et al (2006) Post-transcriptional regulation of microRNA expression. RNA 12:1161–1167
Okamura K, Hagen JW, Duan H et al (2007) The mirtron pathway generates microRNA-class regulatory RNAs in *Drosophila*. Cell 130:89–100
Olsen PH, Ambros V (1999) The lin-4 regulatory RNA controls developmental timing in *Caenorhabditis elegans* by blocking LIN-14 protein synthesis after the initiation of translation. Dev Biol 216:671–680
Peters L, Meister G (2007) Argonaute proteins: mediators of RNA silencing. Mol Cell 26:611–623
Pillai RS, Bhattacharyya SN, Artus CG et al (2005) Inhibition of translational initiation by Let-7 MicroRNA in human cells. Science 309:1573–1576
Poy MN, Eliasson L, Krutzfeldt J et al (2004) A pancreatic islet-specific microRNA regulates insulin secretion. Nature 432:226–230
Prasanth KV, Prasanth SG, Xuan Z et al (2005) Regulating gene expression through RNA nuclear retention. Cell 123:249–263
Rao PK, Kumar RM, Farkhondeh M et al (2006) Myogenic factors that regulate expression of muscle-specific microRNAs. Proc Natl Acad Sci USA 103:8721–8726
Rodriguez A, Griffiths-Jones S, Ashurst JL et al (2004) Identification of mammalian microRNA host genes and transcription units. Genome Res 14:1902–1910
Ruby JG, Jan CH, Bartel DP (2007) Intronic microRNA precursors that bypass Drosha processing. Nature 448:83–86
Sayed D, Hong C, Chen IY et al (2007) MicroRNAs play an essential role in the development of cardiac hypertrophy. Circ Res 100:416–424
Schmitter D, Filkowski J, Sewer A et al (2006) Effects of Dicer and Argonaute down-regulation on mRNA levels in human HEK293 cells. Nucleic Acids Res 34:4801–4815
Sempere LF, Freemantle S, Pitha-Rowe I et al (2004) Expression profiling of mammalian microRNAs uncovers a subset of brain-expressed microRNAs with possible roles in murine and human neuronal differentiation. Genome Biol 5:R13
Sokol NS, Ambros V (2005) Mesodermally expressed *Drosophila* microRNA-1 is regulated by Twist and is required in muscles during larval growth. Genes Dev 19:2343–2354
Stark A, Brennecke J, Bushati N et al (2005) Animal MicroRNAs confer robustness to gene expression and have a significant impact on 3′UTR evolution. Cell 123:1133–1146
Suh MR, Lee Y, Kim JY et al (2004) Human embryonic stem cells express a unique set of microRNAs. Dev Biol 270:488–498
Tang G (2005) siRNA and miRNA: an insight into RISCs. Trends Biochem Sci 30:106–114
Tatsuguchi M, Seok HY, Callis TE et al (2007) Expression of microRNAs is dynamically regulated during cardiomyocyte hypertrophy. J Mol Cell Cardiol 42:1137–1141
Thum T, Galuppo P, Wolf C et al (2007) MicroRNAs in the human heart: a clue to fetal gene reprogramming in heart failure. Circulation 116:258–267

Valadi H, Ekstrom K, Bossios A et al (2007) Exosome-mediated transfer of mRNAs and microRNAs is a novel mechanism of genetic exchange between cells. Nat Cell Biol 9:654–659

van Rooij E, Sutherland LB, Liu N et al (2006) A signature pattern of stress-responsive microRNAs that can evoke cardiac hypertrophy and heart failure. Proc Natl Acad Sci USA 103:18255–18260

van Rooij E, Sutherland LB, Qi X et al (2007) Control of stress-dependent cardiac growth and gene expression by a microRNA. Science 316:575–579

Weiler J, Hunziker J, Hall J (2006) Anti-miRNA oligonucleotides (AMOs): ammunition to target miRNAs implicated in human disease? Gene Ther 13:496–502

Wienholds E, Kloosterman WP, Miska E et al (2005) MicroRNA expression in zebrafish embryonic development. Science 309:310–311

Wightman B, Ha I, Ruvkun G (1993) Posttranscriptional regulation of the heterochronic gene lin-14 by lin-4 mediates temporal pattern formation in *C. elegans*. Cell 75:855–862

Wu L, Fan J, Belasco JG (2006) MicroRNAs direct rapid deadenylation of mRNA. Proc Natl Acad Sci USA 103:4034–4039

Xiao J, Luo X, Lin H et al (2007a) MicroRNA miR-133 Represses HERG K+ channel expression contributing to QT prolongation in diabetic hearts. J Biol Chem 282:12363–12367

Xiao J, Yang B, Lin H et al (2007b) Novel approaches for gene-specific interference via manipulating actions of microRNAs: examination on the pacemaker channel genes HCN2 and HCN4. J Cell Physiol 212:285–292

Xu P, Vernooy SY, Guo M et al (2003) The *Drosophila* microRNA Mir-14 suppresses cell death and is required for normal fat metabolism. Curr Biol 13:790–795

Xu P, Guo M, Hay BA (2004) MicroRNAs and the regulation of cell death. Trends Genet 20:617–624

Yang W, Chendrimada TP, Wang Q et al (2006) Modulation of microRNA processing and expression through RNA editing by ADAR deaminases. Nat Struct Mol Biol 13:13–21

Yang B, Lin H, Xiao J et al (2007) The muscle-specific microRNA miR-1 regulates cardiac arrhythmogenic potential by targeting GJA1 and KCNJ2. Nat Med 13:486–491

Zhao X, Sternsdorf T, Bolger TA et al (2005a) Regulation of MEF2 by histone deacetylase 4- and SIRT1 deacetylase-mediated lysine modifications. Mol Cell Biol 25:8456–8464

Zhao Y, Samal E, Srivastava D (2005b) Serum response factor regulates a muscle-specific microRNA that targets Hand2 during cardiogenesis. Nature 436:214–220

Zhao Y, Ransom JF, Li A et al (2007) Dysregulation of cardiogenesis, cardiac conduction, and cell cycle in mice lacking miRNA-1-2. Cell 129:303–317

MicroRNA Systems Biology

E. Wang

Contents

Abstract In the past few decades, gene regulatory networks and cellular signaling networks have been regarded as the major regulatory systems in a cell. In contrast, RNAs have been thought as molecules which are only transferring genetic information for protein production. Recently, microRNAs (miRNAs) have emerged as another layer of gene regulation. They regulate many key biological processes, including cell growth, death, development and differentiation. This discovery hints that cells have more complicated regulation systems. Genes are working together by forming cellular networks. It has become an emerging concept that miRNAs could intertwine with cellular networks to exert their function. Thus, it is essential to understand how miRNAs take part in cellular processes at a systems-level. In this chapter, I will summarize the most recent progress in understanding of miRNA biology at a systems-level: the principles of miRNA regulation of the major cellular networks including signaling, metabolic, protein interaction and gene regulatory networks. A common miRNA regulatory principle is emerging: miRNAs preferentially regulated the genes that have high regulation complexity. In addition, miRNAs preferentially regulate positive regulatory loops, highly connected

E. Wang
Biotechnology Research Institute, National Research Council of Canada, 6100 Royalmount Avenue, Montreal, Quebec, Canada, and Center for Bioinformatics, McGill University, Montreal, Quebec, Canada
e-mail: Edwin.Wang@cnrc-nrc.gc.ca

V.A. Erdmann et al. (eds.) *RNA Technologies in Cardiovascular Medicine and Research*,
doi: 10.1007/978-3-540-78709-9,

scaffolds and the most network downstream components of cellular signaling networks, while miRNAs selectively regulate the genes which have specific network structural features on metabolic networks.

1 Introduction

In cells, genes are not isolated and do not independently perform a single task; instead, genes are grouped to collaborate and carry out some specific biological function. This collaborative effort of genes indicates that genes are working together in a cell, and we can formulate this conceptually depicting the various interactions as cellular networks. In the past few decades, gene regulatory networks and cellular signaling networks have been regarded as the major regulatory systems in a cell. In contrast, RNAs have been regarded as molecules which are only transferring genetic information for protein production. However, this concept is getting challenged by the recent findings that tiny fragments of noncoding RNA, typically ~22 nucleotides in length, namely microRNAs (miRNAs), are able to negatively regulate protein-coding genes by interfering with mRNA's original instructions. It is currently estimated that miRNAs account for ~1% of predicted genes in higher eukaryotic genomes, and that up to 10–30% of genes might be regulated by miRNAs.

miRNAs have been shown to have biological functions in many aspects. Their targets range from signaling proteins, metabolic enzymes, transcription factors, and so on. The diversity and abundance of miRNA targets offer an enormous level of combinatorial possibilities, and suggest that miRNAs and their targets appear to form a complex regulatory network intertwined with other cellular networks, such as signal transduction networks, metabolic networks, gene regulatory networks and protein interaction networks. It is reasonable to think that miRNAs exert their functions through regulating cellular networks. Thus, it is imperative to understand how miRNAs take part in cellular processes at a systems-level.

Four types of cellular networks have been found in cells: protein interaction networks, metabolic networks, gene regulatory networks, and signaling networks, which can be further classified into two categories. Protein interaction networks encode the information of proteins and their physical interactions. Protein interaction information in the network ranges from basic cellular machinery, such as protein complexes for DNA synthesis, metabolic enzyme complexes, transcription factor complexes, to protein complexes involved in cellular signaling. Simply put, a genome-wide protein interaction network encodes all the protein interaction information across all biological processes in a cell. A gene regulatory network describes regulatory relationships between transcription factors and the protein-coding genes. Similar to protein interaction networks, a gene regulatory network encodes the gene regulatory information for all biological processes and activities in a cell. Therefore, I have classified protein interaction networks and gene regulatory networks into the first category: general network. The second category of networks identified is: specific cellular network, which encompasses metabolic networks and signaling networks, describing specific cellular activities. A cellular metabolic network collects all the metabolic reactions and metabolic flows, while a signaling network encodes signal information flows and biochemical reactions for signal transductions. Traditionally, both types of information are presented using linear pathways, e.g., metabolic pathways and signaling pathways. In a metabolic network, metabolic pathways are intertwined so that metabolic flows are transferable across different pathways. Certain metabolites can be shared and used by many different pathways, while certain end-product metabolites can be produced via bypassing one or several pathways. Signaling networks illustrate inter- and intracellular communications and information processing between signaling proteins. More discussions about the concepts of cellular networks and their biological insights can found in a recent review (Wang et al. 2007).

2 miRNA Regulation of Cellular Signaling Networks

2.1 *Signaling Networks and Computational Analysis*

The components of cellular signaling networks, mainly composed of proteins, are activated or inhibited in response to specific input stimuli and, in turn, serve as stimuli for further downstream proteins. Cellular signaling network is the primary complex cellular system to responding stimuli, signals and messages from other cells and environment. Once a cell receives signals, it processes the information, e.g., signal amplification and noise filtration, and finally the signals reach to transcription factors so that the signaling network triggers the responses of gene regulatory networks. Therefore, a signaling network is the most important complex system in processing the early extra- and intra-cellular signals in a cell (Wang et al. 2007). Cells use signaling networks, as a sophisticated communication system, to perform a series of tasks such as growth and maintenance, cell survival, apoptosis and development.

The relationships of signaling proteins are thought to be critical in determining cell behavior and maintaining cellular homeostasis, therefore, mis-regulation in the expression of genes and their regulators will be reflected on these cellular signaling networks which in turn lead to abnormal end points of development such as cancer and other diseases. miRNAs are posttranscriptional regulators, it is reasonable to think that miRNAs have great potential to regulate signaling networks.

Signaling networks are presented as graphs containing both directed and undirected links. In the networks nodes represent proteins, directed links represent activation or inactivation relationships between proteins, while undirected links represent simply physical interactions between proteins. Network-structural analysis of signaling networks has been limited by the lack of comprehensive datasets for signaling networks. In the past few decades, enormous efforts have been made to study signaling pathways and generated lots of signaling information, especially in mammalian genomes. However, this historically generated information is scatted in literature. Recently, different researchers began to manually curate signaling information and organized them as signaling pathways such as EGFR signaling network and BioCarta signaling pathway database (http://www.biocarta.com) (Oda et al. 2005; Oda and Kitano 2006) or signaling networks (Awan et al. 2007; Ma'ayan et al. 2005). Other researchers used high-throughput technologies to fish new signaling proteins and their interactions based on large-scale experimental studies of protein interactions or genetic interactions between known signaling proteins and other proteins (Lehner et al. 2006). All these efforts offer new possibilities to analyze large and complex cellular signaling networks using the graph and network theory.

So far only a few studies have been conducted for large-scale structural analysis of cellular signaling networks. In 2005, the first network-structural analysis of a literature-mined human cellular signaling network containing ~500 proteins was conducted, and showed that signaling pathways are intertwined in order to manage the numerous cell behavior outputs (Ma'ayan et al. 2005). This work provides a framework for our understanding of how signaling information is processed in cells. In 2006, we conducted an analysis of miRNA regulation of the human signaling network using the same dataset (Cui et al. 2006). In 2007, we collected more signaling information and extended the signaling proteins and their relations to the human signaling network. As a result, the new human signaling network contains more than 1,100 signaling proteins (Awan et al. 2007). Subsequent analysis of cancer-associated genes and cell mobility genes on the signaling network reveals the patterns of oncogenic regulation during tumorigenesis and finds oncogenic hotspots on the human signaling network (Awan et al. 2007). More recently, we further extended the human signaling network to contain more than 1,600 signaling proteins. Our efforts for curating signaling information include recording gene/protein names, molecular types (e.g., growth factor, ligand, adaptor, scaffold, and so on), biochemical reactions (e.g., phosphorylation, acetylation, ubiquitylation, and so on), and interaction types (e.g., activation, inhibition and simply physical interactions of proteins), cellular locations of the proteins, and so on. These data are freely available on our website: http://www.bri.nrc.ca/wang/. We further conducted an integrative analysis of cancer-casually implicated genetic and epigenetic alterations onto the human signaling network compiled from this more comprehensive dataset. Our analysis revealed where the oncogenic stimuli are embedded in the network architecture and illustrated the principles of triggering oncogenic signaling events by genetic and epigenetic alterations. Furthermore, we extracted a human cancer signaling map and showed that different parts/regions of the cancer signaling map are required to coopt during tumorigenesis. Using the same network, we performed a comprehensive analysis of the signaling network in an evolutionary context and underscored new insights into the evolution of cellular signaling networks. Collectively, these efforts indicate that integrative analysis of signaling networks with other datasets would highlight new insights into signaling mechanisms in different biological aspects, e.g., principles of miRNA regulation of signaling networks, cancer development/progression, and evolution.

2.2 Strategies of miRNA Regulation of Cellular Signaling Networks

As miRNAs are able to directly and specifically knock down protein expression, we hypothesized that miRNAs might play an important role in the regulation of the strength and specificity of cellular signaling networks through directly controlling the concentration of network components (proteins) at post-transcriptional and translational levels. We took genome-wide computationally predicted miRNA target genes from two recent studies (Krek et al. 2005; Lewis et al. 2005) and then mapped all the overlapped miRNA targets onto the human signaling network proteins to conduct an network-structural analysis. This analysis revealed several strategies of miRNA regulation of signaling networks (Cui et al. 2006).

We found that miRNAs more frequently target signaling proteins than others, e.g., 29.4 versus 17% of the network proteins and the total genes in human genome, respectively, are miRNA targets. This discovery implies that miRNAs might play a relatively more important role in regulating signaling networks than in other cellular processes. Normally, in signaling networks, cellular signal information flow initiates from extra-cellular space: a ligand binds to a cellular membrane receptor to start the signal, which is then transmitted by intracellular signaling components in cytosol and finally reaches the signaling components in the nucleus. We found that the fraction of miRNA targets increases with the signal information flow from the upstream to the downstream, e.g., from ligands, cell surface receptors, and intracellular signaling proteins to nuclear proteins. For example, only 9.1% of the ligands are miRNA targets, whereas half of the nuclear proteins, most of which are transcription factors, are miRNA targets. In other words, the miRNA targets are enriched more than five times in the most downstream proteins compared to the most upstream proteins. In signaling networks, adaptor proteins recruit downstream signaling components to the vicinity of receptors. They activate, inhibit or relocalize downstream components through direct protein–protein interactions. Adaptors do not have enzyme activity, but physically interact with upstream and downstream signaling proteins. One adaptor is able to recruit distinct downstream components in different cellular conditions. We found that miRNAs preferentially target the downstream components of adaptors, which have potential to recruit more downstream components. For example, the adaptor Grb2 (growth factor receptor-bound protein 2) directly interacts with 14 downstream signaling proteins, half of which are miRNA targets. These downstream components are functionally involved in different signaling pathways that lead to different cellular outputs. For example, SHC (Src homology two domain containing) regulates cell growth and apoptosis through activation of small GTPases of the Ras family, while N-WASP (Wiskott–Aldrich syndrome protein) is involved in the regulation of actin-based cytoskeleton through activation of small GTPases of the Rho family. These two components are targeted by different miRNAs. To accurately respond to extracellular stimuli, adaptors need to selectively recruit downstream components. If an adaptor can recruit more downstream components, these components should have a higher dynamic gene expression behavior. This principle is in agreement with the fact that miRNAs have a high spatio-temporal expression behavior, suggesting that miRNAs might play an important role for precise selection of cellular responses to stimuli by controlling the concentration of adaptors' downstream components.

We further showed that miRNAs more frequently target positively linked network motifs and less frequently target negatively linked network motifs. A complex signaling network can be broken down into distinct regulatory patterns, or network motifs, typically comprised of three to four interacting components capable of signal processing. In our previous work, we showed that mRNA decay plays an important role in motif regulatory behavior (Wang and Purisima 2005). We classified each type of motif into several subgroups based on the number of nodes that are miRNA targets. For example, the three node network may have none of their nodes as a miRNA target (category 0), or may have just one of their nodes as a miRNA target (category 1), or two (category 2) or all three as miRNA targets (category 3). For each motif, we calculated the ratio of positive links to the total directional (positive and negative) links (termed as Ra) in each subgroup and compared it with the average Ra in all the motifs, which is shown as a horizontal line in Fig. 1. For most motifs, the Ra in the subgroup in which none of the nodes are miRNA targets is less than the average Ra of all the motifs (Fig. 1, $P < 4 \times 10^{-3}$, Wilcoxon Ranksum test). This result suggests that miRNAs less frequently target negative regulatory motifs. In contrast, for most motifs, the preponderance of positive links in the subgroups increased as the number of miRNA-targeted components rose (Fig. 1). More significantly, when all nodes are miRNA targets in a motif, all the links in the motif are positive links ($P < 0.01$, Wilcoxon Ranksum test). These results suggest that miRNAs have high potential to target positively linked motifs. For example, AP1 (activator protein 1), CREB (cAMP-responsive element-binding protein) and CBP (CREB-binding protein) form a three-node positive feedback loop. All of the three proteins are miRNA targets.

Positive feedback loops are often used to convert a transient signal into a long-lasting cellular response and to make developmental switches. In the positive feedback loops, noise or fluctuation in any component can be easily amplified, and then drive the system to switch states randomly. In this situation, a negative control would enhance filtering or buffering of such noise or fluctuation amplification. Randomly switching phenotypes has been observed in the yeast galactose network, which contains both positive and negative feedbacks. When the negative feedback was removed from the network, the genes in the network would randomly switch on and off over time (Acar et al. 2005). Compared to transcriptional repressors, miRNAs are likely to tune target protein levels more rapidly at post-transcriptional level. Thus, miRNAs could significantly shorten the response delay. Therefore, by regulating positive regulatory loops, miRNAs might provide fast feedback responses and more effective noise filtering as well as precise definition and maintenance of steady states. In another study, we showed that cross-species gene expression divergences of the miRNA targets are indeed lower than those of others (Cui et al. 2007b). Considering the fact that positive feedback circuits are abundant in genomes (Ferrell 2002), we surmise that miRNAs regulation of positive regulatory

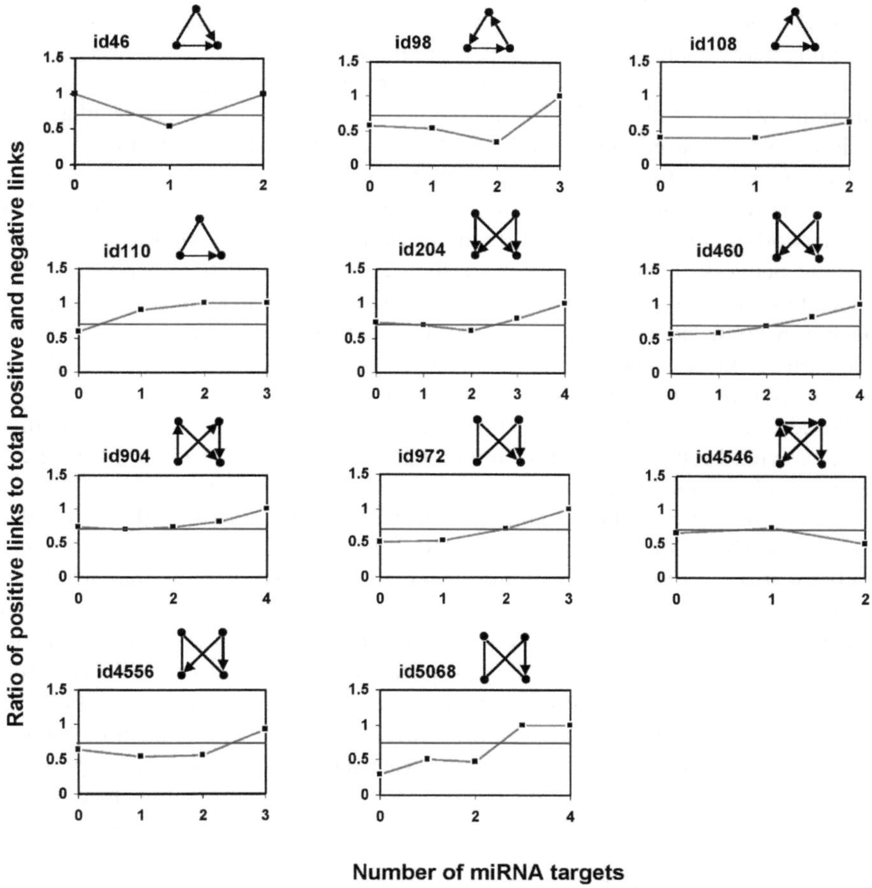

Fig. 1 Relative abundance of positive links in the individual subgroups of each type of network motif. Each type of motif was classified into several subgroups according to the number of nodes which are miRNA targets. For example, a three-node motif can be divided into four subgroups in which the miRNA target numbers are 0, 1, 2 and 3, respectively. The ratio of positive links to total positive and negative links in each subgroup was calculated and plotted as a function of miRNA target numbers per motif. The *horizontal lines* indicate the ratio of positive links to the total positive and negative links in all the respective network motifs. The network motif ID numbering system is from Alon's motif dictionary (http://www.weizmann.ac.il/mcb/UriAlon/NetworkMotifsSW/mfinder/motifDictionary.pdf)

loops might provide a common mechanism for noise filtering and buffering. Compared to transcriptional repressors, miRNAs are likely to tune target protein levels more rapidly at post-transcriptional level.

To explore which cellular machines of the signaling network are regulated by miRNAs, we explored the relationships between network themes and cellular machines. Network motifs are often linked together to form larger subgraphs.

Network themes are examples of such larger subgraphs which are enriched topological patterns containing clusters of overlapping motifs, represent a higher order of regulatory relationships between signaling proteins, and tie to particular biological functions (Zhang et al. 2005). To find the network themes which are regulated by each miRNA, we used the network motifs that contain at least one miRNA target and examined whether some of these motifs could aggregate into clusters. We found that, in general, the network motifs regulated by each miRNA formed one or two network themes. The sizes of the network themes range from 4 to 145 nodes. Most of network themes contain more than 20 nodes. Statistical analysis of the associations between these network themes and cellular machines (transcription machinery, translation machinery, secretion apparatus, motility machinery and electrical response) revealed that nearly 60% of miRNAs in this study could be associated to one or more cellular machines of the signaling network.

We also uncovered that highly linked scaffold proteins have higher probability to be targeted by miRNAs. For example, CRK (v-crk sarcoma virus CT10 oncogene homolog) and SNAP25 (synaptosome-associated protein of 25,000 Da) are targeted by six miRNAs (miR-1, miR-10a, miR-126, miR-133a, miR-20 and miR-93) and five miRNAs (miR-1, miR-128a, miR-130a, miR-153 and miR-27b), respectively. Scaffold proteins are neutrally linked to other two proteins that are either positively or negatively connected. Unlike adaptors, scaffold proteins do not directly activate or inhibit other proteins, but provide regional organization for activation or inhibition between other proteins. Scaffold proteins are able to recruit distinct sets of proteins to different pathways and thus maintain the specificity of signal information flows. Higher linked scaffold proteins can recruit more protein sets and have a higher degree of spatio-temporal expression behavior. The expression of miRNAs is highly specific for tissues and developmental stages, therefore it makes perfect sense that higher linked scaffold proteins are regulated by more miRNAs.

Finally, we discovered that miRNAs avoid targeting common components of cellular machines in the network. In the network, we identified 70 proteins that are shared by all of the five basic cellular machines, e.g., transcription machinery, translation machinery, secretion apparatus, motility machinery and electrical response. We found that only 14.3% of the 70 proteins are miRNA targets, a significant under-representation compared to the fraction of miRNA targets (29.4%) in the network ($P < 2 \times 10^{-4}$). This result suggests that miRNAs avoid disturbing basic cellular processes, because these common proteins are highly shared by basic cellular machines and should be frequently used in various cellular conditions.

These rules or principles uncovered above indicate that miRNAs regulate signaling networks in multiple ways. By selectively regulating positive regulatory motifs, highly connected scaffolds and the most network downstream components, miRNAs may provide a mechanism to terminate the pre-existing messages and facilitate quick and robust transitions for responses to new signals. These functions fit the spatio-temporal behavior of miRNA expression. On the other hand, miRNAs less frequently target negative regulatory motifs, common proteins of basic cellular machines and upstream network components such as ligands.

Although the accuracy of the miRNA target prediction methods has been well demonstrated by experimental validation of randomly selected targets, 12% of them could not be proved as real targets. Therefore, we performed the sensitivity analysis to test the potential effects of the errors on the robustness of the rules we discovered. We mimicked false positives by randomly adding an extra 10 and 20% of network proteins, which are not predicted miRNA targets, to the target list, performed the same analysis and recalculated the *P* values. In addition, we also removed 10 and 20% of miRNA targets to determine the effect of false negatives. The results indicate that the trend remains unchanged by the addition of the false positives or false negatives. Therefore, the principles we obtained in this analysis are robust against substantial errors.

3 miRNA Regulation of Gene Regulatory Networks

Gene regulatory networks describe the regulatory relationships between transcription factors and/or regulatory RNAs and genes. Theoretically, in a cell the entire gene regulatory network encodes a blueprint of gene regulatory relations and a framework for combinatorial mechanisms of using different regulatory relations to perform distinct biological functions. The network reflects the evolutionary selection, e.g., mRNAs of the hub transcription factors decay faster than other transcription factors (Batada et al. 2006; Wang and Purisima 2005). In the past 50 years, *E. coli* and yeast have been used as model organisms to study gene regulation. Therefore, rich information about gene regulatory relations for these organisms has been documented in the literature. RegulonDB, a manually curated database for collecting gene regulatory relations for *E. coli*, is one of the efforts to gather gene regulatory relationships in the literature (Salgado et al. 2004). A genome-wide determination of gene regulatory relations using chromatin immunoprecipitation coupled with DNA microarray in yeast represents an effort to uncover gene regulatory relations via high-throughput approaches (Lee et al. 2002). These efforts made it possible to analyze gene regulatory networks in a large-scale manner. Extensive analyses of gene regulatory networks in *E. coli* and yeast have been conducted. However, gene regulatory information is still less comprehensive in mammalian cells.

To get insights into how miRNA interacts with gene regulatory networks in humans, we took a dataset which represents three transcription factors (TFs), OCT4 (a POU family transcription factor), NANOG (a regulator gene has a great potential to reprogram the cells) and SOX2 (sex determining region Y (sry) -box 2) and their target genes in human embryonic stem cells (Boyer et al. 2005). The regulatory relationships between the transcription factors and their target genes were determined by using chromatin immunoprecipitation coupled with DNA microarray. The three transcription factors totally regulate 2,043 genes, of which 1,314 genes are co-regulated by two of the three TFs and 391 genes are co-regulated by all the transcription factors. Using this dataset, we built a small gene regulatory network in which nodes represent transcription factors or genes and links represent regulatory relations between transcription factors and the regulated genes or transcription factors (Cui et al. 2007a). We mapped the miRNA targets onto the genes of the network. For genes in the network, we divided them into three groups, in which they are regulated by one, two and all three of the transcription factors, respectively, and counted the number of genes that are miRNA targets and the number of genes that are not miRNA targets, respectively, in each group. We revealed that miRNA targets are significantly enriched in the genes that are regulated by more transcription factors. These results tell us that a gene that is regulated by a larger number of transcription factors is also more likely to be regulated by miRNAs.

Because the network is very small, e.g., containing only three transcription factors, we were not confident whether the conclusion we obtained above is robust. To validate and expand the above observation, we examined the relationship between transcription factors and miRNAs for gene regulation at a genome-wide scale (Cui et al. 2007a). Although we were not able to get the datasets for a genome-wide gene regulatory network in human, we could access the datasets, which are computationally determined, of the number of transcription-factor–binding-sites (TFBS) on the promoter region for each gene in human genome. TFBSs or *cis*-regulatory elements are normally located in the promoter region of a gene. Transcription factors regulate a gene through binding to the TFBSs of the gene. The two TFBS datasets were taken from recent publication of Cora et al. (2005) and Xie et al. (2005). Generally speaking, the more TFBSs a gene has, the more transcription factors the gene is regulated by, and the more complex its regulation can be as provided by various possible combinations of transcription factors.

We performed analysis to determine the relationship between the number of the TFBSs and the possibility of being a miRNA target of genes. Toward this end, we grouped genes based on their TFBS numbers (TFBS-count). The TFBS-count is significantly correlated with the miRNA target rate (Pearson's correlation coefficient $r = 0.94$, $P < 3.5 \times 10^{-68}$). For example, the miRNA target rate is doubled from the group of genes that have less than 10 TFBSs to those that have more than 100 TFBSs (from ~35% to ~70%). A similar result was obtained using the TFBS dataset (Xie et al. 2005) ($r = 0.97$, $P < 3.9 \times 10^{-113}$). These results are in agreement with the finding in the human stem cell gene regulation and therefore strongly suggest that miRNAs preferentially target the genes that bear more TFBSs has broad applicability. On the other hand, we analyzed the relationship between the number of miRNAs and the number of TFBSs in the same genes. We found a significant correlation ($r = 0.74$, $P < 6.1 \times 10^{-12}$). A similar result was obtained when using the TFBS dataset of Xie et al. ($r = 0.72$, $P < 9.5 \times 10^{-12}$). These results suggest that the genes that are targeted by more miRNAs have more TFBSs (Cui et al. 2007a).

Collectively, we uncovered a basic rule of miRNA regulation of gene regulatory networks: a gene that is regulated by more transcription factors is also more likely to be regulated by miRNAs. These results indicate that the complexity of gene regulation by miRNAs at the post-transcriptional level is positively related to the complexity of gene regulation by transcription factors at the transcriptional level in human genome. Genes, which are more complexly regulated at transcriptional level, are required to be turned on more frequently and, furthermore, are more likely to be expressed at different temporal and spatial conditions. Therefore, they are also required to be turned off more frequently. miRNAs as negative regulators can exert the turning-off function at post-transcriptional level through repressing mRNA translation and/or mediating cleavage of mRNAs. This is a potentially novel discovery of mechanism for coordinated regulation of gene expression. Such coordinately regulated genes are enriched in certain biological processes and functions, particularly in those involved in developmental processes.

In a seperate study, we showed that miRNAs preferentially regulate positive regulatory loops of signaling networks (Cui et al. 2006). It is also reasonable to hypothesize that miRNAs preferentially regulate positive regulatory loops of gene regulatory networks. Given that positive feedback circuits are abundant in genomes (Ferrell 2002), we surmise that miRNAs frequently regulate gene regulatory networks by targeting positive regulatory loops. Unfortunately, the datasets for positive and negative regulatory relations are currently unavailable in humans/rodents and even in worm and fly, and so we are not able to test this hypothesis at this moment.

4 miRNA Regulation of Metabolic Networks

Metabolites are critical in a cell. Certain metabolites are the basic building blocks of proteins, DNAs and RNAs. Some metabolites such as fatty acids take part in the cellular processes for growth, development and reproduction, while some others are involved in defense mechanisms against parasites and cell signaling. Biochemical characterization of metabolic reactions and enzymes has been conducted for many years. Traditionally, metabolic reactions are organized and illustrated as metabolic pathways. In terms of pathway components, metabolic pathways are so far the clearest and the most comprehensive. Genome sequencing efforts offer comparative genomic analysis of metabolic reactions and enzymes across many species. As a result, the information for metabolic pathways is more enriched than before. This makes it possible to build comprehensive metabolic maps at this time (Feist et al. 2007).

Many metabolites are shared by different metabolic pathways and are further intertwined to form a complex metabolic network. Thus, various cellular activities are accompanied with the changes of metabolism. It is essential to control the rates of metabolic processes in response to changes in the internal or external environment for living cells. Mechanisms that control metabolic networks are complex and involve transcriptional, post-transcriptional and translational regulations. For a long time, we have had reasons to believe that the enzymes of metabolic networks are tightly controlled by transcription factors. Moreover, the principles of transcriptional regulation of metabolic networks by transcription factors have been illustrated through an integrative analysis of gene expression profiles and the yeast metabolic network. Since miRNAs have emerged as an abundant class of negative regulators, it is reasonable to think that miRNAs might extensively regulate metabolic networks.

Indeed, miRNAs have been shown to regulate amino acid catabolism, cholesterol biosynthesis, triglyceride metabolism, insulin secretion, and carbohydrate and lipid metabolism, although the molecular mechanisms of miRNA regulation of metabolism are not clear (Krutzfeldt and Stoffel 2006).

We systematically analyzed the human and *D. melanogaster* metabolic networks by integrating miRNA target genes onto the networks. In both networks, miRNAs selectively regulate certain metabolic processes such as amino acid biosynthesis, and certain sugar and lipid metabolisms, so that they can selectively control metabolite production. When miRNAs regulate specific individual metabolic pathway, they often regulate the last reaction step (LRS) of that pathway. Furthermore, once miRNAs regulate the LRS of a pathway, the cut vertex to the LRS and other enzymes that are in the upstream metabolic flows to the LRS are also enriched with miRNA targets. A cut vertex or a cutpoint is such a bottleneck node that its deletion will disconnect at least one component from the network. Cut vertexes are in crucial network positions and become bottlenecks of the network, and therefore control metabolic flows from a part to another in the network. These results imply that miRNA is strongly involved in coordinated regulation of metabolic processes in metabolic networks.

5 miRNA Regulation of Protein Interaction Networks

Protein interaction networks provide a valuable framework for a better understanding of the functional organization of the proteome and offer a mechanistic basis for most biological processes in organisms. Large-scale determination of interactions between proteins has been conducted in yeast, *E. coli* and other bacteria, worm, fly and human (Wang et al. 2007). Because the datasets for protein interaction networks are relatively more easy of access, extensive analyses of protein interaction networks have been conducted, ranging from pure network structural analysis to the analyses of network motifs, network themes, network communities and evolution.

Recent study showed that miRNAs preferentially regulate the proteins which have more interacting partners in the network. Protein connectivity in a human protein interaction network is positively correlated with the number of miRNA target-site types (Liang and Li 2007). In principle, if a protein has more interacting protein partners, it normally takes part in more biological processes and then its expression is more dynamic (Wang et al. 2007). Therefore, it makes sense that when a protein has more interacting protein partners, it will be regulated by more transcription factors and more miRNAs. This is in agreement with our previous findings that miRNAs preferentially regulate the genes which are regulated by more transcription factors in the gene regulatory network (Cui et al. 2007a). Consistently, genes of two interacting proteins tend to be under similar miRNA regulation, which again is in agreement with these facts that genes encoding interacting proteins tend to have similar mRNA expression profiles (Li et al. 2004). Furthermore, our analysis of microarray profiles of miRNA target genes in different tissues showed a similar trend: broadly expressed mRNAs tend to be regulated by more miRNAs (Yu et al. 2007). Highly linked proteins, e.g., hub proteins, can be divided into two groups based on the clustering coefficient, which is defined as the fraction of the real number of links between a node's neighbors and the maximum possible number of links between them. A hub protein having a high clustering coefficient is likely to be an intra-modular hub, which interacts with most of its partners simultaneously to form a protein complex, and completes a coherent function. On the other hand, a hub protein having a low clustering coefficient tends to be an inter-modular hub, which tends to interact with other proteins in different times and places, and then coordinates different functional modules (Liang and Li 2007). It is understandable that inter-modular hub proteins are more likely to be regulated by miRNAs.

Notably, analyses of miRNA regulation of gene regulatory networks and the protein interaction networks obtained a consistent conclusion, which reflects a general rule of miRNA regulation of whole genome genes, because general networks (gene regulatory networks and the protein interaction networks) collect information for all kinds of activities in cell. On the other hand, analyses of miRNA regulation of metabolic networks and signaling networks obtained different insights. This observation might reflect that cellular specific networks (metabolic networks and signaling networks) have local and specific cellular themes, while general networks (gene regulatory networks and the protein interaction networks) encode global features of cells.

6 miRNA Regulatory Network Motifs

At present, experimental large-scale studies of miRNA regulation of various cellular networks are still not trivial. However, it is experimentally feasible for characterizing miRNA regulatory network motifs or miRNA regulatory circuits. As we mentioned above, network motifs are the statistically significant recurring structural patterns or small subgraphs or sub-networks, which are a result of convergent evolution at the network level, and carry out important functions in cells. Several studies have experimentally explored miRNA regulatory motifs. For example, the secondary vulva cell fate in *C. elegans* is promoted by Notch signaling, which also activates miR-61, which in turn post-transcriptionally represses an inhibitory factor of Notch signaling, thereby stabilizing the secondary vulva fate (Yoo and Greenwald 2005). Similar circuits are also found in the differentiation of neurons in *C. elegans* (Johnston et al. 2005), eye development in *Drosophila* (Li and Carthew 2005; Li et al. 2006), and granulocytic differentiation in human (Fazi et al. 2005). Another example of miRNA regulatory circuits is that miR-17-5p represses E2F1, and both are transcriptionally activated by c-Myc in human cells (O'Donnell et al. 2005). More recently, two groups of researchers published data showing that miR-34 family is a key component of the p53 tumor suppressor network, which controls cellular responses to signals such as DNA damage and oncogene activation (He et al. 2007; Raver-Shapira et al. 2007).

To systematically characterize miRNA regulatory circuits in human and mouse genomes, one computational study tended to take the advantages of intron-based miRNAs, or embedded miRNAs (Tsang et al. 2007). More than 80% of all known miRNAs in human and mouse are embedded in introns of coding or noncoding genes (Kim and Kim 2007). One could hypothesize that the intron-based miRNAs might be co-expressed with their host genes. Indeed, the expression profiles of most embedded miRNAs examined thus far are highly correlated to their host genes at both the tissue and individual cell levels (Li and Carthew 2005), suggesting that they tend to be co-transcribed at identical rates from the same promoters (Kim and Kim 2007). Theses facts led to the assumption that the relative level of host-gene transcription across conditions can accurately serve as a proxy for that of the embedded miRNAs.

An upstream regulatory factor triggers the expression of an embedded miRNA and its targets at the same time, while the transcribed miRNA also regulates its targets. Intuitively, two types of miRNA regulatory motifs, or feedforward loops can be built up (Fig. 2a,b): Type II miRNA feedforward loops, in which an upstream

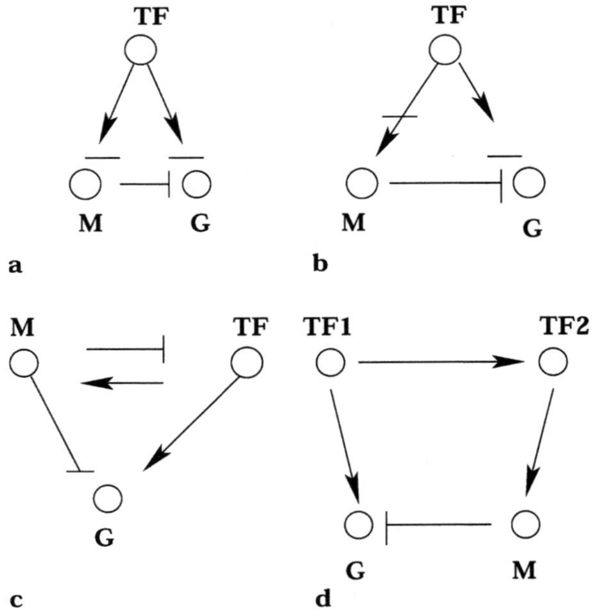

Fig. 2 Types of miRNA regulatory network motifs in human genome. *M* represents miRNA while TF and *G* represent transcription factors and targeted genes, respectively. Links represent regulatory relations. *Arrows* represent activation, while a *bar* represents inhibition. (**a**) Type I miRNA feedforward motif, in which TF sends either activation signal or inhibition signal at the same time to M and G. (**b**) Type II miRNA feedforward motif. In one setting, TF activates M and inhibits G at the same time, while in another setting, TF inhibits M and activates G. (**c**) Composite regulatory motif. (**d**) Indirect feedforward motif

factor could repress the transcription of a target gene and simultaneously activate the transcription of a miRNA that inhibits target gene translation, and Type I miRNA feedforward loops, in which an upstream factor activates the transcription of a target gene and simultaneously activates the transcription of a miRNA that inhibits target gene translation. One example has been characterized experimentally for Type I miRNA feedforward loops, where miR-17-5p represses E2F1, and both are transcriptionally activated by c-Myc in human cells (O'Donnell et al. 2005). This kind of loop has the potential to provide a host of regulatory and signal processing functions (Hornstein and Shomron 2006).

By analyzing the embedded miRNAs in human and mouse using the Novartis human and mouse expression atlas comprising 61 tissues/cell types, Type II miRNA feedforward loops have been found to be prevalent for a significant fraction of the embedded miRNAs. This result is in agreement with previous findings that predicted that target transcripts of several tissue-specific miRNAs tend to be expressed at a lower level in tissues where the miRNAs are expressed (Farh et al. 2005; Stark et al. 2005; Yu et al. 2007). Furthermore, a significant fraction of Type I miRNA feedforward loops were also found, especially prevalent in mature neurons.

In Type II miRNA feedforward loops, a miRNA regulates its targets coherently with transcriptional control, thereby reinforcing transcriptional logic at the post-transcriptional level. As suggested, such circuits can serve as a surveillance mechanism to suppress "leaky" transcription of target genes (Hornstein and Shomron 2006; Stark et al. 2005). It is reasonable to think that these loops would act in concert with other regulators to increase the feedback strength and enhance the robustness of irreversible cellular differentiation. On the other hand, Type I miRNA feedforward loops might prevent proliferation of noise-driven transitions as illustrated in the example of c-Myc/E2F1/miR-17-92 network (O'Donnell et al. 2005).

Another effort to computationally identify miRNA network motifs has been conducted by analyzing the cooperation between transcription factors and miRNAs for regulating the same target genes (Shalgi et al. 2007). The computational techniques for this kind of analysis are similar to those used in identifying transcription factor binding sites. They first looked for transcription factor–miRNA pairs with a high rate of co-occurrence in the promoters and 3 UTR of the regulated genes. Statistical tests showed that transcription factor–miRNA pairs significantly co-occur, which is in agreement with the findings of miRNA regulation of gene regulatory networks. Secondly, they searched miRNA networks using randomization tests. Similar to the above report, Type I and II feedforward loops were also discovered. In addition, two other types of miRNA network motifs were also documented: composite regulatory loops and indirect feedforward loops (Fig. 2c,d). In the former motif, a miRNA (MR) represses a transcription factor (TF) and a target gene (G), while the transcription factor (TF) activates the miRNA (MR) and the target gene (G). In the latter motif, a transcription factor, TF1 activates another transcription factor, TF2 and a target gene, G1, in turn, the TF2 activates a miRNA, MR1, which represses the target gene, G1. However, the functions of these two motif types are not clear yet.

7 Summary

In summary, miRNAs are extensively involved in gene regulation as network motifs in genomes. By analyzing the interactions between miRNAs and general networks (gene regulatory and protein interaction networks), a common miRNA regulatory principle is emerging: miRNAs preferentially regulated the genes that have high regulation complexity. This fact suggests a novel mechanism of coordinated regulation between transcriptional level and post-transcriptional level for gene regulation. In addition, for cellular specific networks (metabolic and signaling networks), miRNAs have different regulatory strategies. For example, miRNAs preferentially regulate positive regulatory motifs, highly connected scaffolds and most network downstream components of cellular signaling networks, which might provide a mechanism to terminate the pre-existing messages and facilitate quick and robust transitions for responses to new signals. On the other hand, miRNAs less frequently target negative regulatory motifs, common proteins of basic cellular machines and upstream network components such as ligands in signaling networks. In metabolic networks, miRNAs selectively regulate the genes which have specific network structural features on the network, which might provide effective and selective regulation of cellular metabolism.

Acknowledgements This work is partially supported by Genome and Health Imitative. I thank our team members, Dr. Y. Deng and Mrs. M. Mistry for comments.

References

Acar M, Becskei A, van OA (2005) Enhancement of cellular memory by reducing stochastic transitions. Nature 435:228–232

Awan A, Bari H, Yan F et al (2007) Regulatory network motifs and hotspots of cancer genes in a mammalian cellular signaling network. IET Syst Biol 1:292–297

Batada NN, Hurst LD, Tyers M (2006) Evolutionary and physiological importance of hub proteins. PLoS Comput Biol 2:e88

Boyer LA, Lee TI, Cole MF et al (2005) Core transcriptional regulatory circuitry in human embryonic stem cells. Cell 122:947–956

Cora D, Herrmann C, Dieterich C et al (2005) Ab initio identification of putative human transcription factor binding sites by comparative genomics. BMC Bioinformatics 6:110

Cui Q, Yu Z, Purisima EO et al (2006) Principles of microRNA regulation of a human cellular signaling network. Mol Syst Biol 2:46

Cui Q, Yu Z, Pan Y et al (2007a) MicroRNAs preferentially target the genes with high transcriptional regulation complexity. Biochem Biophys Res Commun 352:733–738

Cui Q, Yu Z, Purisima EO et al (2007b) MicroRNA regulation and interspecific variation of gene expression. Trends Genet 23:372–375

Farh KK, Grimson A, Jan C et al (2005) The widespread impact of mammalian microRNAs on mRNA repression and evolution. Science 310:1817–1821

Fazi F, Rosa A, Fatica A et al (2005) A minicircuitry comprised of microRNA-223 and transcription factors NFI-A and C/EBPalpha regulates human granulopoiesis. Cell 123:819–831

Feist AM, Henry CS, Reed JL et al (2007) A genome-scale metabolic reconstruction for *Escherichia coli* K-12 MG1655 that accounts for 1260 ORFs and thermodynamic information. Mol Syst Biol 3:121

Ferrell JE (2002) Self-perpetuating states in signal transduction: positive feedback, double-negative feedback and bistability. Curr Opin Cell Biol 14:140–148

He L, He X, Lim LP et al (2007) A microRNA component of the p53 tumour suppressor network. Nature 447:1130–1134

Hornstein E, Shomron N (2006) Canalization of development by microRNAs. Nat Genet 38(Suppl):S20–S24

Johnston RJ Jr, Chang S, Etchberger JF et al (2005) MicroRNAs acting in a double-negative feedback loop to control a neuronal cell fate decision. Proc Natl Acad Sci USA 102:12449–12454

Kim YK, Kim VN (2007) Processing of intronic microRNAs. EMBO J 26:775–783

Krek A, Grun D, Poy MN et al (2005) Combinatorial microRNA target predictions. Nat Genet 37:495–500

Krutzfeldt J, Stoffel M (2006) MicroRNAs: a new class of regulatory genes affecting metabolism. Cell Metab 4:9–12

Lee TI, Rinaldi NJ, Robert F et al (2002) Transcriptional regulatory networks in *Saccharomyces cerevisiae*. Science 298:799–804

Lehner B, Crombie C, Tischler J et al (2006) Systematic mapping of genetic interactions in *Caenorhabditis elegans* identifies common modifiers of diverse signaling pathways. Nat Genet 38:896–903

Lewis BP, Burge CB, Bartel DP (2005) Conserved seed pairing, often flanked by adenosines, indicates that thousands of human genes are microRNA targets. Cell 120:15–20

Li X, Carthew RW (2005) A microRNA mediates EGF receptor signaling and promotes photoreceptor differentiation in the *Drosophila* eye. Cell 123:1267–1277

Li S, Armstrong CM, Bertin N et al (2004) A map of the interactome network of the metazoan *C. elegans*. Science 303:540–543

Li Y, Wang F, Lee JA et al (2006) MicroRNA-9a ensures the precise specification of sensory organ precursors in *Drosophila*. Genes Dev 20:2793–2805

Liang H, Li WH (2007) MicroRNA regulation of human protein protein interaction network. RNA 13:1402–1408

Ma'ayan A, Jenkins SL, Neves S et al (2005) Formation of regulatory patterns during signal propagation in a mammalian cellular network. Science 309:1078–1083

Oda K, Kitano H (2006) A comprehensive map of the toll-like receptor signaling network. Mol Syst Biol 2:2006.0015

Oda K, Matsuoka Y, Funahashi A et al (2005) A comprehensive pathway map of epidermal growth factor receptor signaling. Mol Syst Biol 1:2005.0010

O'Donnell KA, Wentzel EA, Zeller KI et al (2005) c-Myc-regulated microRNAs modulate E2F1 expression. Nature 435:839–843

Raver-Shapira N, Marciano E, Meiri E et al (2007) Transcriptional activation of miR-34a contributes to p53-mediated apoptosis. Mol Cell 26:731–743

Salgado H, Gama-Castro S, Martinez-Antonio A et al (2004) RegulonDB (version 4.0): transcriptional regulation, operon organization and growth conditions in *Escherichia coli* K-12. Nucleic Acids Res 32:D303–D306

Shalgi R, Lieber D, Oren M et al (2007) Global and local architecture of the mammalian microRNA-transcription factor regulatory network. PLoS Comput Biol 3:e131

Stark A, Brennecke J, Bushati N et al (2005) Animal microRNAs confer robustness to gene expression and have a significant impact on 3 UTR evolution. Cell 123:1133–1146

Tsang J, Zhu J, van OA (2007) MicroRNA-mediated feedback and feedforward loops are recurrent network motifs in mammals. Mol Cell 26:753–767

Wang E, Purisima E (2005) Network motifs are enriched with transcription factors whose transcripts have short half-lives. Trends Genet 21:492–495

Wang E, Lenferink A, O'Connor-McCourt M (2007) Cancer systems biology: exploring cancer-associated genes on cellular networks. Cell Mol Life Sci 64:1752–1762

Xie X, Lu J, Kulbokas EJ et al (2005) Systematic discovery of regulatory motifs in human promoters and 3′ UTRs by comparison of several mammals. Nature 434:338–345

Yoo AS, Greenwald I (2005) LIN-12/Notch activation leads to microRNA-mediated down-regulation of Vav in *C. elegans*. Science 310:1330–1333

Yu Z, Jian Z, Shen SH et al (2007) Global analysis of microRNA target gene expression reveals that miRNA targets are lower expressed in mature mouse and *Drosophila* tissues than in the embryos. Nucleic Acids Res 35:152–164

Zhang LV, King OD, Wong SL et al (2005) Motifs, themes and thematic maps of an integrated *Saccharomyces cerevisiae* interaction network. J Biol 4:6

Part II
RNA Interference

Targeting Viral Heart Disease by RNA Interference

S. Merl and R. Wessely(✉)

Contents

Abstract Viral heart disease (VHD) is an important clinical disease entity both in pediatric as well as adult cardiology. Coxsackieviruses (CVBs) are considered an important cause for VHD in both populations. VHD may lead to dilated cardiomyopathy and heart failure which can ultimately require heart transplantation. However, no specific treatment modality is so far available. We and others have shown that coxsackieviral replication and cytotoxicity can be successfully targeted by RNA interference, thus leading to increased cell viability and even prolongation of survival in vivo. However, considerable limitations have to be solved before this novel therapeutic approach may enter the clinical trials arena.

1 Introduction

Picornaviridae comprise one of the largest families of human infectious pathogens of major clinical significance. Among the different genera, particularly rhinoviruses and enteroviruses such as coxsackievirus and poliovirus are of particular medical and economical interest. Coxsackievirus B3 (CVB3) has been identified as the most causative agent for the pathogenesis of viral heart disease both in adult and

R. Wessely
Deutsches Herzzentrum and 1 Medizinische Klinik, Technische Universität, Lazarettstrasse 36, 80636 München, Germany
e-mail: rwessely@web.de

V.A. Erdmann et al. (eds.) *RNA Technologies in Cardiovascular Medicine and Research*,
doi: 10.1007/978-3-540-78709-9,

pediatric patients (Woodruff 1980). Coxsackieviral heart disease encompasses acute as well as chronic forms of myocarditis (Feldman and McNamara 2000) and may lead to arrhythmias, heart failure, and sudden death. In addition, the coxsackieviral genome can establish a persistent infection of the myocardium without clinical signs of myocarditis and thus may induce viral cardiomyopathy (Kandolf et al. 1993; Wessely et al. 1998b). Viral cardiomyopathy may evolve to a life-threatening progressive disease, characterized by enlargement of the heart chambers, myocardial and ventricular dysfunction and finally heart failure that may ultimately lead to heart transplantation. Besides its relevance for the pathogenesis of viral heart disease, extracardiac diseases including hepatitis (Wessely et al. 2001), pancreatitis (Zaragoza et al. 1999), aseptic meningitis (Feuer et al. 2003) as well as encephalomyelitis (Bauer et al. 2002), have been attributed to CVB3. At present, treatment with ß-interferon is clinically under investigation to attenuate clinical symptoms and improve cardiac function. However, a specific antiviral drug therapy for coxsackievirus-mediated disease is currently not available.

RNA interference (RNAi) is an evolutionarily highly conserved endogenous mechanism for the regulation of gene expression by sequence-specific RNA degradation. First described by Fire et al. (1998) as an unexpected anomaly during antisense inhibition of gene expression in *Caenorhabditis elegans*, the process of RNA interference proved to be conserved also in many higher organisms (Billy et al. 2001; Elbashir et al. 2001a,b). The post-transcriptional gene silencing is triggered by long, double-stranded RNA molecules, which are cleaved by a cytoplasmatic dsRNA specific endonuclease known as Dicer into 21–23 nucleotides long, double-stranded short interfering RNA (siRNA) molecules (Bernstein et al. 2001; Elbashir et al. 2001a; Hammond et al. 2000). The siRNA molecules associate with helicase and nuclease molecules to form the targeting complex, known as RNA-induced silencing complex (RISC). Within RISC, the double-stranded siRNA molecule is unwound by helicase activity and the sense strand released. The antisense strand of the siRNA directs the complex to homologous complementary target mRNA sequences and hybridizes by Watson-Crick base pairing. Finally, the endonuclease Argonaute 2, also a component of the multiprotein complex RISC, promotes the precise and highly sequence-specific degradation of target RNA. Since RNA interference has probably evolved besides its role in basic cellular processes determining cell fate and differentiation (Abrahante et al. 2003; Brennecke et al. 2003) as an endogenous defence mechanism for protection from invading genetic elements like transposons and particularly viruses (Downward 2004), this ubiquitous pathway has opened exciting possibilities for exploiting it experimentally for functional genomics and in particular therapeutically in the fight against infections.

Initially, the attempt to provoke a specific knockdown of an individual gene by treating mammalian cells with long double-stranded RNAs failed. In mammalian cells, duplex RNA molecules longer than 30 nucleotides induce an interferon response, which triggers nonspecific degradation of RNA and inhibition of protein synthesis (Stark et al. 1998), leading ultimately to cell death. The crucial finding of 21–23 bp long siRNA molecules mediating the sequence-specific degradation of homologous RNA without affecting cellular functions (Elbashir et al. 2001a) alleviated this problem. By entering the RNA interference pathway equally to Dicer products downstream of the interferon pathway, these molecules mostly circumvent the stimulation of a stress response. In this process, gene silencing may either be achieved by exogenous delivery of chemically synthesized siRNAs (Elbashir et al. 2001a) or by vector-derived expression of either siRNAs or double-stranded short hairpin RNAs (shRNA), which are subsequently processed by Dicer in functional siRNAs (Brummelkamp et al. 2002; Miyagishi and Taira 2002; Paul et al. 2002).

The technique of siRNA-mediated silencing rapidly turned out to be an excellent experimental tool to analyze the function of individual genes, resulting in several studies demonstrating the impact of siRNAs to interfere with cellular processes such as apoptosis (Kartasheva et al. 2002; Lassus et al. 2002), or cell cycle regulation (Chen et al. 2002). Notably, the finding of synthetic siRNA mediating highly sequence specific RNA degradation provided a novel therapeutic tool to target distinct RNAs involved in the pathogenesis of human disease, such as cancer, autoimmune disease, dominant genetic disorders and in particular viral diseases. In this context, potential targets for RNAi-based antiviral therapies constitute of viral genes that are essential for virus replication or, on the other hand, cellular genes that are essential for virus entry or involved in the virus life cycle. For RNA viruses, theoretically any region of the viral genome, even a non-coding sequence, may serve as a potential target site for nucleic-acid based silencing. Presently, RNAi targeted pathogens of clinical importance encompass the human immunodeficiency virus (Boden et al. 2003; Coburn and Cullen 2002; Das et al. 2004; Hu et al. 2002; Jacque et al. 2002; Ji et al. 2003; Lee et al. 2002, 2003; Novina et al. 2002; Park et al. 2003), influenza virus (Ge et al. 2003), herpes simplex virus (Palliser et al. 2006), respiratory syncytial virus (Bitko and Barik 2001), hepatitis B virus (Hamasaki et al. 2003; McCaffrey et al. 2003; Shlomai and Shaul 2003; Ying et al. 2003) and poliovirus (Gitlin et al. 2002). In vivo approaches demonstrating RNAi mediated intracellular immunity encompass hepatitis B (Giladi et al. 2003; McCaffrey et al. 2003; Morrissey et al. 2005b), hepatitis C (McCaffrey et al. 2002; Wang et al. 2005), influenza (Ge et al. 2004; Tompkins et al. 2004), and respiratory syncytial virus (RSV) (Bitko et al. 2005; Zhang et al. 2005). Ongoing clinical trials utilize RNAi-based immunity to protect cells from infection with viruses such as respiratory syncytial virus infection (Alnylam Pharmaceuticals, Cambridge, Mass., USA) and trials protecting against infection with the human immunodeficiency virus (Benitec, Melbourne, Vic, Australia).

Our laboratory has focused on the investigation of coxsackievirus induced pathophysiologic processes in the context of viral heart diseases, particularly in consideration of potential therapeutic implications. Coxsackievirus B3 belongs to the picornaviridae virus family and is a close relative of poliovirus. Its genome is a messenger-like positive strand RNA molecule with a single open reading frame encoding a monocistronic polyprotein that is processed post-translational. Upon CVB3 infection, cardiac tissue injury occurs either by a direct viral cytopathic effect (Badorff et al. 1999; Wessely et al. 1998a) and/or by immunomodulatory mechanisms precipitated by viral infection (Knowlton and Badorff 1999). Therefore, a therapeutic drug concept promoting virus elimination can be considered as a key therapeutic approach to cure or attenuate CVB3-related disease. The

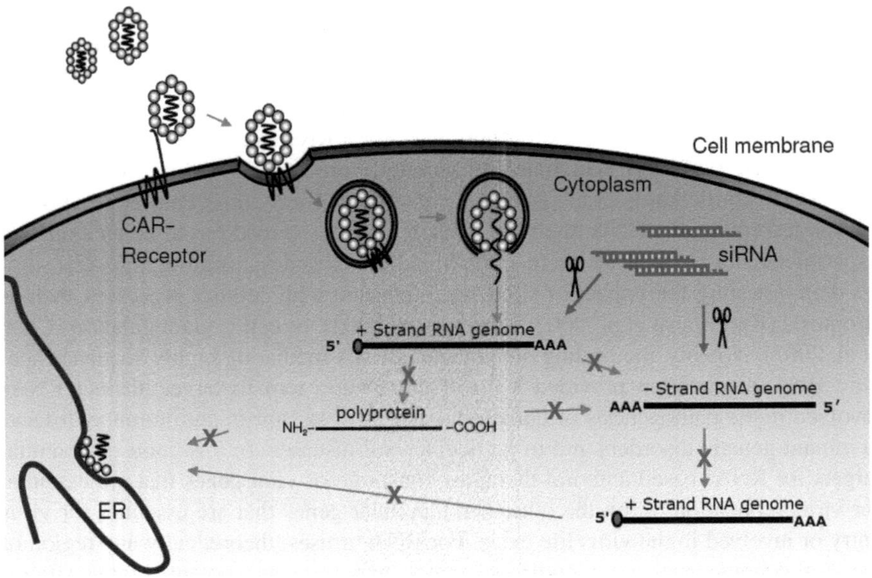

Fig. 1 Targeting coxsackieviral infection cycle by RNA interference. The coxsackieviral life cycle begins with the entry of the host cells, which express the coxsackievirus-adenovirus receptor (*CAR*). Upon cell entry, coxsackievirus releases its messenger like plus-stranded RNA genome. Subsequently, the viral genome mediates its own translation into a monocistronic polyprotein, which is processed post-translationally by viral proteases. The viral RNA dependent RNA polymerase is essential for virus replication via a negative-stranded intermediate. Ultimately the viral RNA assembles with viral proteins at the endoplasmatic reticulum (*ER*) to form infectious virions that are released from the host cell. Therapeutic antiviral RNA interference should result in the degradation of positive-stranded viral RNA as well as negative-stranded intermediates, resulting in the complete abrogation of the subsequent viral maturation process

pivotal observation that siRNAs may facilitate viral clearance and intracellular immunity without apparent side effects during infections with pathogenic human viruses such as the human immunodeficiency virus (HIV) (Rossi 2006) has raised exciting possibilities. The use of highly specific siRNAs targeting distinct regions of the viral genome (Fig. 1) as well as host genes that are relevant for virus entry and maturation represents a novel therapeutic strategy to cure or attenuate in particular coxsackievirus-mediated diseases.

2 Preclinical Results of Therapeutic Antiviral RNA Interference: Focus on CVB3-Associated Disease

In recent years, the evidence of siRNA conferring genomic immunity via the endogenous RNA interference pathway generated vigorous enthusiasm to exploit the RNAi mechanism therapeutically in the fight against a broad range of intractable diseases, including neurodegenerative diseases (Rodriguez-Lebron and Gonzalez-Alegre 2006) such as Alzheimer, Huntington and Parkinson's disease, as well as macular degeneration and cancer (Alisky and Davidson 2004; Barik 2005; Dorsett and Tuschl 2004; Shankar et al. 2005). The RNAi-mediated silencing of various oncogenes and cancer-associated genes proved to be efficient for the attenuation of tumor growth, tumor survival, cellular invasion and metastasis both in vitro and in vivo (Friedrich et al. 2004; Gartel and Kandel 2006). Given its primordial antiviral function, the RNA interference pathway soon became important for antiviral drug development (Table 1).

The interest to identify RNAi-based therapeutic strategies has been particularly strong for the reverse-transcribing human immunodeficiency virus (HIV). Several laboratories obtained significant inhibition of the HIV-1 replication applying both synthetic and vector-derived siRNAs/shRNAs directed against the viral genome and HIV-encoded RNAs, such as the TAR element, tat, rev, gag, env, vif, nef and reverse transcriptase (Boden et al. 2003; Coburn and Cullen 2002; Das et al. 2004; Hu et al. 2002; Jacque et al. 2002; Lee et al. 2002, 2003; Novina et al. 2002; Park et al. 2003). Novina et al. (2002) used RNAi to substantially suppress the viral *gag* gene, which encodes a precursor protein being a key component for uncoating as well as maturation processes during the HIV infection cycle. Due to its enormous genetic diversity as well as long-term siRNA treatment evolving mutational variants presenting nucleotide substitutions and deletions in their siRNA target sequences (Boden et al. 2003; Das et al. 2004), HIV can escape the antiviral RNAi. Indeed, the antiviral effectiveness was retained by simultaneously targeting different regions of the viral RNA (Boden et al. 2003; Gitlin et al. 2002). As an alternative strategy, cellular cofactors essential for the virus entry have also been successfully silenced, resulting in a profound inhibition of HIV entry and subsequent replication, encompassing the viral receptor CD4 (Novina et al. 2002) and the co-receptors CCR5 (Lee et al. 2003; Qin et al. 2003) as well as CXCR4 (Ji et al. 2003). In a previous study, the therapeutic effectiveness was significantly enhanced by the lentiviral-based delivery of a combination of shRNAs targeting the HIV tat and rev RNAs, a ribozyme directed against the host co-receptor CCR5, and a nucleolar-localizing TAR RNA decoy, each blocking HIV by a distinct mechanism (Li et al. 2005). At present, a clinical trial using simultaneous application of multiple siRNAs is in progress (Benitec, Melbourne, Vic, Australia).

The first evidence for the in vivo antiviral effectiveness of RNAi was provided by targeting hepatitis B virus (Giladi et al. 2003; McCaffrey et al. 2003; Morrissey et al. 2005a,b). The expression of anti-HBV shRNAs efficiently suppressed HBV expression plasmids in the liver upon hydrodynamic co-transfection via the tail vein (McCaffrey et al. 2002), resulting in a substantial reduction of HBV RNA and replicated HBV genomes as well as a significant decrease of secreted hepatitis B surface antigen levels (McCaffrey et al. 2003; Morrissey et al. 2005a).

Additional proof for the antiviral potential of RNAi has come from studies with poliovirus. Poliovirus is an enterovirus such as CVB3 and belongs to the family of Picornaviridae, utilizing a messenger-like positive-sense single-stranded RNA genome. The transfection of short interfering RNAs directed against genomic

Table 1 RNAi in vitro and in vivo approaches targeting viral diseases

Genome		Virus family	Virus	RNAi target	In vitro/ in vivo	Reference
DNA	dsDNA	Herpesviridae	Herpes simplex virus	Helicase/primase associated protein UL5	In vitro	Palliser et al. (2006)
				Glycoprotein B (UL27)	In vitro/ in vivo	Palliser et al. (2006)
				Single-stranded DNA binding protein UL29	In vitro/ in vivo	Palliser et al. (2006)
				Glycoprotein E (US8)	In vitro	Bhuyan et al. (2004)
	dsDNA reverse transcribing	Hepadnaviridae	Hepatitis B virus	Pre-genomic RNA	In vitro/ in vivo	Hamasaki et al. (2003), McCaffrey et al. (2003), Morrissey et al. (2005a)
				Core protein HBcAG	In vitro/ in vivo	Hamasaki et al. (2003), McCaffrey et al. (2003), Shlomai and Shaul (2003), Ying et al. (2003)
				Non-structural transactivator protein HBx	In vitro	Shlomai and Shaul (2003)
RNA	Plus-stranded	Flaviviridae	Hepatitis C virus	5′ untranslated region (UTR)	In vitro	Krönke et al. (2004), Wang et al. (2005), Yokota et al. (2003)
				Core protein	In vitro	Randall and Rice (2004)
				Helicase/protease NS3	In vitro	Kapadia et al. (2003)
				Non-structural protein NS4B	In vitro	Randall and Rice (2004)
				Polymerase NS5B	In vitro	Kapadia et al. (2003), Wilson et al. (2003)
				Polymerase NS5B	In vivo	McCaffrey et al. (2002)
			West Nile virus	Capsid protein	In vitro	McCown et al. (2003)
				Polymerase NS5	In vitro	McCown et al. (2003)
		Picornaviridae	Coxsackievirus B3	Capsid protein VP1	In vitro/ in vivo	Ahn et al. (2005), Kim et al. (2007), Merl and Wessely (2007), Yuan et al. (2005)
				Capsid protein VP2	In vitro	Merl and Wessely (2007)
				Protease 2A	In vitro/ in vivo	Merl et al. (2005), Yuan et al. (2005)

			Protease 3C	In vitro	Merl and Wessely (2007)
			Polymerase 3D	In vitro/ in vivo	Ahn et al. (2005), Kim et al. (2007), Merl and Wessely (2007), Schubert et al. (2005), Werk et al. (2005), Yuan et al. (2005)
		Poliovirus	Capsid protein	In vitro	Gitlin et al. (2002)
			Polymerase 3D	In vitro	Gitlin et al. (2002)
Minus-stranded	Paramyxoviridae	Respiratory syncytial virus	Fusion protein F	In vitro	Bitko and Barik (2001)
			Phosphoprotein P	In vitro	Bitko and Barik (2001)
			Phosphoprotein P	In vivo	Bitko et al. (2005)
			Non-structural protein NS1	In vivo	Zhang et al. (2005)
	Orthomyxoviridae	Influenza virus	Nucleocapsid protein NP	In vitro/ in vivo	Ge et al. (2003, 2004), Tompkins et al. (2004)
			RNA transcriptase PA	In vitro/ in vivo	Ge et al. (2003, 2004), Tompkins et al. (2004)
Reverse transcribing	Retroviridae	Human immune deficiency virus	Group-specific antigen (gag)	In vitro	Hu et al. (2002), Novina et al. (2002)
			Cellular receptor CD4	In vitro	Novina et al. (2002)
			Cellular co-receptor CCR5	In vitro	Lee et al. (2003), Qin et al. (2003)
			Cellular co-receptor CXCR4	In vitro	Ji et al. (2003)
			Transactivator of transcription (tat)	In vitro	Boden et al. (2003), Coburn and Cullen (2002), Lee et al. (2003)
			Regulator of virion (rev)	In vitro	Coburn and Cullen (2002), Lee et al. (2002)
			Integrase	In vitro	Hu et al. (2002)
			Long terminal repeat (LTR)	In vitro	Jacque et al. (2002)
			Viral infectivity factor (vif)	In vitro	Jacque et al. (2002)
			Negative regulatory factor (nef)	In vitro	Das et al. (2004), Jacque et al. (2002)
			Envelope protein (env)	In vitro	Park et al. (2003)

regions encoding either a viral capsid-protein or the RNA dependent RNA polymerase effectively protected mammalian cells against polioviral infection, promoting viral clearance from most infected cells (Gitlin et al. 2002, 2005). In this context it was also observed that poliovirus can escape siRNA challenge due to emerging mutational variants harboring unique point mutations within their siRNA binding motif.

The development of an RNAi-based therapeutic strategy to combat coxsackievirus-induced heart diseases is due to its medical as well as economic impact and the lack of a specific anti-coxsackieviral drug therapy under focus of intense research. A number of independent laboratories, including ourselves, showed as a proof of concept that intracellular immunity against coxsackieviral infection could be successfully conferred by RNAi in vitro and in vivo.

Several groups achieved significant attenuation of viral replication and cytopathogenicity in cell culture by directly targeting the coxsackieviral genome. The process of target site validation revealed numerous effective siRNA binding motifs in genomic regions coding for the viral capsid protein VP1 (Ahn et al. 2005; Merl and Wessely 2007; Yuan et al. 2005), the capsid protein VP2 (Merl and Wessely 2007), the viral protease 2A (Merl et al. 2005; Yuan et al. 2005), the viral protease 3C (Merl and Wessely 2007) and the RNA dependent RNA polymerase 3D (Ahn et al. 2005; Kim et al. 2007; Merl and Wessely 2007; Schubert et al. 2005; Werk et al. 2005; Yuan et al. 2005). Of note, siRNAs targeting the 5′ as well as 3′ UTR had no substantial effect on viral cytopathogenicity and replication (Kim et al. 2007; Merl and Wessely 2007; Werk et al. 2005). The silencing was achieved either by treatment with synthetic siRNAs or by vector derived expression of shRNAs.

Yuan and co-workers reported a 92% inhibition of CVB3 replication in human HeLa cells and murine immortalized cardiomyocytes pre-treated with siRNA directed against the viral protease 2A (Yuan et al. 2005). In this study, the administration of siRNA post-infection proved to be effective in limiting even ongoing coxsackieviral infections, a fact that is of major importance for the clinical applicability of antiviral RNAi. Using fusion constructs composed of a reporter gene and subgenomic fragments of viral RNA, Werk et al. identified effective target sites in the genomic 3D region, which exhibited therapeutic potential to inhibit the virus propagation up to 80–90% (Werk et al. 2005). In our previous work, the application of the most effective siRNA directed against the RNA dependent RNA polymerase 3D resulted in an approximately fourfold prolonged survival of coxsackievirus-infected cells and an inhibition of viral replication by more than 10^5-fold compared to control siRNAs (Merl and Wessely 2007). Indeed, it has been demonstrated that therapeutic anti-coxsackieviral RNAi effectiveness can be hampered by the emergence of viral escape mutants harboring single point mutations in the central part of their siRNA target sites. Yet, the appearance of mutated coxsackievirus progeny can be sufficiently suppressed by the simultaneous application of at least three siRNAs targeting distinct genomic regions, thereby improving therapeutic outcome in terms of prolongation of cell survival by a factor of more than 6 compared to infected, control-siRNA treated cells (Merl and Wessely 2007). Similarly, Schubert et al. used a vector simultaneously expressing two different short hairpin RNAs to preserve therapeutic effectiveness when targeting viral RNA harboring subtle mutations in its target sites (Schubert et al. 2005). Upon administration of a dual shRNA cocktail directed against the genomic 3D region, the virus propagation in HeLa cells was efficiently inhibited resulting in an 80–90% decrease of the virus titer. As an alternative strategy, the coxsackievirus–adenovirus receptor (CAR), which is a host gene essential for the cell entry of group B coxsackieviruses and various adenovirus subtypes (Bergelson et al. 1997), has also been successfully silenced by RNAi. The suppression of CAR led to an attenuation of virus propagation by approximately 60% in HeLa cells (Werk et al. 2005). Similar results were obtained by Fechner et al. reporting shRNA-mediated silencing of CAR resulting in the inhibition of CVB3 infection for up to 97% in human immortalized cardiomyocytes (Fechner et al. 2007).

First evidence for the in vivo anti-coxsackieviral effectiveness of RNAi was provided in Type I IFNR–knockout mice (Muller et al. 1994), a knockout mice strain highly susceptible to CVB3 infection with increased mortality even when infected with low viral titers (Wessely et al. 2001). The application of siRNA directed against the viral protease 2A was carried out via hydrodynamic tail vein injection. siRNA-2A had a profound therapeutic impact promoting significant inhibition of viral replication, attenuated organ damage and consequently prolonged survival (Merl et al. 2005). Repeated siRNA administration further improved survival in this animal model compared to single application. Kim et al. examined the antiviral effectiveness of siRNA directed against the viral capsid protein VP1 as well as the RNA dependent RNA polymerase following hydrodynamic transfection of vector-derived shRNAs in Balb/c mice (Kim et al. 2007). Irrespective of the time point of the shRNA application prior or subsequent to viral infection, this therapeutic strategy led to a reduction of viral replication and tissue damage.

3 Factors That Limit Early Efficacy of Antiviral RNA Interference

Considering that enteroviruses undergo high-level replication predominantly at initial stages of infection, early therapeutic effectiveness is considered of critical importance for the antiviral impact of RNAi. The early antiviral efficacy of RNAi depends on various important factors that include optimized siRNA design as well as target site accessibility and off-target effects. Though numerous determinants for optimal silencing potency still remain to be elucidated, many features conferring enhanced siRNA functionality have been unraveled. siRNA design criteria being crucial for the success of the RNAi approach encompass structural features such as moderate to low G/C content (30–50%) and a lack of siRNA intrinsic secondary structures as well as internal repeats. Internal secondary structures may potentially hamper the association of the siRNA molecules with the components of the targeting complex. Another decisive feature determining enhanced effectiveness is the low thermodynamic stability at the 5′ terminus of the antisense strand. The thermodynamic 5′ end stability of the antisense strand versus the sense strand determines the rate of asymmetric antisense strand incorporation in the targeting complex (Khvorova et al. 2003; Schwarz et al. 2003). In addition, position specific determinants, such as base preferences at certain positions of the siRNAs, are of major importance to ensure high silencing activity of the siRNA. To impede any off-target effect, which is the unintended silencing of distinct genes, bioinformatics database queries for sequence homologies between the potential siRNA and the transcriptome of the targeted organism such as a Blast search should be conducted. Yet, several guidelines on siRNA design have been published including recommendations of fundamental importance by Reynolds et al. (2004). In addition, several academic and commercial entities provide online algorithms, to design siRNAs with enhanced target specificity according to thermodynamic, structural as well as position specific criteria (http://www.dharmacon.com; http://www.ambion.com/techlib/misc/siRNA_finder.html; http://www1.qiagen.com/Products/GeneSilencing/CustomSiRna/SiRna Designer.aspx).

The development of algorithms for rational siRNA design in consideration of thermodynamic and structural determinants led to a widespread use of siRNA-mediated gene silencing in research and development in life sciences. Yet, occasionally, siRNAs designed in conformity with theses algorithms exhibit no silencing activity, suggesting that additional factors impact on silencing activity. Several studies demonstrated that siRNA efficacy is at least in part influenced by structural properties of the targeted RNA, such as local intramolecular folding (Bohula et al. 2003; Heale et al. 2005; Vickers et al. 2003). In a previous study, Luo and Chang (2004) attributed this positional effect of different siRNAs mostly to the local secondary structure of the mRNA at the target site. They showed that the target site accessibility can be characterized by a single parameter, the "hydrogen bond (H-b) index." The gene-silencing effect inversely depends on the (H-b) index, which is the average number of hydrogen bonds formed between nucleotides in the target region and the rest of the mRNA. Ding et al. (2004) predicted the target site accessibility using a statistical probability profile of single-stranded regions generated for the entire target RNA. In this context, Yoshinari et al. could demonstrate that the silencing effectiveness of distinct siRNAs directed against the HIV-1 transactivation response element (TAR) was significantly influenced by the tight stem–loop structure of TAR (Yoshinari et al. 2004). Brown et al. supported the findings of the inaccessibility of the highly structured TAR, showing that silencing activity can be recovered upon enhancing the target site accessibility by disrupting the secondary structure of TAR using 2′-*O*-methyl oligonucleotides complementary to regions 5′ or 3′ of the siRNA binding motif (Brown et al. 2005).

Albeit the messenger-like positive stranded RNA genome of coxsackievirus suggests RNAi susceptibility for the complete genomic sequence, it has been shown frequently that genomic target selection has a major impact on therapeutic outcome. In general, siRNAs directed against the coding regions of CVB3 conferred considerable antiviral immunity. However, genomic targets located in the 5′ and 3′ untranslated, noncoding regions were less effective compared to targets in the protein coding region, despite designing siRNAs under careful consideration of all recommended criteria (Kim et al. 2007; Merl and Wessely 2007; Werk et al. 2005;

Yuan et al. 2005). The loss of antiviral effectiveness might be at least in part be due to steric hindrance by secondary and tertiary structures of the viral target RNA, as well as to proteins binding to regulatory genomic regions rendering the siRNA binding sites not fully accessible for the RNAi targeting complex. Notably, the highly folded structure of the 5′ UTR region of the coxsackieviral genome with its pronounced stem–loop motifs may hinder siRNA binding significantly. Even though previous studies reported efficient suppression of hepatitis C virus replication by siRNAs targeting single-stranded regions inbetween two stem–loop motifs of the viral 5′ UTR (Yokota et al. 2003) or even double-stranded regions in a stem–loop motif (Kanda et al. 2007; Prabhu et al. 2006), the overall findings implicate that RNAi efficacy can be optimized by selection of sequences located in viral protein coding regions, at least in the context of an enterovirus.

4 Factors That Limit Subsequent Efficacy of Antiviral RNA Interference

Several organisms such as plants (Palauqui and Vaucheret 1998; Voinnet et al. 1998) and nematodes (Grishok and Mello 2002) have developed mechanisms to amplify the dsRNA-induced RNA degradation signal resulting in a robust and long-lasting silencing even through cell division (Fire et al. 1998; Kennerdell and Carthew 1998). The local triggered gene silencing can spread to untreated cells throughout the organism and is often inherited by the next generation when spreading into germ line cells (Grishok and Mello 2002; Hammond et al. 2001). However, mammalian cells lack an RNA-dependent RNA polymerase to generate new siRNAs (Chi et al. 2003; Stein et al. 2003). As a result, gene silencing and thus antiviral efficacy is temporarily restricted to progressive degradation of siRNA by cellular nucleases and dilution to non-efficient levels due to cell division. Indeed, the silencing effect may last in terminally differentiated and cell cycle-arrested cells, such as neurons, for 3 weeks or more (Omi et al. 2004). However, in proliferating cells, RNAi activity induced by synthetic siRNA duplexes persists in dependence of the cell proliferation status about 3–7 days (Holen et al. 2002; Yang et al. 2001).

Since the coxsackieviral genome can persist chronically in the myocardium, the long-term efficacy is of significant importance for the antiviral RNAi approach. As a result, a number of groups have focused either on the use of chemical modifications to improve the siRNA intrinsic pharmacokinetic properties (deFougerolles et al. 2005) or the use of vector derived siRNA/shRNA. Chemical modifications are conducted to increase siRNA stability and to protect against degradation by endogenous nucleases while retaining full silencing activity. These modifications include RNA phosphate backbone modifications, replacing some phosphates of the siRNA duplex by RNase-protecting phosphorothioate (Braasch et al. 2003; Harborth et al. 2003), introduction of 2′-fluoro (2′-F) pyrimidines into the siRNA duplex (Layzer et al. 2004), or the modification of the ribose (Braasch et al. 2003), such as the introduction of 2′-*O*-methyl, 2′-deoxy-2′-fluorouridine. The incorporation of a number of synthetic RNA-like high affinity nucleotide analogues (LNA) substantially improved the siRNA molecule bio-stability in serum significantly. Furthermore, therapeutic application of siRNA targeting the genomic RNA of the pathogen SARS corona virus provided evidence that LNA is compatible with the intracellular targeting machinery and actually showed enhanced silencing effectiveness (Elmen et al. 2005). Similarly, Allerson et al. demonstrated that the introduction of 2′-*O*-methyl and 2′-fluoro nucleotides conferred greatly enhanced plasma stability and additionally a more than 500-fold increase of in vitro potency (Allerson et al. 2005). A previous work reported that a combination of diverse chemical modifications can dramatically augment siRNA serum stability while maintaining high silencing activity. In this context, the siRNA duplex consisted of a sense strand with all pyrimidines substituted by 2′-fluoro pyrimidines, all purines substituted by deoxyriboses, as well as inverted abasic caps at the 5′ and 3′ termini. Within the antisense strand, all pyrimidines were replaced by 2′-fluoro pyrimidines and all purines by 2′-*O*-methyl purines. The 3′ end of the antisense strand exhibited a phosphorothioate linkage (Morrissey et al. 2005a).

An alternative approach to achieve long-term silencing effects suggests the use of vector-based systems for endogenous siRNA expression in target cells (Tuschl and Borkhardt 2002; Yu et al. 2002). The expression cassettes contain RNA polymerase III promoters which either express separately sense and antisense strands that subsequently constitute the active siRNA duplex, or express short hairpin RNAs as stem–loop structures that are processed by Dicer into functional siRNAs. Upon transfection, the plasmid expresses large amounts of shRNAs resulting in a profound and long lasting suppression (McCaffrey et al. 2002) in contrast to chemically synthesized siRNA. Kim et al. could demonstrate the anti-coxsackieviral impact of plasmid-derived short hairpin RNAs in Cos-7 cells and in mice (Kim et al. 2007). Potential adverse side effects of the long-term silencing still remain unknown; however, inducible expression systems with the transcription being under tight control of a specific inducer or repressor are available.

Indeed, the use of shRNA-expressing plasmids is often limited due to the low plasmid transfection efficiencies of the majority of mammalian cells, in particular primary cells (Dykxhoorn et al. 2003). Therefore, efficient delivery as well as highly stable expression of shRNA can currently only be achieved in a broad range of proliferating and non-proliferating mammalian cells, stem cells and transgenic mice using retroviral (Brummelkamp et al. 2002), lentiviral (Rubinson et al. 2003), as well as adenoviral (Arts et al. 2003) expression systems. Viral gene transfer provides high transduction efficiencies, yet, albeit significant progress has been achieved, important safety issues remain (Thomas et al. 2003). Besides safety concerns, involving immunogenic as well as inflammatory host responses and the possibility of emerging replication competent viruses, insertional mutagenesis due to non-specific retroviral integration into the host genome potentially triggering oncogenic transformation is an important risk during viral gene therapy. By contrast, synthetic, chemically modified siRNA molecules facilitate indeed a timely restricted, though robust, silencing effect avoiding the risks of insertional mutagenesis, immunogenicity or the potential toxic effects of long-term RNAi expression.

Another critical factor that strongly influences the efficacy of RNAi-based antiviral strategies at subsequent stages in a clinical setting is the highly error-prone nature of the virally encoded RNA-dependent RNA polymerase. Incomplete silencing may enable the RNA-dependent RNA polymerase to generate several mutational variants, including viruses harboring subtle mutations within the respective siRNA target sequences. These so-called viral escape mutants can evade the siRNA recognition and augment the viral replication, thereby neutralizing the therapeutic siRNA treatment. The ability to counteract therapeutic RNAi by emergence of escape mutants has been observed in a number of viruses (Boden et al. 2003; Das et al. 2004; Gitlin et al. 2002, 2005). During HIV-1 inhibition, the appearance of a siRNA resistant, mutated virus progeny was observed after several weeks of culture. These RNAi-resistant viruses presented nucleotide substitutions or deletions in the siRNA binding motif (Boden et al. 2003; Das et al. 2004). Emergence of viral escape mutants from siRNA challenge has also been described for members of the enteroviruses family. During the therapeutic application of siRNA targeting the polioviral genome, a siRNA resistant virus progeny emerged after infection at a high multiplicity of infection (Gitlin et al. 2002, 2005). These escape mutants harbored single nucleotide alterations in their siRNA target sequences. Equally, the anti-coxsackieviral efficacy of RNAi proved to be highly dependent on the emergence of a siRNA resistant virus progeny. The resistant viruses exhibited single point mutations in the central part of their respective siRNA target site that allowed for a rapid escape from siRNA-mediated viral genome degradation, thus resulting in the complete loss of therapeutic efficacy (Merl and Wessely 2007). Mutational diversity within the virus progeny was mainly restricted to silent transversions on wobble positions as well as transversions leading to conservative amino acid changes retaining the physicochemical properties thereby ensuring the propagation of viable mutants.

The emergence of escape mutants might be hampered partly by targeting highly conserved genomic regions, since nucleotide substitutions or deletions occurring in these regions often affect viral viability. However, these regions are often not fully accessible to the RISC complex due to proteins binding to regulatory genomic regions as well as steric hindrance by highly ordered secondary and tertiary structures. For example, the pronounced stem–loop motifs render the 5′ UTR of the coxsackieviral genome less susceptible to siRNA-induced degradation as noted below (Kim et al. 2007; Merl and Wessely 2007; Werk et al. 2005; Yuan et al. 2005). In addition, under selective pressure of RNAi, the emergence of a siRNA-resistant virus progeny harboring subtle point mutations in their respective siRNA target sequences was observed despite targeting the conserved genomic regions encoding the viral RNA dependent RNA polymerase 3D and the viral protease 3C (Merl and Wessely 2007).

An additional option to overcome the hindrance of viral escape is to target simultaneously different genomic regions thereby reducing the probability of emerging escape mutants limited by the mutation rate of the viral polymerases. An enhanced antiviral effectiveness was achieved by the application of siRNA pools simultaneously targeting multiple regions of the viral RNA during HIV-1 as well as polioviral infection (Boden et al. 2003; Gitlin et al. 2002; Ji et al. 2003). Similarly, viral escape was minimized during coxsackieviral infections upon simultaneous administration of siRNA directed against distinct genomic regions (Merl and Wessely 2007). The occurrence of mutated viruses was reduced by more than half following simultaneous treatment with two independent antiviral siRNAs compared to single siRNA treatment, from more than 90% to 40%. However, only the combined application of three different siRNAs could achieve an almost complete suppression of mutated virus progeny, suggesting the combined treatment with at least three different siRNA molecules to be sufficient to suppress significantly viral escape. Schubert et al. used a vector simultaneously expressing two different short hairpin RNAs to retain high therapeutic effectiveness actually when targeting viral RNA harboring subtle mutations in its target sites (Schubert et al. 2005). In summary, exogenous or endogenous delivery of cocktails of siRNAs that target multiple viral sequences may be the best option to prevent viral escape that represents the major reason for the subsequent limitation of antiviral RNAi efficacy.

5 Conclusions and Future Prospects

In recent years, the discovery of the endogenous RNA interference pathway conferring genomic immunity has opened exciting possibilities for experimental exploitation for functional genomics and in particular therapeutically in the fight against a broad range of yet intractable diseases. In this perspective, RNAi-based antiviral therapy has great potential to cure or attenuate in particular viral diseases that are not efficiently treatable to date.

In the context of viral heart disease, several independent laboratories, including ourselves, were recently able to demonstrate that intracellular immunity can be successfully conferred by siRNA-mediated targeting of the coxsackieviral genome or the coxsackievirus-adenovirus receptor (CAR), thus leading to attenuated coxsackieviral replication and cytopathogenicity indicating a potential therapeutic role for RNAi against CVB3-related diseases. The in vivo application of siRNA directed against the viral genome had profound therapeutic impact promoting significant inhibition of viral replication, attenuated organ damage and consequently prolonged survival in highly susceptible mice. Despite these encouraging findings, major obstacles must still be overcome to enhance therapeutic efficacy before the value of therapeutic antiviral RNAi can be investigated clinically. These obstacles include the complete suppression of viral escape, the prevention of unwanted off-target effects, security aspects, tissue-specific delivery, siRNA stability, in vivo delivery systems and prolongation of therapeutic effectiveness. Yet, the development of non-viral delivery methods in terms of liposomal siRNA encapsulation as well as conjugation with specific ligands, antibodies, and aptamers have been first promising steps towards the clinical applicability of antiviral RNAi that may be available in the not too distant future to provide a novel therapeutic strategy to attenuate or cure life-threatening viral heart diseases.

References

Abrahante J, Daul A, Li M et al (2003) The Caenorhabditis elegans hunchback-like gene lin-57/hbl-1 controls developmental time and is regulated by microRNAs. Dev Cell 4:625–637

Ahn J, Jun E, Lee H et al (2005) A small Interfering RNA targeting coxsackievirus B3 protects permissive HeLa cells from viral challenge. J Virol 79:8620–8624

Alisky J, Davidson B (2004) Towards therapy using RNA interference. Am J Pharmacogenomics 4:45–51

Allerson C, Sioufi N, Jarres R et al (2005) Fully 2′-modified oligonucleotide duplexes with improved in vitro potency and stability compared to unmodified small interfering RNA. J Med Chem 48:901–904

Arts G, Langemeijer E, Tissingh R et al (2003) Adenoviral vectors expressing siRNAs for discovery and validation of gene function. Genom Res 13:2325–2332

Badorff C, Lee G, Lamphear B et al (1999) Enteroviral protease 2A cleaves dystrophin: evidence of cytoskeletal disruption in an aquired cardiomyopathy. Nat Med 5:320–326

Barik S (2005) Silence of the transcripts: RNA interference in medicine. J Mol Med 83:764–773

Bauer S, Gottesman G, Sirota L et al (2002) Severe coxsackie virus B infection in preterm newborns treated with pleconaril. Eur J Pediatr 161:491–493

Bergelson J, Cunningham J, Droguett G et al (1997) Isolation of a common receptor for coxsackie B viruses and adenoviruses 2 and 5. Science 275:1320–1323

Bernstein E, Caudy A, Hammond S et al (2001) Role for a bidentate ribonuclease in the initiation step of RNA interference. Nature 409:363–366

Bhuyan P, Kariko K, Capodici J et al (2004) Short interfering RNA-mediated inhibition of herpes simplex virus type 1 gene expression and function during infection of human keratinocytes. J Virol 78:10276–10281

Billy E, Brondani V, Zhang H et al (2001) Specific interference with gene expression induced by long, double-stranded RNA in mouse embryonal teratocarcinoma cell lines. Proc Natl Acad Sci USA 98:14428–14433

Bitko V, Barik S (2001) Phenotypic silencing of cytoplasmic genes using sequence-specific double-stranded short interfering RNA and its application in the reverse genetics of wild type negative-strand RNA viruses. BMC Microbiol 1:34

Bitko V, Musiyenko A, Shulyayeva O et al. (2005) Inhibition of repiratory viruses by nasally administered siRNA. Nat Med 11:50–55

Boden D, Pusch O, Lee F et al (2003) Human immunodeficiency virus type 1 escape from RNA interference. J Virol 77:11531–11535

Bohula E, Salisbury A, Sohail M et al (2003) The efficacy of small interfering RNAs targeted to the type 1 insulin-like growth factor receptor (IGF1R) is influenced by secondary structure in the IGF1R transcript. J Biol Chem 278:15991–15997

Braasch D, Jensen S, Liu Y et al (2003) RNA interference in mammalian cells by chemically-modified RNA. Biochemistry 42:7967–7975

Brennecke J, Hipfner D, Stark A et al (2003) Bantam encodes a developmentally regulated microRNA that controls cell proliferation and regulates the proapoptotic gene hid in Drosophila. Cell 113:25–36

Brown K, Chu C, Rana T (2005) Target accessibility dictates the potency of human RISC. Nat Struct Mol Biol 12:469–470

Brummelkamp T, Bernards R, Agami R (2002) A system for stable expression of short interfering RNAs in mammalian cells. Science 296:550–553

Chen Z, Indjeian V, McManus M et al (2002) CP110, a cell cycle-dependent CDK substrate, regulates centrosome duplication in human cells. Dev Cell 3:339–350

Chi J, Chang H, Wang N et al (2003) Genomewide view of gene silencing by small interfering RNAs. Proc Natl Acad Sci USA 100:6343–6346

Coburn G, Cullen B (2002) Potent and specific inhibition of human immunodeficiency virus type 1 replication by RNA interference. J Virol 76:9225–9231

Das AT, Brummelkamp T, Westerhout E et al (2004) Human immunodeficiency virus type 1 escapes from RNA interference-mediated inhibition. J Virol 78:2601–2605
deFougerolles A, Manoharan M, Meyers R et al (2005) RNA interference in vivo: towards synthetic small inhibitory RNA-based therapeutics. Methods Enzymol 392:278–296
Ding Y, Chan C, Lawrence C (2004) Sfold web server for statistical folding and rational design of nucleic acids. Nucleic Acids Res 32:135–141
Dorsett Y, Tuschl T (2004) siRNAs: application in functional genomics and potential as therapeutics. Nat Rev Drug Discov 3:318–329
Downward J (2004) RNA interference. BMJ 328:1245–1248
Dykxhoorn D, Novina C, Sharp P (2003) Killing the messenger: short RNAs that silence gene expression. Nat Rev Mol Cell Biol 4:457–467
Elbashir S, Harborth J, Lendeckel W et al (2001a) Duplexes of 21-nucleotide RNAs mediate RNA interference in cultured mammalian cells. Nature 411:494–498
Elbashir S, Lendeckel W, Tuschl T (2001b) RNA interference is mediated by 21- and 22-nucleotide RNAs. Genes Dev 15:188–200
Elmen J, Thonberg H, Ljungberg K et al (2005) Locked nucleic acid (LNA) mediated improvements in siRNA stability and functionality. Nucleic Acids Res 33:439–447
Fechner H, Pinkert S, Wang X et al (2007) Coxsackievirus B3 and adenovirus infections of cardiac cells are efficiently inhibited by vector-mediated RNA interference targeting their common receptor. Gene Ther 14:960–971
Feldman A, McNamara D (2000) Myocarditis. N Engl J Med 343:1388–1398
Feuer R, Mena I, Pagarigan R et al (2003) Coxsackievirus B3 and the neonatal CNS: the roles of stem cells, developing neurons, and apoptosis in infection, viral dissemination, and disease. Am J Pathol 163:1379–1393
Fire A, Xu S, Montgomery M et al (1998) Potent and specific genetic interference by double-stranded RNA in *Caenorhabditis elegans*. Nature 391:806–811
Friedrich I, Shir A, Klein S et al (2004) RNA molecules as anti-cancer agents. Semin Cancer Biol 14(4):223–230
Gartel A, Kandel E (2006) RNA interference in cancer. Biomol Eng 23(1):17–34
Ge Q, McManus M, Nguyen T et al (2003) RNA interference of influenza virus production by directly targeting mRNA for degradation and indirectly inhibiting all viral RNA transcription. Proc Natl Acad Sci USA 100:2718–2723
Ge Q, Filip L, Bai A et al (2004) Inhibition of influenza virus production in virus infected mice by RNA interference. Proc Natl Acad Sci USA 101:8676–8681
Giladi H, Ketzinel-Gilad M, Rivkin L et al (2003) Small interfering RNA inhibits hepatitis B virus replication in mice. Mol Ther 8:769–776
Gitlin L, Karelsky S, Andino R (2002) Short interfering RNA confers intracellular antiviral immunity in human cells. Nature 418:430–434
Gitlin L, Stone J, Andino R (2005) Poliovirus Escape from RNA Interference: short interfering RNA-target recognition and implications for therapeutic approaches. J Virol 79:1027–1035
Grishok A, Mello C (2002) RNAi (Nematodes: *Caenorhabditis elegans*). Adv Genet 46:339–360
Hamasaki K, Nakao K, Matsumoto K et al (2003) Short interfering RNA-directed inhibition of hepatitis B virus replication. FEBS Lett 543:51–54
Hammond S, Bernstein E, Beach D et al (2000) An RNA-directed nuclease mediates post-transcriptional gene silencing in Drosophila cells. Nature 404:293–296
Hammond S, Caudy A, Hannon G (2001) Post-transcriptional gene silencing by double-stranded RNA. Nat Rev Genet 2:110–119
Harborth J, Elbashir S, Vandenburgh K et al (2003) Sequence, chemical, and structural variation of small interfering RNAs and short hairpin RNAs and the effect on mammalian gene silencing. Antisense Nucleic Acid Drug Dev 13:83–105
Heale B, Soifer H, Bowers C et al (2005) siRNA target site secondary structure predictions using local stable substructures. Nucleic Acids Res 33:e30

Holen T, Amarzguioui M, Wiiger M et al (2002) Positional effects of short interfering RNAs targeting the human coagulation trigger tissue factor. Nucleic Acids Res 30(8):1757–1766
Hu W, Myers C, Kilzer J et al (2002) Inhibition of retroviral pathogenesis by RNA interference. Curr Biol 12:1301–1311
Jacque J, Triques K, Stevenson M (2002) Modulation of HIV-1 replication by RNA interference. Nature 418:435–438
Ji J, Wernli M, Klimkait T et al (2003) Enhanced gene silencing by the application of multiple specific small interfering RNAs. FEBS Lett 552:247–252
Kanda T, Steele R, Ray R et al (2007) Small interfering RNA targeted to hepatitis C virus 5′ nontranslated region exerts potent antiviral effect. J Virol 81:669–676
Kandolf R, Klingel K, Zell R et al (1993) Molecular pathogenesis of enterovirus-induced myocarditis: virus persistence and chronic inflammation. Intervirology 35:140–151
Kapadia S, Brideau-Andersen A, Chisari F (2003) Interference of hepatitis C virus RNA replication by short interfering RNAs. Proc Natl Acad Sci USA 100:2014–2018
Kartasheva N, Contente A, Lenz-Stoppler C et al (2002) P53 induces the expression of its antagonist p73 delta N, establishing an autoregulatory feedback loop. Oncogene 21:4715–4727
Kennerdell J, Carthew R (1998) Use of dsRNA-mediated genetic interference to demonstrate that frizzled and frizzled 2 act in the wingless pathway. Cell 95:1017–1026
Khvorova A, Reynolds A, Jayasena S (2003) Functional siRNAs and miRNAs exhibit strand bias. Cell 115:209–216
Kim J, Chung S, Hwang H et al (2007) Expression of short hairpin RNAs against the coxsackievirus B3 exerts potential antiviral effects in COS-7 cells and in mice. Virus Res 125:9–13
Knowlton K, Badorff C (1999) The immune system in viral myocarditis: maintaining the balance. Circ Res 85:559–561
Krönke J, Kittler R, Buchholz F et al (2004) Alternative approaches for efficient inhibition of hepatitis C virus replication by small interfering RNAs. J Virol 78:3436–3446
Lassus P, Opitz-Araya X, Lazebnik Y (2002) Requirement for caspase-2 in stress-induced apoptosis before mitochondrial permeabilization. Science 297:1352–1354
Layzer J, McCaffrey A, Tanner A et al (2004) In vivo activity of nuclease-resistant siRNAs. RNA 10:766–771
Lee NS, Dohjima T, Bauer G et al (2002) Expression of small interfering RNAs targeted against HIV-1 rev transcripts in human cells. Nat Biotechnol 20:500–505
Lee MT, Coburn G, McClure M et al (2003) Inhibition of human immunodeficiency virus type 1 replication in primary macrophages by using Tat- or CCR5-specific small interfering RNAs expressed from a lentivirus virus. J Virol 77:11964–11972
Li M, Kim J, Li S et al (2005) Long-term inhibition of HIV-1 infection in primary hematopoietic cells by lentiviral vector delivery of a triple combination of anti-HIV shRNA, anti-CCR5 ribozyme, and a nucleolar-localizing TAR decoy. Mol Ther 12:900–909
Luo K, Chang D (2004) The gene-silencing efficiency of siRNA is strongly dependent on the local structure of mRNA at the targeted region. Biochem Biophys Res Commun 318:303–310
McCaffrey A, Meuse L, Pham T et al (2002) RNA interference in adult mice. Nature 418:38–39
McCaffrey A, Nakai H, Pandey K et al (2003) Inhibition of hepatitis B virus in mice by RNA interference. Nat Biotechnol 21:639–644
McCown M, Diamond M, Pekosz A (2003) The utility of siRNA transcripts produced by RNA polymerase I in down regulating viral gene expression and replication of negative- and positive-strand RNA viruses. Virology 313:514–524
Merl S, Wessely R (2007) Anti-coxsackieviral efficacy of RNA interference is highly dependent on genomic target selection and emergence of escape mutants. Oligonucleotides 17:44–53
Merl S, Michaelis C, Jaschke B et al (2005) Targeting 2A protease by RNA interference attenuates coxsackievirale cytopathogenicity and promotes survival in highly susceptible mice. Circulation 111:1583–1592
Miyagishi M, Taira K (2002) U6 promoter-driven siRNAs with four uridine 3′ overhangs efficiently suppress targeted gene expression in mammalian cells. Nat Biotechnol 20:497–500

Morrissey D, Blanchard K, Shaw L et al (2005a) Activity of stabilized short interfering RNA in a mouse model of hepatitis B virus replication. Hepatology 41:1349–1356
Morrissey D, Lockridge J, Shaw L et al (2005b) Potent and persistent in vivo anti-HBV activity of chemically modified siRNAs. Nat Biotechnol 23:1002–1007
Muller U, Steinhoff U, Reis L et al (1994) Functional role of type I and type II interferons in antiviral defense. Science 264:1918–1921
Novina C, Murray M, Dykxhoorn D et al (2002) siRNA-directed inhibition of HIV-1 infection. Nat Med 8:681–686
Omi K, Tokunaga K, Hohjoh H (2004) Long-lasting RNAi activity in mammalian neurons. FEBS Lett 558:89–95.
Palauqui J, Vaucheret H (1998) Transgenes are dispensable for the RNA degradation step of cosuppression. Proc Natl Acad Sci USA 95:9675–9680
Palliser D, Chowdhury D, Wang Q et al (2006) An siRNA-based microbicide protects mice from lethal herpes simplex virus 2 infection. Nature 439:89–94
Park W, Hayafune M, Miyano-Kurosaki N et al (2003) Specific HIV-1 env gene silencing by small interfering RNAs in human peripheral blood mononuclear cells. Gene Ther 10:2046–2050
Paul C, Good P, Winer I et al (2002) Effective expression of small interfering RNA in human cells. Nat Biotechnol 20:505–508
Prabhu R, Garry R, Dash S (2006) Small interfering RNA targeted to stem–loop II of the 5′ untranslated region effectively inhibits expression of six HCV genotypes. Virol J 3:100
Qin X, An D, Chen I et al (2003) Inhibiting HIV-1 infection in human T cells by lentiviral-mediated delivery of small interfering RNA against CCR5. Proc Natl Acad Sci USA 100:183–188
Randall G, Rice C (2004) Interfering with hepatitis C virus RNA replication. Virus Res 102:19–25
Reynolds A, Leake D, Boese Q et al (2004) Rational siRNA design for RNA interference. Nat Biotechnol 22:326–330
Rodriguez-Lebron E, Gonzalez-Alegre P (2006) Silencing neurodegenerative disease: bringing RNA interference to the clinic. Expert Rev Neurother 6:223–233
Rossi J (2006) RNAi as a treatment for HIV-1 infection. Biotechniques 40(Suppl):25–29
Rubinson D, Dillon C, Kwiatkowski A et al (2003) A lentivirus-based system to functionally silence genes in primary mammalian cells, stem cells and transgenic mice by RNA interference. Nat Genet 33:401–406
Schubert S, Grunert H, Zeichhardt H et al (2005) Maintaining inhibition: siRNA double expression vectors against coxsackieviral RNAs. J Mol Biol 346:457–465
Schwarz D, Hutvanger G, Du T et al (2003) Asymmetry in the assembly of the RNAi enzyme complex. Cell 115:199–208
Shankar P, Manjunath N, Lieberman J (2005) The prospect of silencing disease using RNA interference. JAMA 293:1367–1373
Shlomai A, Shaul Y (2003) Inhibition of hepatitis B virus expression and replication by RNA interference. Hepatology 37:764–770
Stark G, Kerr I, Williams B, Silvermann R et al (1998) How cells respond to interferons. Annu Rev Biochem 67:227–264
Stein P, Svoboda P, Anger M et al (2003) RNAi: mammalian oocytes do it without RNA-dependent RNA polymerase. RNA 9:187–192
Thomas C, Ehrhardt A, Kay M (2003) Progress and problems with the use of viral vectors for gene therapy. Nat Rev Genet 4:346–358
Tompkins S, Lo C, Tumpey T et al (2004) Protection against lethal influenza virus challenge by RNA interference in vivo. Proc Natl Acad Sci USA 101:8682–8686
Tuschl T, Borkhardt A (2002) Small interfering RNAs: a revolutionary tool for the analysis of gene function and gene therapy. Mol Interv 2:158–167
Vickers T, Koo S, Bennett C et al (2003) Efficient reduction of target RNAs by small interfering RNA and RNase H-dependent antisense agent. J Biol Chem 278:7108–7118

Voinnet O, Vain P, Angell S et al (1998) Systemic spread of sequence-specific transgene RNA degradation in plants is initiated by localized introduction of ectopic promoterless DNA. Cell 95:177–187

Wang Q, Contag C, Ilves H et al (2005) Small hairpin RNAs efficiently inhibit hepatitis C IRES-mediated gene expression in human tissue culture cells and a mouse model. Mol Ther 12:562–568

Werk D, Schubert S, Lindig V et al (2005) Developing an effective RNA interference strategy against a plus-strand RNA virus: silencing of coxsackievirus B3 and its cognate coxsackievirus-adenovirus receptor. Biol Chem 386:857–863

Wessely R, Henke A, Zell R et al (1998a) Low-level expression of a mutant coxsackieviral cDNA induces a myocytopathic effect in culture: an approach to the study of enteroviral persistence in cardiac myocytes. Circulation 98:450–457

Wessely R, Klingel K, Santana L et al (1998b) Transgenic expression of replication-restricted enteroviral genomes in heart muscle induces defective excitation-contraction coupling and dilated cardiomyopathy. J Clin Invest 102:1444–1453

Wessely R, Klingel K, Knowlton K et al (2001) Cardioselective infection with coxsackievirus B3 requires intact type I interferon signaling: implications for mortality and early viral replication. Circulation 103:756–761

Wilson J, Jayasena S, Khvorova A et al (2003) RNA interference blocks gene expression and RNA synthesis from hepatitis C replicons propagated in human liver cells. Proc Natl Acad Sci USA 100:2783–2788

Woodruff J (1980) Viral myocarditis. A review. Am J Pathol 101:425–484

Yang S, Tutton S, Pierce E et al (2001) Specific double-stranded RNA interference in undifferentiated mouse embryonic stem cells. Mol Cell Biol 21:7807–7816

Ying C, DeClercq E, Neyts J (2003) Selective inhibition of hepatitis B virus replication by RNA interference. Biochem Biophys Res Commun 309:482–484

Yokota T, Sakamoto N, Enomoto N et al (2003) Inhibition of intracellular hepatitis C virus replication by synthetic and vector-derived small interfering RNAs. EMBO Rep 4:602–608

Yoshinari K, Miyagishi M, Taira K (2004) Effects on RNAi of the tight structure, sequence and position of the targeted region. Nucleic Acids Res 32:691–699

Yu J, DeRuiter S, Turner D (2002) RNA interference by expression of short-interfering RNAs and hairpin RNAs in mammalian cells. Proc Natl Acad Sci USA 99:6047–6052

Yuan J, Cheung P, Zhang H et al (2005) Inhibition of coxsackievirus B3 replication by small interfering RNAs requires perfect sequence match in the central region of the viral positive strand. J Virol 79:2151–2159

Zaragoza C, Ocampo C, Saura M et al (1999) Inducible nitric oxide synthase protection against coxsackievirus pancreatitis. J Immunol 163:5497–5504

Zhang W, Yang H, Kong X et al (2005) Inhibition of respiratory syncytial virus infection with intranasal nanoparticles targeting the viral NS1 gene. Nat Med 11:56–62

Design of siRNAs and shRNAs for RNA Interference: Possible Clinical Applications

V. Pekarik

Contents

Abstract A random observation of so-called co-suppression in Petunia plants about 18 years ago turned out to be a foundation of dynamically expanding technology of RNAi, with the impact comparable to the discovery of PCR or mouse knockout.

The original discovery was based on an observation that an introduction of double stranded RNA (dsRNA) into cells can trigger a silencing mechanism, which turned out to be sequence specific. The dsRNA is processed into short ~22-nt-long RNA fragments, which became used instead of dsRNA. Replacing dsRNA with the siRNA raised a new challenge in the design of highly specific siRNAs. Sequence analysis and comparison of efficiency in gene knockdown resulted in formulation of basic rules to guide the design of efficient siRNAs. A challenge to produce

V. Pekarik
Center for Anatomy, Institute for Cell- and Neurobiology, Charité University Hospital, Phillippstrasse 12, 10115 Berlin, Germany
Centre de Medicina Regenerativa de Barcelona (CMR[B]), Parc de Recerca Biomèdica de Barcelona (PRBB), c/ Dr. Aiguader 88, 7 planta, 08003 Barcelona
e-mail: vpekarik@cmrb.eu

V.A. Erdmann et al. (eds.) *RNA Technologies in Cardiovascular Medicine and Research*,
doi: 10.1007/978-3-540-78709-9,

siRNAs in vivo was met by designing vectors expressing short hairpins RNAs (shRNAs) from RNA polymerase III promoters (U6, H1, 7SK), which were in cells processed into active siRNAs.

In parallel with the study of RNAi, progress in understanding the cellular processing and function of microRNAs (miRNAs) was achieved. RNAi and miRNA systems share most if not all components of the silencing pathway. A direct consequence was an incorporation of siRNA sequence into a context of native miRNA, which enhanced the RNAi efficacy and allowed production of siRNAs from tissue specific RNA polymerase II promoters.

In this chapter, I will summarize general rules for the design of efficient siRNA duplexes and their conversion into shRNAs. I will summarize the means of shRNA production in vitro and in vivo including the use of various expression vectors and delivery strategies and compare it to the miRNA processing. I will also discuss critical steps of experimental design to knockdown a specific mRNA in cell culture or in a tissue specific manner. At the end, I will discuss the possibilities this technology offers to treat, or at least ameliorate, the progression of several inherited or viral diseases.

Abbreviations ADAR: Adenosine deaminase acting on RNA; ALS: Amyotrophic lateral sclerosis; cDNA: Copy DNA; cRNA: Copy RNA; dsRNA: Double stranded RNA; DYT1: Primary dystonia; EGFP: Enhanced green fluorescent protein; eIF2α: Alpha-subunit of eukaryotic translation initiation factor 2; FANA: 2′-deoxy-2′-fluoro-β-D-arabinonucleic acid; LNA: Locked nucleic acid; miRNA: MicroRNA; miRNA*: Passenger strand of microRNA; nt: Nucleotide; OAS: Oligoadenylate synthase; ORF: Open reading frame; PKR: Protein kinase R; polyQ: Polyglutamine; pre-miRNA: Precursor miRNA; Prnp: Prion protein gene; PrP: Prion protein; RISC: RNA interference silencing complex or RNA-induced silencing complex; rRNA: Ribosomal RNA; SCA: Spinocerebellar ataxia; shRNA: Short hairpin RNA; siRNA: Short interfering RNA; siRNA*: Passenger strand of siRNA duplex; SOD1: Cu/Zn superoxide dismutase; ssRNA: Single stranded RNA; tetO: Tetracycline operator; TLR3: Toll-like receptor 3; TRBP: Transactivating response (TAR) RNA-binding protein; TSE: Transmissible spongiform encephalopathy; UTR: Untranslated region

1 Introduction

RNA interference (RNAi) has undergone a progressive development since the first observation of an obscure phenomenon of discolored petals in petunia overexpressing an extra copy of chalcone synthase gene (Napoli et al. 1990; van der Krol et al. 1990) about 18 years ago. The mechanism of gene silencing in plants was characterized several years later and was found to be induced by RNA-mediated RNA degradation (Metzlaff et al. 1997). A similar mechanism operates in invertebrates and vertebrates including mammals.

The technique of RNAi was exploited to study gene function in *Caenorhabditis elegans* and *Drosophila* (Fire et al. 1998). Later on, it was found that the technique is applicable to other eukaryotes including vertebrates. During the initial phase, a long double-stranded RNA (dsRNA) was used. This approach is working very well for *C. elegans* and *Drosophila* but has met severe obstacles in vertebrate systems. During the early period, the mechanism of RNAi was thoroughly studied and a picture of the molecular machinery emerged. The main discovery was that the effectors of RNAi are short cleavage products of long dsRNA used in initial transfection experiments. Because these small fragments of uniform length of about 21 nucleotides (nt) interfere with gene expression they were named short interfering RNAs (siRNAs). These fragments were found to be more effective in inducing gene silencing, and their use was not accompanied by the adverse effects previously observed when long dsRNA was used. Further research led to a discovery of protein machinery, mediating sequence specific mRNA degradation and named the RNA interference silencing complex (RISC). Dissecting the RNAi pathway ultimately led to one of the major discoveries in recent years. The research revealed a completely new and unexpected gene expression regulatory network, based on the presence of very small regulatory RNAs. These are capable of repressing translation of certain mRNAs. For their tiny size they were named microRNAs (miRNAs). To date, a whole score of miRNAs have been discovered in many organisms. miRNAs utilize the same machinery as RNAi, but the miRNA binding sites are mostly located in 3′ untranslated regions (UTR) of mRNAs while siRNAs are usually targeting open reading frames (ORF). miRNAs seldom possess 100% complementarity to the target sequence and inhibit mRNA translation. siRNAs have 100% complementarity to the target sequence and initiate cleavage of the target mRNA mediated by Argonaute proteins contained in RISC. The discovery of miRNAs and their cellular expression and processing allowed the design of new ways of how to produce si-RNAs in vivo from expression constructs. siRNA and miRNA machinery are closely intertwined and share most if not all cellular factors.

siRNAs are produced and used as annealed RNA strands. In cells, the double strand RNA dissociates and only one strand (guide) gets integrated into RISC. Design of efficient siRNAs requires knowledge of the cellular machinery responsible for selection of a guide strand and its loading into RISC. Additionally, conversion of siRNAs into hairpins requires knowledge of the available expression systems based on RNA polymerase II and III promoters. As the production of active siRNAs from hairpins in cell utilizes cellular miRNA processing machinery, its understanding is a prerequisite for shRNA design.

In this review, I will focus on the basics of miRNA expression and processing, summarize basic rules and highlight specific points important for the design of siRNAs. Furthermore, I will outline the important principles for conversion of siRNAs into shRNAs and briefly depict expression systems for the production of shRNAs in vivo.

2 Adverse Effects of Long dsRNAs

In original experiments carried out in worms, the active siRNAs were generated intracellularly from long dsRNAs by Dicer (Billy et al. 2001), an intracellular RNase III. However, application of long dsRNAs in vertebrate experiments led to many undesired side effects (Karpala et al. 2005). Later on, it was discovered that mammalian cells have orchestrated defense mechanisms responding to dsRNA. Originally, the purpose seemed to be to respond against virus replication or transposone movement (Fig. 1).

Extracellularly applied dsRNA (in cell culture media during transfection) is detected by pattern recognition receptors such as toll-like receptor 3 (TLR3). dsRNA is responsible for stimulating important protective responses, such as the activation of dicer-related antiviral pathways, induction of type 1 IFN, and stimulation

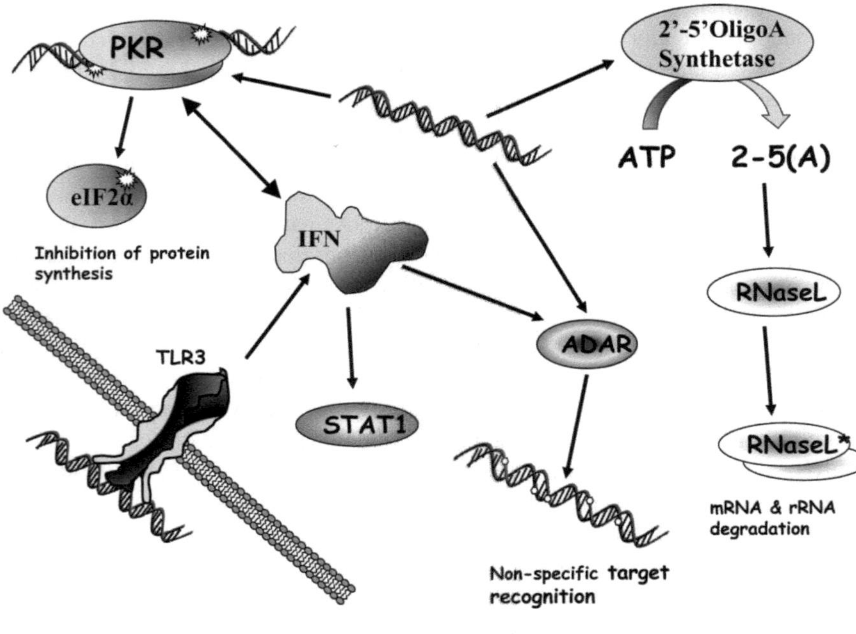

Fig. 1 Nonspecific effects caused by long dsRNA. dsRNA induced antiviral response can be activated by extracellular stimulation of Toll-like receptor 3 (*TLR3*) resulting in activation of interferon pathway. Transfected intracellular dsRNA activates protein kinase R (*PKR*), which in turn phosphorylates *eIF2*α and also activates interferon (*IFN*) response. Phosphorylated *eIF2*α inhibits protein synthesis. dsRNA can be modified by *ADAR*, which deaminates dsRNA converting adenosine into inosine. siRNA derived from such modified dsRNA will have altered target specificity. Mechanism leading to a degradation of mRNA and rRNA is triggered by activation of oligoadenylate synthetase (*2′-5′oligoA synthetase*). 2′-5′ linked adenylate oligomers activate the endoribonuclease *RNaseL*, which begins destroying diverse RNAs further inhibiting protein synthesis

of dsRNA-activated protein kinase (protein kinase R; PKR) and oligoadenylate synthetase (OAS) (Karpala et al. 2005). Activated PKR phosphorylates the alpha-subunit of eukaryotic translation initiation factor 2 (eIF2α) which inhibits protein synthesis and may eventually lead to the initiation of apoptosis (Scheuner et al. 2006). 2′-5′ oligoadenylate synthetase is stimulated by dsRNA to produce 5′-phosphorylated, 2′-5′-linked oligoadenylates (2-5A) from ATP, which in turn activates RNase L. Activated RNase L cleaves various cellular RNAs including ribosomal RNAs (rRNA). Off-target effects, often reported when dsRNA was used, are probably mediated through interferon activation of ADAR (adenosine deaminase acting on RNA) converting adenosine of dsRNA into inosine. siRNAs where inosine replaces adenine will have different sequence specificity then original siRNAs.

Recent discoveries suggest that unexpected experimental results can be caused by the activation of genes whose promoters are targeted by short dsRNAs produced by cleavage of long dsRNAs. Short dsRNA can induce transcriptional activation (Li et al. 2006; Janowski et al. 2007) or transcriptional gene silencing (Morris et al. 2004), which are accompanied by changes in histone acetylation and methylation.

A number of experimental artifacts caused by nonspecific activation of antiviral response led to replacement of long dsRNA with specific short interfering RNAs in mammalian experiments, which do not provoke interferon activation. The use of siRNAs to a great extent eliminates or decreases the negative effects of dsRNAs, but it is not 100% efficient. Overexpression of 21-nt shRNAs from lentiviral vectors can activate oligoadenylate synthetase (OAS1) (Fish and Kruithof 2004; Pebernard and Iggo 2004), one of interferon responsive genes. Lowering the lentiviral vector titer reduces both expression of shRNA and induction of OAS1, without a major impact on the efficacy of gene silencing. Preserving the wild-type sequence around the transcription start site of U6 or H1 promoters, in particular a C/G sequence at positions −1/+1, helps to avoid interferon induction (Pebernard and Iggo 2004).

3 siRNA Structure and Design

siRNAs are short RNAs incorporated in a protein complex guiding the cleavage of mRNAs complementary to the siRNA sequence. They are derived from a duplex of two complementary 21-nt-long RNA strands (siRNA/siRNA*, where * denotes passenger strand) where only 19 nt are paired and two nucleotides at 3′ ends form protruding termini (Fig. 2).

3.1 siRNA Design Parameters

Original design parameters were derived from early experiments in *Drosophila* (Elbashir et al. 2001) and provided a good base for the development of more powerful algorithms. Years of research have led to the formulation of several principal rules,

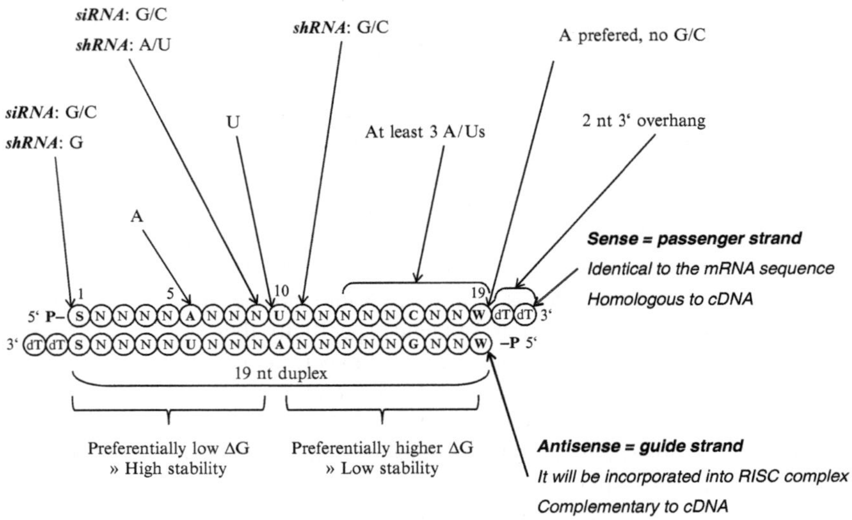

Fig. 2 siRNA anatomy. siRNA is formed by annealing of two complementary RNA strands with two nucleotide overhangs. These overhangs are not important for the sequence recognition and can be replaced with deoxynucleotides. The deoxythymidine (*dT*) is used most often to reduce the cost of chemical synthesis. siRNA should be designed to fulfill several basic rules. Khvorova et al. (2003) have found that an asymmetry in ΔG (Gibbs energy) is important factor for preferential loading of guide strand into RISC. Certain sequence preferences can improve efficiency of silencing. Reynolds et al. (2004) identified few sequence characteristics among the most efficient siRNAs. Presence of at least 3 A/U bases at positions 15–19, an A in position 19, while G/C at the position 19 is avoided improved efficiency to knockdown a specific transcript. Functional shRNAs prefer A/U in position 9 and G/C in position 11 while siRNAs prefer G/C at position 9 (Li et al. 2007). *S* Strong base, *W* weak base

which were later incorporated into many prediction programs. Many of these rules were discovered while studying the processing of natural miRNAs.

The first step in assembly of active RISC is the incorporation of a guide strand (siRNA) into the protein complex, which is likely facilitated by a putative RNA helicase. The assembly of RISC is asymmetrical with a clear strand bias, meaning that only one strand of the duplex is preferentially loaded (Schwarz et al. 2003) while the other strand (siRNA*) is destroyed. The asymmetry is driven by thermodynamic stability of each end of the siRNA duplex ends (Khvorova et al. 2003). The strand whose 5′ end has higher ΔG (free energy), i.e., lower internal stability, is preferentially incorporated into RISC. Antisense (guide) strand of chemically synthesized siRNAs should have lower internal stability at its 5′ end. This is further supported by the work of Reynolds et al. (2004) who found few sequence specific characteristics. There should be at least 3 A/U bases at positions 15–19 (sense strand), an A in position 19 (sense strand) while G/C at the position 19 should be avoided. A U as another weak base is, in position 19, well tolerated. When the design does not allow following the A19 rule or the sequence has homogeneous internal stability through the whole 19 nt region, a single nucleotide mismatch at the position 19 (sense strand) or replacement of guanosine with inosine in a sense strand can help to prioritize loading of the antisense strand into RISC (Schwarz et al. 2003). New insights into the role of sequence of siRNAs on the efficiency was provided by Katoh and Suzuki (2007) who screened all siRNAs corresponding to every single position of EGFP (enhanced green fluorescent protein). Their data clearly showed that efficacy of siRNAs is associated with the thermodynamic instability of the guide strand 5′ end as proposed. Another interesting finding is a discovery that every third position of highly effective siRNA contains an A/U nucleotide with a specific emphasis on a U in position 10 (sense strand), which positively correlates with a previously published finding (Jagla et al. 2005). The ideal sequence should have the following sequence: 5′-NNAANAANNUNAANAWNAA-3′ (Katoh and Suzuki 2007). The effectivity of siRNAs with the three nucleotide periodicity ($3n + 1$) correlates with the binding affinity to TRBP (transactivating response (TAR) RNA-binding protein), which was identified as a partner protein for human Dicer. However, these results will require further experimental validation. Characteristic and rules for a "perfect" siRNA are summarized in Fig. 2.

3.2 Thermodynamics Parameters

In previous parts of the text I have mentioned that a thermodynamical stability, or rather instability, of the 5′ end of the guide strand is an important parameter for designing siRNAs. It is also a parameter often used to infer a probability score that a specific siRNA will be efficient against its target.

Many biologists are not entirely familiar with the concept of free energy and how to interpret ΔG values provided or required by various prediction programs.

Gibbs energy is the chemical potential that is minimized when a system reaches equilibrium. The change in Gibbs free energy (ΔG) during a hybridization reaction between Watson-Crick complementary single stranded oligonucleotides provides a measure for the stability of the resulting complex. The interaction between siRNA guide strand and mRNA is a chemical reaction where siRNA and its target mRNA (or siRNA* strand in the siRNA/siRNA* duplex) form a double-stranded RNA (dsRNA). Formation of such complex will lead to a release of energy, which will change the energy of the whole system. ΔG is the net exchange of energy between the system and its environment. Free energy is closely related to other thermodynamics parameters, namely enthalpy and entropy, and can be described by the following equation.

$$\Delta G = \Delta H - T\Delta S,$$

H, Enthalpy – the heat content; ΔH, Enthalpy change; S, Entropy – a measure of chaos or disorganization of system; ΔS, energy spent by the system to organize itself; T, absolute temperature (Kelvin) (37°C = 310.15 K).

A negative value indicates that during the formation of siRNA/mRNA (or siRNA/siRNA*) duplex more energy is released and therefore formation of a product (dsRNA) is favored. More negative value also means that to dissociate a dsRNA duplex is more difficult than to dissociate a duplex with more positive value. The end of siRNA duplex with lower ΔG is therefore more likely to start to dissociate.

However, the dissociation of a siRNA duplex for its incorporation into RISC is only one side of the problem. The hybridization between the siRNA strand incorporated in RISC and the target mRNA has to be taken into consideration. siRNAs (miRNAs) with low ΔG are therefore more likely to interact with its respective target mRNA and probably will be more effective than siRNAs with less negative values. Therefore, siRNAs with overall lower ΔG should be chosen.

The equation has another important implication. The ΔG is temperature dependent and will change at different temperatures; therefore ΔG calculated for experiments carried out in *Drosophila* at 25°C will be different from the same siRNA/mRNA target ΔG calculated for mouse in vivo or cell culture experiments at 37°C. It is also important to mention that there is no real correlation between ΔG and melting temperature (T_m) of the oligonucleotide.

3.3 siRNA Design Algorithms

Most of the rules for siRNA design were implemented in various software packages or programs available through the internet. Many of them incorporate the original Tuschl algorithm (Elbashir et al. 2001). Thermodynamics parameters were taken into account in the Khvorova et al. (2003) algorithm while sense strand base preferences (positions 3, 10, 13 and 19) were incorporated into Reynolds et al. (2004) algorithm. Due to its small size, a siRNA duplex can cause cross-reactivity resulting in silencing off-target genes. An algorithm cross-checking for off-target sequences was incorporated into *Deqor* program (Henschel et al. 2004). Specific position base preferences and a penalty against GC stretches were used in Ui-Tei prediction algorithm (Ui-Tei et al. 2004). The three nucleotide periodicity rule ($3n + 1$) discovered by Katoh and Suzuki (2007) was implemented in *siExplorer*. Further parameters derived from correlation of specific primary sequence motifs with functionality were added by Amarzguioui and Prydz (2004). A radial basis function neuronal network and a supervised learning technique to predict functional siRNAs has been used by Takasaki et al. (2006). Neuronal network based prediction are part of the *BioPredsi* software (Huesken et al. 2005). Shabalina et al. (2006) used neuronal network to derive more sequence preferences and avoidances in active siRNAs and the findings are implemented in *ThermoComposition* software. Vert et al. (2006) in their *DSIR* program assigns weight values to individual bases in each position of siRNA strand. Another freeware named *Optirna* is presented by Ladunga (2007).

All approaches can, to some degree, discriminate between efficient and inefficient siRNAs. Detailed study of individual parameters and comparison of published algorithms and software has shown that *BioPredsi*, *ThermoComposition*, and *DSIR* perform better than the others (Matveeva et al. 2007). Linear regression that uses local duplex stability, nucleotide position-dependent preferences and total G/C content as input parameters is implemented in *siRNA scales* program (Matveeva et al. 2007).

Few more algorithms were compared by Saetrom and Snove (2004) who have shown that Amarzguioui and Prydz (2004) and Reynolds et al. (2004) algorithms with Ui-Tei algorithm (*siDirect*) (Ui-Tei et al. 2004) and GPboost (Saetrom and Snove 2004) outperform most of the others and have both stable and high performance.

Many companies specialized on production of chemical synthesized siRNAs provide design tools based on algorithms mentioned above or they use their proprietary algorithms. siDESIGN Center interface provides a high flexibility in design parameters and utilizes a feature-weighted selection algorithm with internal filtering. Several algorithms and websites where the siRNA design service is provided are summarized in Table 1.

Table 1 siRNAs design algorithms

Software or algorithm name/Reference	Website	Note
OptiRNA/Ladunga (2007)	http://optirna.unl.edu/	Support vector machine-based selection
siExplorer/Katoh and Suzuki (2007)	http://rna.chem.t.u-tokyo.ac.jp/cgi/siexplorer.htm	(3*n* + 1) A/U periodicity
siRNA-scales/Matveeva et al. (2007)	http://gesteland.genetics.utah.edu/siRNA_scales/	Local duplex stability, nucleotide position, G/C content
BioPredsi/Huesken et al. (2005)	http://www.biopredsi.org/start.html	Artificial intelligence algorithm
DSIR/Vert et al. (2006)	http://cbio.ensmp.fr/dsir/	Linear regression-based model
siDirect/Ui-Tei et al. (2004)	http://alps3.gi.k.u-tokyo.ac.jp/~yamada/sidirect2/index.php	Sequence features
siRNA Design Software	http://i.cs.hku.hk/~sirna/software/sirna.php	Combines several design tools to create list of candidates
Deqor/Henschel et al. (2004)	http://cluster-1.mpi-cbg.de/Deqor/deqor.html	Cross-reactivity considered
ThermoComposition/Shabalina et al. (2006)	ftp://ftp.ncbi.nih.gov/pub/shabalin/siRNA/	Neuronal network based
Amarzguioui and Prydz (2004)	Available upon request	Sequence features
Reynolds et al. (2004)	Available upon request	Sequence features
GPboost/Saetrom and Snove (2004)	Available upon request	Weighted sum of sequence motifs/patterns
siDESIGN center	http://www.dharmacon.com/designcenter	Feature-weighted selection algorithm
BLOCK-iT RNAi designer	https://rnaidesigner.invitrogen.com/rnaiexpress/setOption.do?designOption=sirna	Unspecified selection algorithm with specificity check
siRNA target finder	http://www.ambion.com/techlib/misc/siRNA_finder.html	Basic selection

Summary of the most common algorithms used to design siRNAs. Many of the algorithms were implemented into various software packages. siRNA design software provides either a gateway to other programs or makes use of existing design tools to create a set of candidates from an input mRNA. BioPredsi and ThermoComposition use artificial intelligence algorithm to identify active siRNAs. Other programs use individual algorithms or their combinations. The level of flexibility in design parameters varies among programs. A few proprietary design tools are included

Several highest scoring siRNA sequences should be analyzed for their specificity through NCBI's online BLAST interface (http://www.ncbi.nlm.nih.gov/BLAST/). A step by step protocol was recently published (Birmingham et al. 2007)

Still, even the best prediction programs have discrimination ability ranging at around 90%. As a rule of thumb, several siRNAs targeting the same mRNA should be designed and tested for their effectiveness.

3.4 siRNA Production

Enzymatic cleavage of long dsRNA precursors with recombinant Dicer (Myers et al. 2003) or with bacterial RNase III (Yang et al. 2002) will produce a mix of active siRNAs. Another possibility is chemical synthesis of individual siRNAs.

The enzymatic processing of dsRNA is relatively easy, cheap, and fast to perform. dsRNAs can be produced from previously cloned cDNA constructs used for preparation of in situ hybridization probes by in vitro transcription. Instead of synthesizing DIG-labeled cRNA, unlabeled strands (sense and antisense) can be produced, annealed together, and enzymatically cleaved producing a siRNA cocktail. This technique has many advantages but also some drawbacks. The cocktail of random 21-mers provides a robustness to the system as it is likely that while containing some ineffective siRNAs it will also contain a few powerfully silencing siRNAs. With this method, there is no need for selective design of specific siRNAs, and screens of many genes can be accomplished in just a few days. On the other hand, the system lacks in specificity as whole gene families can be targeted at the same time. The use of many siRNAs in the same experiment can lead to the overloading of the intracellular machinery, thus resulting in many nonspecific effects (Grimm et al. 2006) arising from miRNA deficiency. Certain siRNAs due to their short length can have homology with other unrelated genes which can be accidentally knocked-down together with the target gene.

Nowadays, the predominantly used technique is chemical synthesis of target specific siRNAs. This technique provides an excellent specificity, when even a single allele can be targeted, but it is more expensive and demands elaborate sequence design. Yet another advantage is the possibility to convert desired siRNA into a hairpin, which can be delivered into cells in a form of expression construct, allowing for long-term expression of the silencing siRNA.

A siRNA duplex is formed by two chemically synthesized RNA strands. The cost of chemical synthesis of RNA oligonucleotide is much higher than of DNA oligonucleotide of the same length. RNAs are unstable in biological systems due to presence of ubiquitous RNases. To lower the price of chemical synthesis the 3′ overhangs can be replaced with deoxynucleotides (Elbashir et al. 2001). As these two nucleotides do not contribute to sequence recognition, usually two deoxythymidine nucleotides are used (Fig. 2).

3.5 Chemical Modifications of siRNAs

To improve biological stability of siRNA molecules, a few chemical modifications were tested. The first modification, the replacement of unpaired 3′ ribonucleotide overhangs for deoxyribonucleotides, has already been mentioned. Modifications to increase half-life of RNA oligonucleotides can affect the sugar phosphate backbone, the ribose ring, or the RNA bases (Fig. 3). As a general rule, the sense strand tolerates chemical modifications better. Phosphorothioates are very often used in antisense knockdown strategies. Their introduction into either the sense or antisense strand of siRNA duplex does not prohibit RNAi-mediated RNA degradation but moderately affects the efficiency of RNAi (Chiu and Rana 2003). Replacement of sugar-phosphate backbone with peptidonucleic acid proved to be effective in antisense approaches, but to my knowledge it was never tested in RNAi. Another modification which is popular in antisense strategies relies on the replacement of the ribose ring with morpholine ring. Unfortunately, this approach was not tested in siRNAs either. A chemical modification of uracil in its five position is well tolerated in the antisense strand as proved with the introduction of 5-bromo-uridine or 5-iodo-uridine. On the other hand, the introduction of *N*(3)-methyl-uridine into antisense strand completely abolished RNAi (Chiu and Rana 2003). The most promising strategy seems to be a modification of the ribose ring.

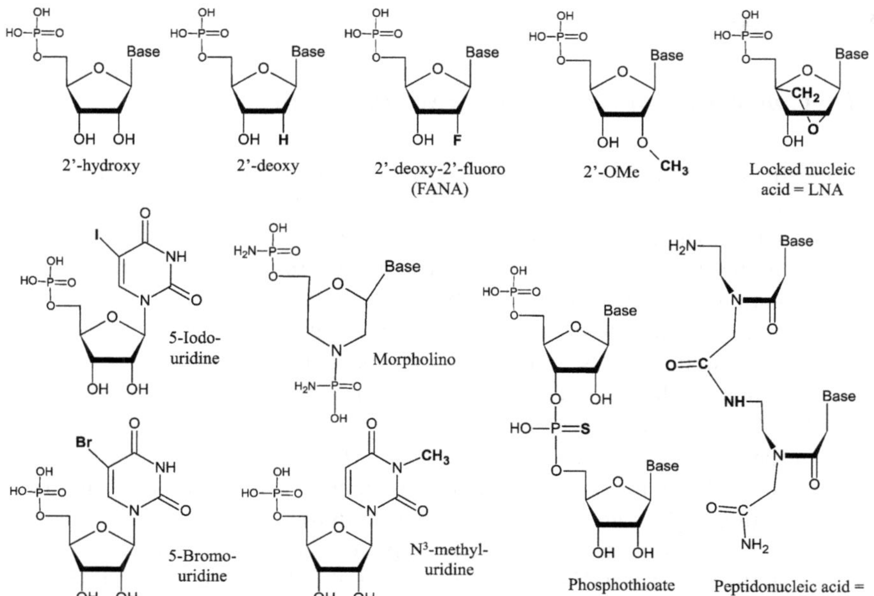

Fig. 3 Chemical modification of siRNAs. The most commonly used chemical modification of the ribose ring, sugar-phosphate backbone or the bases themselves are depicted

Even though replacement of ribose with 2′-deoxy-ribose is not tolerated within the duplex (except the overhangs) other modifications were found promising. Introduction of 2′-deoxy-2′-fluoro-β-D-arabinonucleic acid (FANA) nucleotides is tolerated in both strands of siRNA duplex, and fully modified sense strand was shown to be fourfold more potent and had longer half-life in serum compared to unmodified siRNA (Dowler et al. 2006). Surprisingly, the modification of a ribose ring with 2′-OMe groups significantly reduced RNAi (Chiu and Rana 2003). Another alteration, which involves 2′ position of ribose ring, is locked nucleic acid (LNA). LNA contains a methylene bridge connecting the 2′-oxygen with the 4′carbon of the ribose ring. This bridge locks the ribose ring in the *3′-endo* conformation characteristic of RNA. When used in antisense strategies, LNA has been reported to have unequalled affinity and a very high nuclease resistance. The very high affinity and specificity of LNA has a great potential for design of single allele specific RNAi. Introduction of a single LNA oligonucleotide into an antisense strand is well tolerated with the exception of position 1 (antisense strand), which completely abolishes activity. Replacement of several ribonucleotides in the sense strand with LNA is well tolerated (Elmen et al. 2005). However, the information available is not sufficient to draw a final conclusion and further experiments will be needed to fully understand the rules governing efficiency of chemical modifications in RNAi experiments.

4 shRNAs

4.1 Conversion of siRNA Duplexes into Hairpins

Chemically synthesized siRNAs are a useful tool to validate a target sequence or perform short term knockdown experiments in cell culture systems. For studies of proteins with a long half-life, functional genomics studies, cell type specific experiments or tissue specific knockdowns the use of siRNAs is clearly compromised.

The solution of these problems is stable expression of RNAi effectors molecules from plasmid or viral vectors. Active siRNAs are generated by cellular machinery, under normal circumstances, responsible for processing of miRNAs.

Cells contain miRNAs transcribed as a primary precursor transcripts (pri-pre-miRNA) which are then processed through a series of enzymatic cleavages taking place in the cell nucleus and in the cytoplasm (Fig. 4). Most miRNA genes are transcribed by RNA polymerase II either as mono- or polycistronic transcripts. miRNA coding regions are also often present in introns of other genes (Lee et al. 2002). Primary transcripts forming long imperfect hairpins are processed in the nucleus by the Drosha-DGCR8 complex (Lee et al. 2003; Zeng et al. 2005). This RNase III enzyme cleaves near the base of the hairpin releasing pre-miRNA, which is then transported out of the nucleus by the Exportin-5-Ran-GTP complex (Lund et al. 2004; Yi et al. 2003). In the cytoplasm, the pre-miRNA is processed by

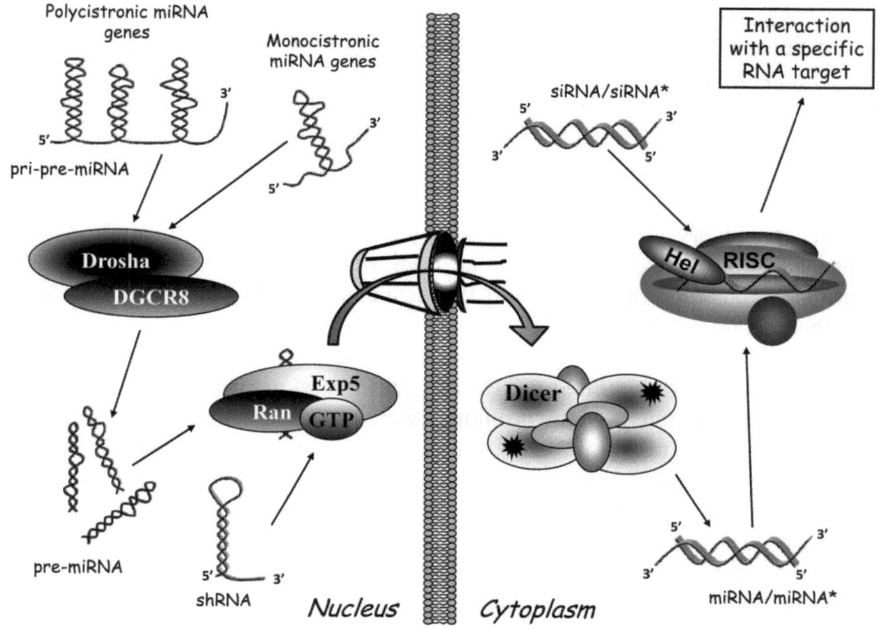

Fig. 4 miRNA/siRNA processing pathway. miRNAs are transcribed as pimary-pre-miRNAs either as mono- or polycistronic transcripts, which are cleaved by a complex of ribonuclease III *Drosha* and *DGCR8* into *pre-miRNAs*. *Pre-miRNAs* are transported to the cytoplasm by nuclear karyoptherin Exportin-5 in complex with *Ran-GTP*. The same complex exports *shRNAs* produced from U6 or H1 promoters. Translocated *shRNAs* or *pre-miRNAs* are in cytoplasm processed by another RNase III *Dicer* into active *siRNAs*, which are dissociated and the guide strand is incorporated into RISC

another RNase III enzyme Dicer. The enzyme cleaves near the loop and releases a miRNA:miRNA* duplex formed by a guide strand (miRNA) complementary to a miRNA binding site in mRNA 3′UTR, guiding RISC to a specific mRNA, and a passenger strand (miRNA*) which is released during the RISC holoenzyme assembly and destroyed. Details of RISC formation are not completely clear.

In *Drosophila* experiments it was shown that a putative RNA helicase Armitage (Armi) is required for RISC maturation (Tomari et al. 2004). This theory is supported by findings that RISC associates with a multiprotein complex containing MOV10, a human homolog of Armitage (Chendrimada et al. 2007), and identification of another RNA helicase A (RHA) that functions as an siRNA-loading factor (Robb and Rana 2007). On the other hand, the Argonaute 2 (Ago2), a RISC protein, was found to be solely responsible for loading siRNAs into RISC. Ago2 directly receives the double-stranded siRNA from the RISC assembly machinery and then cleaves the siRNA passenger strand, thereby liberating the single-stranded guide (Matranga et al. 2005). This theory, however, does not explain an asymmetry in loading selectively the guide strand into RISC (Khvorova et al. 2003; Schwarz et al. 2003). Binding of mRNA to RISC with 100% complementarity to the guide strand activates RNase activity of Ago2 and the mRNA is destroyed. Micro RNAs are seldom completely complementary to the miRNA binding site and most often they recognize only a short stretch of nucleotides complementary to the 5′ end of the miRNA called "seed" (nucleotides 2–7 of miRNA) (Lewis et al. 2005). In such cases, the mRNA is transported to so-called P (processing) bodies where it is stalled from translation or destroyed.

This machinery can be effectively utilized for production of siRNAs from expression constructs. siRNAs can be converted into three principal types of expression constructs, a shRNA with 19-nt stem, a shRNA with long (29-nt) stem, or imbedded into miRNA scaffold.

The shRNAs precursors are most often expressed from ubiquitously active RNA polymerase III promoters (H1, U6, 7SK) or viral or tissue specific RNA polymerase II promoters. Many promoters were modified by introduction of tetracycline operator (tetO) allowing for tetracycline-responsive expression. An expression cassette is then incorporated into a plasmid or viral vectors. The use of viral vectors, such as lentiviruses, allows easy infection of broad range of cells in vitro and in vivo, and generation of transgenic animals through infection of oocytes. Plasmids are easy to use in cell culture systems, cheap to prepare, and do not posses significant biosafety risk.

The most common strategy to convert siRNAs into coding sequence for expression from RNA polymerase III promoters is to connect the 3′ end of sense strand with the 5′ end of antisense strand by a short loop. The sequence of the loop does not seem to be important for efficient processing of shRNA but the length seems to be a crucial factor for the export of the shRNA from the nucleus (Zeng and Cullen 2004). Artificial loop sequences [5′-TTCAAGAGA-3′ (Brummelkamp et al. 2002)] or miRNA derived loops [miR-30 5′-CUGUGAAGCCACAGAUGGG-3′ (Boden et al. 2004) or miR-23 5′-CTTCCTGTCA-3′ (Kawasaki and Taira 2003)] were successfully used in previous experiments. Shortening of miR-30 loop down to nine nucleotides (5′-AAGCCACAG-3′) still preserves full silencing activity of shRNA. Further shortening of the loop down to six or four nucleotides impairs interaction of the shRNA with the nuclear export complex (Exportin-5/Ran/GTP) and effectively inactivates such shRNA (Zeng and Cullen 2004).

The use of RNA polymerase III promoters imposes another sequence restriction to the design of a short hairpin. All RNA polymerase III uniformly initiate transcription with a guanosine, which has to be present in position of +1. These polymerases also terminate transcription after a stretch of 4–6 uridines, consequently sequences of four and more consecutive uridines have to be avoided.

This approach was further modified by increasing the length of the stem region up to 29 nucleotides (Siolas et al. 2005). For unknown reasons and in contradiction to a study analyzing the effect of the loop on shRNA efficiency, the authors chose a 4-nt loop.

The third way of producing siRNAs in cells is to imbed the siRNA sequence into a miRNA (miR-30) scaffold (Silva et al. 2005) mimicking cellular processing of pre-pri-miRNA transcript.

Authors of the two new conversion strategies claim substantial improvements over the other methods. Until recently a systematic comparison of these strategies was missing. Li et al. (2007) compared various approaches by converting of the same siRNAs into various shRNAs expression systems. The authors demonstrated that U6-expressed shRNAs with 19-nt stem and 9-nt loop region (5′-UUCAAGAGA-3′) are more efficient than miR-30 based embedded siRNAs for target knockdown. The shRNA with 29-nt stem were more efficient than shRNAs with a 19-nt stem in the context of a 4-nt loop but shRNAs with a 19-nt stem are superior to shRNAs with a 29-nt stem for target knockdown in the context of a 9-nt loop. This is in agreement with work of Zeng and Cullen (2004) who have shown that shortening of the loop down to 4-nt severely impairs the efficiency of shRNA.

Li et al. (2007) have also found that functional shRNAs and siRNAs exhibit similar but not identical sequence preferences. The main difference is that functional siRNAs prefer G/C at position 9 (sense strand) and have a weak preference for G/C at position 11 while functional shRNAs prefer A/U in position 9 and G/C in position 11. These differences are included in Fig. 2.

Under some circumstances, a tissue specific expression of siRNAs is desired. In such a situation an expression of miRNA-embedded siRNA from tissue specific RNA polymerase II promoter is advisable.

Many experimental approaches are focused on improving and increasing expression level of siRNAs in transfected or infected cells. The pursuit of the highest expression level might impose additional problems. Grimm et al. (2006) have shown that a high expression level of siRNA results in saturation of the intracellular miRNA processing machinery, which in turn leads to a downregulation of cell specific miRNAs (Grimm et al. 2006). It seems that a rate limiting factor in miRNA processing cascade is the nuclear karyopherin exportin-5. Moderate siRNA expression level can provide more relevant results which are less likely biased by interference with cellular miRNAs levels. This is further supported by a finding that the use of a high concentration (~100 nM) of siRNAs in transfection experiments induced a significant number of genes often involved in stress response. Such activation of stress response genes was eliminated when 20 nM siRNAs were used (Semizarov et al. 2003).

4.2 *Cloning and Expression of shRNAs*

The cloning of shRNAs into U6 promoter driven expression constructs has a few pitfalls. Introduction of a restriction site to the region of transcription initiation would be desired for cloning of the shRNAs, but the sequence cannot be modified as every alteration will lead to inactivation of the promoter. An elegant way to circumvent this problem is to use a restriction enzyme with different recognition and cleavage sequence. The process is schematically drawn in Fig. 5. We have introduced inverted BbsI restriction site downstream of the U6 promoter. BbsI recognizes

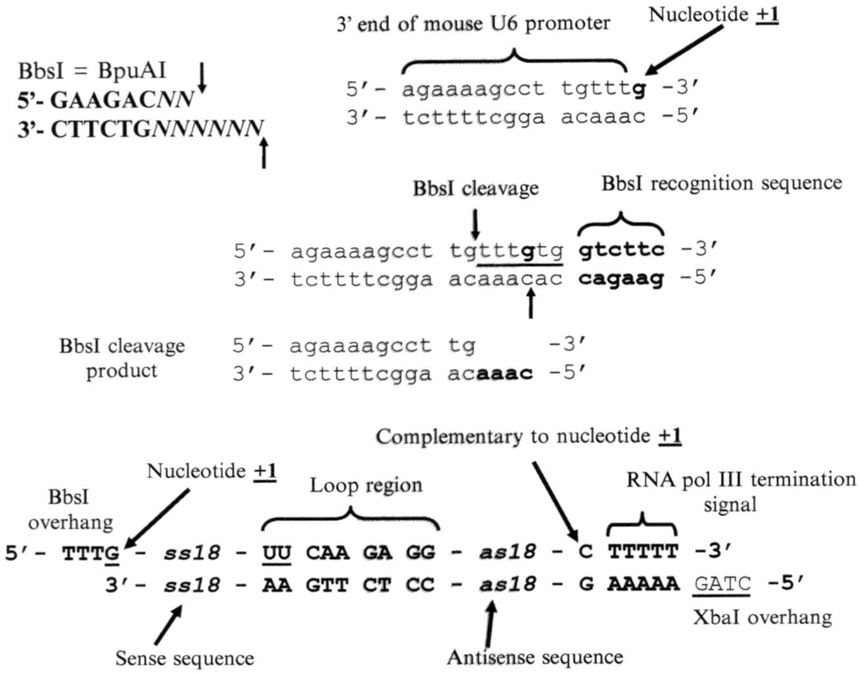

Fig. 5 Cloning of shRNA into BbsI restriction site. BbsI restriction enzyme recognizes six nucleotide sequence but cleaves at juxtaposed site. BbsI site can be introduced into antisense strand in the proximity of U6 (or H1) promoter. BbsI cleavage results in overhang comprising +1 G nucleotide and the last three nucleotides of the promoter. Compatible overhang is then implemented into the synthesized oligos together with another overhang compatible to the second restriction endonuclease cleavage site

6-nt DNA motif but cleaves two, respective four nucleotides from the recognition site leaving 4-nt 5′ overhang. Several nucleotides apart from BbsI site we have introduced the XbaI site. The shRNA insert is formed by annealing of two oligonucleotides that will leave a 3′ overhang compatible to nucleotides –3 to –1 of the 3′ end of U6 promoter and nucleotide +1 of the shRNA transcript. The G nucleotide in position +1 is required for RNA polymerase III initiation of transcription and impose another sequence constraint.

Oligonucleotides used for preparation of the insert should be HPLC-purified to reduce the number of synthesis errors. Other restriction endonucleases recognizing degenerate sequence (SapI, EarI, BsmI) can be used instead of BbsI.

Another factor influencing the cloning of shRNAs is a bacterial strain used for propagation of plasmids containing shRNAs. shRNAs expression constructs will contain inverted repeats, which are often recombined out by most commonly used *E. coli* strains. Several strains were created to combat such events. SURE strain (recB *recJ sbcC umuC*::Tn5 (KanR) *uvrC*) or GT116 (*recA1 ΔsbcC-sbcD*) are deficient in several recombination pathways especially the SbcCD. The *sbcC* and *sbcD* genes encode a dimeric protein called SbcCD that recognizes and cleaves hairpins rendering plasmids with hairpin structures particularly unstable in *E. coli*. Deletion of the sbcCD operon significantly improves the number of recombinant clones in plasmids containing hairpin structures. Deletion or mutation in *recA*, *recJ* or *uvrC* further improves the stability of inverted repeats or other non-B DNA conformation often found in mammalian DNA. Another possibility to increase the stability of shRNA constructs in bacteria lies in targeted mutagenesis of the sense strand. Double-stranded RNA can well accommodate G:U wobble pairs replacing Watson-Crick pairs (Varani and McClain 2000). The wobble G:U pair is not well tolerated in a complex of siRNA guide strand with mRNA and inhibits silencing by RISC; therefore, only the passenger strand should be altered. Besides stabilizing the shRNA in bacteria this approach can be used to improve thermodynamical parameters of siRNA by increasing ΔG of 3′ end of the passenger siRNA strand.

Some plasmids, especially coding for siRNAs against tumor suppressors genes, can posses a biosafety risk. To reduce the chance of spreading such plasmids into the environment, plasmids with a conditional origin of replication R6Kγ (gamma) can be used (Silva et al. 2005). Such plasmids can be propagated in bacteria expressing *pir* gene coding for R6K specific initiator protein π (pi).

Alternatively, shRNAs coding sequence can be cloned downstream of T7 or other bacteriophage RNA polymerase promoter (SP6, T3) and expressed in vitro. Transcribed shRNAs can be transfected into cells where they will be processed by Dicer into siRNAs.

5 Clinical Application

Besides becoming an excellent experimental tool the RNAi becomes a clinically important technique. The ability to downregulate not only a protein but also a specific message makes it a useful strategy to combat not only sporadic or inherited dominant diseases but also certain virus infections, e.g., HIV being one of the most prominent. RNAi holds a potential to treat chronic hepatitis B virus (HBV) and hepatitis C virus (HCV) infection. Globally, these hepatotrophic viruses are the most important causes of cirrhosis and liver cancer. Available treatments have their limitations, which makes development of novel effective RNAi-based therapies for HBV and HCV especially significant.

RNAi can find application in cell or tissue transplantation therapies. The capability to downregulate components of histocompatibility systems can improve success of allograft transplantations. Administration of dendritic cells modified by RNA interference prolongs cardiac allograft survival (Xiang et al. 2007).

A superior specificity to recognize a well-defined sequence opens new avenues to develop allele-specific RNAi protocols targeting the mutant allele only, allowing treatment of dominant mono allelic syndromes and diseases.

5.1 Antiviral Applications

RNA interference (RNAi) targeting viral mRNAs is widely used to block virus replication in mammalian cells. The specific antiviral RNAi response can be induced via transfection of synthetic small interfering RNAs (siRNAs) or via intracellular expression of short hairpin RNAs (shRNAs). For HIV-1, both approaches resulted in profound inhibition of virus replication. However, the therapeutic use of a single siRNA/shRNA appears limited due to the rapid emergence of RNAi-resistant escape viruses. This is the most common problem of most of the anti-HIV therapeutical approaches. Selective introduction of multiple shRNAs targeting several crucial genes involved in HIV replication and infectivity can lead to effective virus control.

RNAi can be used as antiviral therapy against other viruses having RNA genomes or RNA replication intermediates.

5.2 Allele Specific RNAi

Many severe predominantly neurological diseases are caused by expression of a protein with dominant negative function coded by one allele. Most of these proteins are vitally important for cellular functions; therefore, overall downregulation of both alleles would create even more problems.

Primary dystonia (DYT1) is the most common form of inherited dystonia. DYT1 is caused by a deletion in the *TOR1A* gene that eliminates a glutamic acid residue from the protein torsinA (*TOR1AΔE*). Currently, there are no preventive or curative therapies. Delivery of shRNA targeting torsinA(ΔE) by a recombinant feline immunodeficiency virus effectively silenced torsinA(ΔE) in a neural model of the disease (Gonzalez-Alegre et al. 2005). The RNAi technology is potentially effective even in situations when the biological role of the affected protein is unclear, such as in the case of torsinA.

A single nucleotide polymorphism (SNP) in the sickle β-globin gene (β^S) leads to sickle cell anemia. A specific siRNA with position 10 of the guide strand designed to align with the targeted SNP specifically silences β^S gene expression without affecting the expression of the γ-globin or normal β-globin genes. Silencing was increased by altering the 5′ end of the siRNA antisense (guide) strand to enhance its binding to the RNA-induced silencing complex (RISC) (Dykxhoorn et al. 2006).

Another example of allele specific interference is targeting the mutated allele of Cu/Zn superoxide dismutase (SOD1) causing devastating neurodegenerative disease, the amyotrophic lateral sclerosis (ALS). The disease is characterized by motor neuron degeneration, paralysis and death. As mutant SOD1 acquires a toxic property that kills motor neurons, by reducing the mutant protein the disease progression may be slowed or prevented. shRNAs selectively targeting mutant allele in a transgenic mouse model were developed and have been shown to significantly delay ALS onset and extend survival (Xia et al. 2006). ALS is caused by many different mutations and not all of them can be selectively targeted. A way to address this problem is a use of shRNA downregulating both wild type and mutant alleles with simultaneous delivery of modified gene allele resistant to shRNAs (Xia et al. 2005).

Other diseases where RNAi can be clinically applied are Alzheimer disease or Parkinson's disease.

Huntington's diseases and several types of spinocerebellar ataxias (SCA) are caused by accumulations of proteins containing extended polyglutamine (polyQ) stretches. Polyglutamine stretches are encoded predominantly by a CAG triplet forming a CAG repetitive sequence with increased tendency to expand. In such cases, it is not possible to design siRNA selectively targeting mutant genes because the repeated sequence is present in the wild-type allele, too. Naturally occurring single nucleotide polymorphisms (SNP) are often present in many genes and differ between maternal and paternal alleles. An interesting approach is to target such a polymorphism linked to the allele carrying expansion instead of targeting the mutation itself.

Prion diseases are fatal neurodegenerative diseases characterized by the accumulation of PrP(Sc), the infectious and protease-resistant form of the cellular prion protein [PrP(C)]. Clinical application of RNAi holds a promise of treatment of prion protein (PrP) caused transmissible spongiform encephalopathies (TSE). Pfeifer et al. (2006) have generated lentivectors expressing PrP(C)-specific short hairpin RNAs (shRNAs) that efficiently silenced expression of the prion protein gene (Prnp) in primary neuronal cells. After intracranial injection, lentiviral shRNA reduced PrP(C) expression in transgenic mice carrying multiple copies of Prnp. Such reduction in PrP level was shown to reduce infection rate and prolong survival of animals infected with PrP(Sc).

Nonphysiological activation of many onçogenes is often caused by a point mutation or by genomic translocation resulting in fusion of oncogenes with other genes. shRNAs targeting the junction in chimeric protein can be an attractive therapeutical target in anti tumor therapies.

6 Concluding Remark

Use of siRNAs or interference with miRNA system can be compared to a butterfly effect. A single butterfly flapping its wings in Brazil sets off a tornado in Texas. The phrase describes a part of chaos theory stating that even small variations of the initial condition may produce large variations in the long-term behavior of the nonlinear dynamic system. All biological processes are subject to these fundamental laws. The RNAi is clearly changing the state of initial conditions. In well-controlled experiments the RNAi can provide very specific results. However, one should be aware of the fact that sometimes beating of RNAi wings can cause a tornado in the experimental system.

References

Amarzguioui M, Prydz H (2004) An algorithm for selection of functional siRNA sequences. Biochem Biophys Res Commun 316:1050–1058

Billy E, Brondani V, Zhang H et al (2001) Specific interference with gene expression induced by long, double- stranded RNA in mouse embryonal teratocarcinoma cell lines. Proc Natl Acad Sci USA 98:14428–14433

Birmingham A, Anderson E, Sullivan K et al (2007) A protocol for designing siRNAs with high functionality and specificity. Nat Prot 2:2068–2078

Boden D, Pusch O, Silbermann R et al (2004) Enhanced gene silencing of HIV-1 specific siRNA using microRNA designed hairpins. Nucleic Acids Res 32:1154–1158

Brummelkamp TR, Bernards R, Agami R (2002) A system for stable expression of short interfering RNAs in mammalian cells. Science 296:550–553

Dowler T, Bergeron D, Tedeschi AL et al (2006) Improvements in siRNA properties mediated by 2′-deoxy-2′-fluoro-beta-D-arabinonucleic acid (FANA). Nucleic Acids Res 34:1669–1675

Dykxhoorn DM, Schlehuber LD, London IM et al (2006) Determinants of specific RNA interference-mediated silencing of human beta-globin alleles differing by a single nucleotide polymorphism. Proc Natl Acad Sci USA 103:5953–5958

Elbashir SM, Martinez J, Patkaniowska A et al (2001) Functional anatomy of siRNAs for mediating efficient RNAi in *Drosophila melanogaster* embryo lysate. EMBO J 20:6877–6888

Elmen J, Thonberg H, Ljungberg K et al (2005) Locked nucleic acid (LNA) mediated improvements in siRNA stability and functionality. Nucleic Acids Res 33:439–447

Fire A, Xu S, Montgomery MK et al (1998) Potent and specific genetic interference by double-stranded RNA in *Caenorhabditis elegans*. Nature 391:806–811

Fish RJ, Kruithof EK (2004) Short-term cytotoxic effects and long-term instability of RNAi delivered using lentiviral vectors. BMC Mol Biol 5:9

Gonzalez-Alegre P, Bode N, Davidson BL et al (2005) Silencing primary dystonia: lentiviral-mediated RNA interference therapy for DYT1 dystonia. J Neurosci 25:10502–10509

Grimm D, Streetz KL, Jopling CL et al (2006) Fatality in mice due to oversaturation of cellular microRNA/short hairpin RNA pathways. Nature 441:537–541

Henschel A, Buchholz F, Habermann B (2004) DEQOR: a web-based tool for the design and quality control of siRNAs. Nucleic Acids Res 32:W113–W120

Huesken D, Lange J, Mickanin C et al (2005) Design of a genome-wide siRNA library using an artificial neural network. Nat Biotechnol 23:995–1001

Check E (2007) RNA interference: hitting the on switch. Nature 448:855–858

Chendrimada TP, Finn, KJ, Ji X et al (2007) MicroRNA silencing through RISC recruitment of eIF6. Nature 447:823–828

Chiu YL, Rana TM (2003) siRNA function in RNAi: a chemical modification analysis. RNA 9:1034–1048

Jagla B, Aulner N, Kelly PD et al (2005) Sequence characteristics of functional siRNAs. RNA 11:864–872

Janowski BA, Younger ST, Hardy DB et al (2007) Activating gene expression in mammalian cells with promoter-targeted duplex RNAs. Nat Chem Biol 3:166–173

Karpala AJ, Doran TJ, Bean AG (2005) Immune responses to dsRNA: implications for gene silencing technologies. Immunol Cell Biol 83:211–216

Katoh T, Suzuki T (2007) Specific residues at every third position of siRNA shape its efficient RNAi activity. Nucleic Acids Res 35:e27

Kawasaki H, Taira K (2003) Short hairpin type of dsRNAs that are controlled by tRNA(Val) promoter significantly induce RNAi-mediated gene silencing in the cytoplasm of human cells. Nucleic Acids Res 31:700–707

Khvorova A, Reynolds A, Jayasena SD (2003) Functional siRNAs and miRNAs exhibit strand bias. Cell 115:209–216

Ladunga I (2007) More complete gene silencing by fewer siRNAs: transparent optimized design and biophysical signature. Nucleic Acids Res 35:433–440

Lee Y, Jeon K, Lee JT et al (2002) MicroRNA maturation: stepwise processing and subcellular localization. EMBO J 21:4663–4670

Lee Y, Ahn C, Han J et al (2003) The nuclear RNase III Drosha initiates microRNA processing. Nature 425:415–419

Lewis BP, Burge CB, Bartel DP (2005) Conserved seed pairing, often flanked by adenosines, indicates that thousands of human genes are microRNA targets. Cell 120:15–20

Li LC, Okino ST, Zhao H et al (2006) Small dsRNAs induce transcriptional activation in human cells. Proc Natl Acad Sci USA 103:17337–17342

Li L, Lin X, Khvorova A et al (2007) Defining the optimal parameters for hairpin-based knockdown constructs. RNA 13:1765–1774 (Epub ahead of print)

Lund E, Guttinger S, Calado A et al (2004) Nuclear export of microRNA precursors. Science 303:95–98

Matranga C, Tomari Y, Shin C et al (2005) Passenger-strand cleavage facilitates assembly of siRNA into Ago2-containing RNAi enzyme complexes. Cell 123:607–620

Matveeva O, Nechipurenko Y, Rossi L et al (2007) Comparison of approaches for rational siRNA design leading to a new efficient and transparent method. Nucleic Acids Res 35:e63

Metzlaff M, O'Dell M, Cluster PD et al (1997) RNA-mediated RNA degradation and chalcone synthase A silencing in petunia. Cell 88:845–854

Morris KV, Chan SW, Jacobsen SE et al (2004) Small interfering RNA-induced transcriptional gene silencing in human cells. Science 305:1289–1292

Myers JW, Jones JT, Meyer T et al (2003) Recombinant dicer efficiently converts large dsRNAs into siRNAs suitable for gene silencing. Nat Biotechnol 21:324–328

Napoli C, Lemieux C, Jorgensen R (1990) Introduction of a chimeric chalcone synthase gene into petunia results in reversible co-suppression of homologous genes in trans. Plant Cell 2:279–289

Pebernard S, Iggo RD (2004) Determinants of interferon-stimulated gene induction by RNAi vectors. Differentiation 72:103–111

Pfeifer A, Eigenbrod S, Al-Khadra S et al (2006) Lentivector-mediated RNAi efficiently suppresses prion protein and prolongs survival of scrapie-infected mice. J Clin Invest 116:3204–3210

Reynolds A, Leake D, Boese Q et al (2004) Rational siRNA design for RNA interference. Nat Biotechnol 22:326–330

Robb GB, Rana TM (2007) RNA helicase A interacts with RISC in human cells and functions in RISC loading. Mol Cell 26:523–537

Saetrom P, Snove O Jr (2004) A comparison of siRNA efficacy predictors. Biochem Biophys Res Commun 321:247–253

Semizarov D, Frost L, Sarthy A et al (2003) Specificity of short interfering RNA determined through gene expression signatures. Proc Natl Acad Sci USA 100:6347–6352

Shabalina SA, Spiridonov AN, Ogurtsov AY (2006) Computational models with thermodynamic and composition features improve siRNA design. BMC Bioinformatics 7:65

Scheuner D, Patel R, Wang F et al (2006) Double-stranded RNA-dependent protein kinase phosphorylation of the alpha-subunit of eukaryotic translation initiation factor 2 mediates apoptosis. J Biol Chem 281:21458–21468

Schwarz DS, Hutvagner G, Du T et al (2003) Asymmetry in the assembly of the RNAi enzyme complex. Cell 115:199–208

Silva JM, Li MZ, Chang K et al (2005) Second-generation shRNA libraries covering the mouse and human genomes. Nat Genet 37:1281–1288

Siolas D, Lerner C, Burchard J, Ge W et al (2005) Synthetic shRNAs as potent RNAi triggers. Nat Biotechnol 23:227–231

Takasaki S, Kawamura Y, Konagaya A (2006) Selecting effective siRNA sequences by using radial basis function network and decision tree learning. BMC Bioinformatics 7(Suppl 5):S22

Tomari Y, Du T, Haley B et al (2004) RISC assembly defects in the *Drosophila* RNAi mutant armitage. Cell 116:831–841
Ui-Tei K, Naito Y, Takahashi F et al (2004) Guidelines for the selection of highly effective siRNA sequences for mammalian and chick RNA interference. Nucleic Acids Res 32:936–948
van der Krol AR, Mur LA, Beld M et al (1990) Flavonoid genes in petunia: addition of a limited number of gene copies may lead to a suppression of gene expression. Plant Cell 2:291–299
Varani G, McClain WH (2000) The G × U wobble base pair. A fundamental building block of RNA structure crucial to RNA function in diverse biological systems. EMBO Rep 1:18–23
Vert JP, Foveau N, Lajaunie C et al (2006) An accurate and interpretable model for siRNA efficacy prediction. BMC Bioinform 7:520
Xia XG, Zhou H, Zhou S et al (2005) An RNAi strategy for treatment of amyotrophic lateral sclerosis caused by mutant Cu, Zn superoxide dismutase. J Neurochem 92:362–367
Xia X, Zhou H, Huang Y et al (2006) Allele-specific RNAi selectively silences mutant SOD1 and achieves significant therapeutic benefit in vivo. Neurobiol Dis 23:578–586
Xiang J, Gu X, Zhou Y et al (2007) Administration of dendritic cells modified by RNA interference prolongs cardiac allograft survival. Microsurgery 27:320–323
Yang D, Buchholz F, Huang Z et al (2002) Short RNA duplexes produced by hydrolysis with *Escherichia coli* RNase III mediate effective RNA interference in mammalian cells. Proc Natl Acad Sci USA 2:2
Yi R, Qin Y, Macara IG et al (2003) Exportin-5 mediates the nuclear export of pre-microRNAs and short hairpin RNAs. Genes Dev 17:3011–3016
Zeng Y, Cullen BR (2004) Structural requirements for pre-microRNA binding and nuclear export by Exportin 5. Nucleic Acids Res 32:4776–4785
Zeng Y, Yi R, Cullen BR (2005) Recognition and cleavage of primary microRNA precursors by the nuclear processing enzyme Drosha. EMBO J 24:138–148

RNA Interference and MicroRNA Modulation for the Treatment of Cardiac Disorders

Status and Challenges

W. Poller(✉), L. Suckau, S. Pinkert, and H. Fechner

Contents

Abstract The current status and challenges of RNA interference (RNAi) and microRNA modulation strategies for the treatment of myocardial disorders are discussed and related to the classical gene therapeutic approaches of the past decade. Section 2 summarizes the key issues of current vector technologies which determine if they may be suitable for clinical translation of experimental

W. Poller
Department of Cardiology and Pneumology, Campus-Benjamin Franklin, Charite University Medicine Berlin, Germany
e-mail: wolfgang.poller@charite.de

V.A. Erdmann et al. (eds.) *RNA Technologies in Cardiovascular Medicine and Research*,
doi: 10.1007/978-3-540-78709-9,

RNAi or microRNA therapeutic protocols. We then present and discuss examples dealing with the potential of cardiac RNAi therapy. First, an approach to block a key early step in the pathogenesis of a virus-induced cardiomyopathy by RNAi targeting of a cellular receptor for cardiopathogenic viruses (Section 3). Second, an approach to improve cardiac function by RNAi targeting of late pathway of heart failure pathogenesis common to myocardial disorders of multiple etiologies. This strategy is directed at myocardial Ca^{2+} homeostasis which is disturbed in heart failure due to coronary heart disease, heart valve dysfunction, cardiac inflammation, or genetic defects (Section 4). Whereas the first type of strategies (directed at early pathogenesis) need to be tailor-made for each different type of pathomechanism, the second type (targeting late common pathways) has a much broader range of application. This advantage of the second type of approaches is of key importance since enormous efforts need to be undertaken before any regulatory RNA therapy enters the stage of possible clinical translation. If then the number of patients eligible for this protocol is large, the actual transformation of the experimental therapy into a new therapeutic option of clinical importance is far more likely to occur.

Abbreviations AAV: Adeno-associated virus; AdV: Adenovirus; CAR: Coxsackievirus–adenovirus-receptor; DCM: Dilated cardiomyopathy; DCMi: Inflammatory cardiomyopathy; HF: Heart failure; miRNA: microRNA; misiRNA: miRNA-based shRNA; PLB: Phospholamban; RNAi: RNA interference; shRNA: Short hairpin RNA

1 Introduction to RNA-Based and Gene Therapeutic Strategies

In this chapter we will discuss the current status and challenges of using RNA interference (RNAi) and microRNA modulation strategies for the treatment of myocardial disorders and relate them to the classical gene therapeutic approaches of the past decade. The introductory Sect. 2 summarizes the key issues of current vector technologies which critically determine if they may be suitable for clinical translation of experimental RNAi or microRNA therapeutic protocols. We will then present and discuss two examples dealing with the potential of cardiac RNAi therapy. The first example deals with an attempt to block a very early step in the pathogenesis of a virus-induced cardiomyopathy and the resulting heart failure by RNAi targeting of a cellular receptor for cardiopathogenic viruses (Sect. 3). The second examples deals with an approach to improve cardiac function by RNAi targeting a late, common pathway of heart failure pathogenesis which is common to myocardial disorders of multiple etiologies. This strategy is directed at myocardiac Ca^{2+} homeostasis which is a major component of heart failure due to coronary heart disease (ischemic cardiomyopathy), heart valve dysfunction (valvular cardiomoypathy), cardiac inflammation (inflammatory cardiomyopathy), or genetic defects (dilated cardiomyopathy) (Sect. 4). Whereas the first type of strategies (directed at early pathogenic steps) needs to be tailor-made for each different type of pathogenesis, the second type (which target late common pathways) has a much broader range of application. This advantage of the second type of approach is of major importance since enormous efforts need to be undertaken before an RNAi or micro RNA-based therapeutic protocol enters the stage of possible clinical translation. If, then, the number of patients eligible for this protocol is large, the actual transformation of the experimental therapy into a new therapeutic option of major clinical importance is far more likely to occur.

2 Key Issues of Current Vector Technologies

The classical first concept of gene therapy which has been developed for monogenic disorders caused by deficiency of a single gene is gene substitution therapy. In the field of cardiology, several monogenic disorders primarily manifested in the myocardium could theoretically be cured by this approach. Among them are the various long-QT syndromes, the arrhythmogenic right ventricular dysplasia, the Brugada syndrome, the hypertrophic cardiomyopathies caused by a host of mutations in various sarcomeric proteins, and the dilated cardiomyopathies. A second concept of gene therapy is the enhancement of gene functions to cells, tissues, and organs by overexpression of primarily endogenous genes using gene transfer vectors. Under appropriate conditions, a therapeutic effect may be achieved not only in genetically determined but also in acquired diseases. A third concept is the addition of truly novel gene functions to the target organ by vector-based expression of foreign genes. Novel concepts of far more recent origin involve the use of non-protein-coding, small regulatory RNA molecules for therapeutic purposes. They include the fourth concept of short hairpin RNA transcription from vectors for the suppression of gene functions in the diseased target organ or organism via the mechanism of RNA interference, and the fifth concept of using microRNA generation from vectors to achieve therapeutic modulation of cell and organ functions on the level of the cellular microRNA system (see Part I of this book). Naturally, all of the following key issues of classical gene therapy (concepts 1–3) also apply similarly to the RNA-based therapies (concepts 4 and 5). Figure 1 illustrates the fundamental difference in action between classical gene therapy vs. RNA therapy, but at the same time points to an important partial overlap between the two otherwise distinct strategies. Common to both is the need to deliver the therapeutic structure (cDNA-encoded protein vs. regulatory RNA) to the right place, at an appropriate concentration for a sufficient period of time, and with adequate safety features. This partial overlap allows to exploit advanced vector technologies, initially developed in the gene therapy field, for most types of RNA therapy, too.

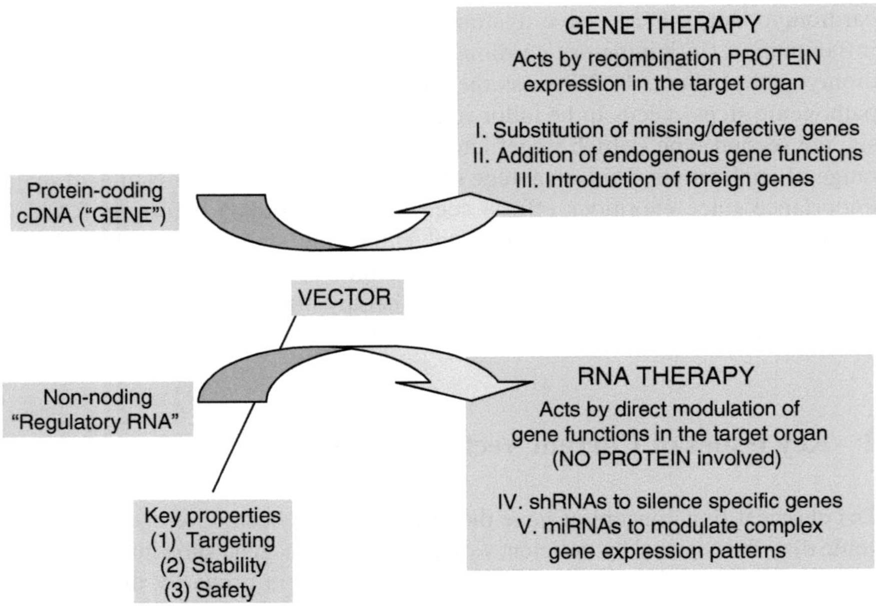

Fig. 1 Strategic Concepts of Gene Therapy *versus* Regulatory RNA Therapy

2.1 *Targeting of Vectors for Regulatory RNAs or Genes*

The first key issue of vector technology is the mere physical steering (targeting) of the vector (irrespective of expressing a protein-encoding cDNA, a short hairpin RNA or a pre-microRNA) to the target tissue, e.g., the myocardium, which proved to be a difficult task. Various approaches (Hoshijima et al. 2002; Rockman et al. 1998; White et al. 2000; Kypson et al. 1999; Koch et al. 2000; Ikeda et al. 2002) including intracoronary application (Giordano et al. 1996; Maurice et al. 1999) have been employed to achieve this goal, including already in clinical trials (Ylä-Herttuala and Alitalo 2003; Simons et al. 2000; Grines et al. 2002; Hedman et al. 2003). A better understanding of the molecular and cellular determinants of vector targeting in the cardiovascular system has evolved during the past few years based on studies of the expression patterns of vector receptors (i.e. the receptors primarily mediating the cellular uptake of the virus from which the vector is derived) in animals (Fechner et al. 1999, 2003a) and humans (Noutsias et al. 2001; Poller et al. 2002a,b), and on the discovery of anatomical barriers such as the vascular endothelium or intracellular matrix (Fechner et al. 1999) inhibiting or preventing vector transfer to particular parts of the heart and other organs and tumors. Interestingly, the expression of a particularly important vector receptor, the Coxsackievirus–

Adenovirus-Receptor (CAR), was found to be highly variable in human hearts (generalized induction in dilated cardiomyopathy) (Noutsias et al. 2001; Communal et al. 2003) and in animals (local induction after myocardial infarction) (Fechner et al. 2003a). Receptor-directed strategies (Curiel 1999) for the improvement of myocardial vector targeting have been evaluated but without significant success in vivo. If perfect physical vector targeting to the diseased tissue cannot be achieved, additional transcriptional confinement of the transgene may be achieved by using cardiac-specific promotors (Franz et al. 1997; Henderson et al. 1989; Reynolds et al. 2001). Recognition of the strong receptor-dependency of vector targeting has prompted investigation dealing with the possible therapeutic potential of tropism modifications by altering the receptor-interaction domains of the viruses used as vectors. A number of studies on the vector-susceptibility of different primary cell types for tropism modification have been conducted both for adenoviral (Wickham 2000; Wickham et al. 1997) and AAV vectors (Nicklin and Baker 2002; Rabinowitz et al. 2002; Ponnazhagan et al. 2002; Ponnazhagan and Hoover 2004; Shi and Bartlett 2003; Perabo et al. 2003; Büning et al. 2003). Basically, these approaches employ alterations of viral surface structures mediating cellular receptor binding and internalization of the vector. This has been achieved by genetic engineering of vector genes encoding vector surface components (Ogorelkova et al. 2006; Poller et al. 2002b), by bi-functional antibodies binding both to a vector surface epitope and a cellular structure specific for the desired target cells (Noutsias et al. 2001), or by bi-specific targeting proteins conjugated to the vector surface via avidin–biotin complexes (Perabo t al. 2003). However, none of these approaches has achieved similar success in vivo as the most recent generation of pseudotyped AAV vectors (in particular AAV8 and AAV9 for cardiac targeting), although use of these vectors may be combined with further tropism optimization by introducing additional alterations in the pseudotyped AAV capsids, and with tissue-specific promoters.

It should be emphasized that in order for the above tropism modifications to become useful in vivo the vectors first need to get direct access to the desired target cells, which may be prevented, however, by anatomical barriers (vascular endothelium, basal membrane, extracellular matrix). In addition to studies addressing the receptor issue, a number of approaches to overcome the anatomical barriers have been evaluated (Vale et al. 1999; Boekstegers et al. 2000; Beeri et al. 2002; Davidson et al. 2001; Price et al. 1998; Donahue et al. 1997; Hajjar et al. 1998), but none of these methods is simple or as yet appropriate for application in the clinical setting. In myocardial gene therapy, the final target cells – beyond the anatomical barriers – are particularly difficult targets as compared to other cell types and, at the current state of the art, viral vectors are the only option to achieve transfer rates sufficient for therapeutic efficacy. So far, no report on successful in vivo treatment of cardiac diseases using non-viral systems (including nanoparticles) has been published. Given the obstacles against efficient and targeted vector delivery, the recent development of imaging techniques for the in vivo detection of transgene expression may greatly facilitate the further development, monitoring, and assessment of cardiac gene therapeutic procedures (Wu et al. 2002; Inubushi et al. 2003; Auricchio et al. 2003).

2.2 Stability of Regulatory RNA or Transgene Expression

The second key issue of vector technology is the stability of transgene function (be it transgenic protein expression or short hairpin or pre-microRNA transcription). If genetic therapy for heart failure is not only meant to serve as a bridge-to-transplant or bridge-to-recovery, then long-term stability of the therapeutic vector plus transgene is required. Among the virus-based vector systems used for myocardial gene transfer, the currently most promising systems to achieve long-term stability are derived from adeno-associated viruses (AAVs). AAV vectors – although still difficult to produce at the high titers needed for in vivo applications – have shown stability for than a year in fully immunocompetent hemophilic dogs treated intramuscularly with a coagulation factor IX-producing AAV vector (Monahan et al. 1998; Herzog et al. 2002). Based on the experimental data a factor IX AAV vector has been evaluated in a clinical phase I trial (Kay et al. 2000). The high stability appears to be a consequence of inherent fundamental biological properties of the AAV genome (Afione et al. 1996). Although it is not always clear whether transgene expression occurs from integrated or episomal vector genomes, sequencing of vector–genome junctions has demonstrated the presence of integrated AAV genomes in cultured human cells and in mice after treatment with AAV vectors (Miller et al. 2002; Rutledge and Russell 1997; Nakai et al. 1999, 2001, 2003). AAV vectors have only recently been employed for the cardiac genetic treatment of heart failure (Hoshijima et al. 2002), for the systemic treatment of Fabry disease (Takahashi et al. 2002), and for substitution gene therapy of an inherited cardiomyopathy due to a δ-sarcoglycan gene defect (Kawada et al. 2002), with transgene expression over several months in all cases.

2.3 Safety of RNA-Based and Gene Therapeutic Approaches

The third key of vector technology – of paramount importance before any clinical application may be considered – is the safety of the gene transfer protocols. One safety feature under intense investigation is approaches towards regulation of the transgene after injection into the host. This may not be required for any application but is necessary in situations when adaptation of transgene activity to the physiologically required level is desired. An option to shut down the transgene completely at any time in the case of serious adverse effects is highly desirable under safety considerations (Bliznakov 2002). Vectors systems that can be shut off by withdrawal of an inducer drug have been described, most of them using doxycycline as inducer (Fechner et al. 2003b; Srour et al. 2003; Chtarto et al. 2003).

Safety issues inherent to specific vector types are, e.g., the inflammatory responses against adenoviral vectors which were transient, however, and appeared to be of no clinical significance in two recent cardiological clinical trials in 2002 and 2003 employing intracoronary injection of adenoviral vectors expressing VEGF (Grines et al. 2002; Hedman et al. 2003). In a previous trial in the year 1999

(Raper et al. 2002); a liver-targeted gene therapy protocol aiming at the treatment of ornithin transcarboxylase deficiency had used vector doses three orders of magnitude higher than those in the recent cardiological trials. The highest dose used therein was 3.3 × 1,010 vector particles (Grines et al. 2002), whereas in the former liver-directed study the multiorgan failure and death occurred in one patient who had received the highest dose of 3.8 × 1,013 particles (Raper et al. 2002). That tragic event prompted thorough additional investigations (Lehrman 1999; Marshall 2000) into the risks of adenovector-mediated gene therapy including serious adverse events even if they may occur only very rarely, in predisposed individuals, or at very high vector doses. A survey of clinical trials in 100 cancer patients employing intravascular adenoviral vectors reports that doses up to 2.5 × 1,013 had an acceptable safety profile (Reid et al. 2002). Even very rare serious side effects will be considered unacceptable, however, for non-malignant diseases.

The safety of AAV vector-based cardiac gene therapy is currently under investigation in a Phase I safety trial at the Mount Sinai Hospital in New York (unpublished data). With respect to possible risks specific to the AAV vectors their immunogenicity is low as compared to adenovectors (Sun et al. 2002, 2003). Another possible risk has been deduced from the capacity of wild-type AAV for chromosomal integration which appears to be lost, however, in recombinant AAV vectors such as used for gene therapy (Rutledge and Russell 1997; Rutledge et al. 1998; Miller et al. 2002, 2004). A recent study has searched for possible chromosomal effects of AAV vector integration (Miller et al. 2002) which deserve particular attention after a recent report on the late and unexpected occurrence of leukemia in children treated with retroviral vectors for severe combined immmunodeficiency (SCID), after successful correction of the primary genetic defect (Kohn et al. 2003). Whereas retroviral vectors such as used in the SCID study are integrating into the human genome at random sites, wild-type AAV shows a preference for chromosome 19 (Miller et al. 2002; Nakai et al. 1999). The possible consequences of genomic integration of AAV vectors deserves close attention over years, since the retrovector-associated leukemia occurred only years after successful ex vivo gene therapy for SCID by retrovirally mediated transfer of the γc gene into CD34+ cells (Kohn et al. 2003; Check 2003; Hacein-Bey-Abina et al. 2002, 2003).

3 RNA Interference Targeting Early Pathogenic Steps in Cardiomyopathies

3.1 Overview

Since coxsackievirus B3 (CoxB3) and adenoviruses may cause acute myocarditis and inflammatory cardiomyopathy, isolation of the common Coxsackievirus–Adenovirus-Receptor (CAR) has provided an interesting new target for molecular antiviral therapy. Whereas many viruses show high mutation rates enabling them to develop escape mutants, mutations of their cellular virus receptors are far less likely. We report on antiviral efficacies of CAR gene silencing by short hairpin (sh)RNAs in the cardiac-derived HL-1 cell line and in primary neonatal rat cardiomyocytes (PNCMs). Treatment with CAR-shRNA-generating vectors resulted in almost complete silencing of CAR expression both in HL-1 cells and PNCMs. While in HL-1 cells, CAR was already silenced 24 h after transduction of CAR-shRNA expressing vector in PNCMs CAR downregulation becomes visible only at day 6 and thereafter. CAR knockout resulted in strong inhibition of CoxB3 infections by up to 97% in HL-1 cells and by up to 90% in PNCMs, while adenovirus infections were inhibited by only 75% in HL-1 cells but up to 92% in PNCMs. We conclude that CAR knockout by shRNA vectors is promising against CoxB3 and adenovirus infections, but cell type specific CAR silencing by vector expressed shRNAs needs to be considered for the antiviral approaches.

3.2 *Introduction to Viral Cardiomyopathies*

Initially, cardiac viral infections were documented in the clinical context of acute myocarditis. Coxsackievirus B3 (CoxB3) was the first virus detected in this condition in humans (Bowles et al. 1986), and adenoviruses of serotypes 2 and 5 were later described as common agents of myocarditis in children (Bowles et al. 2003). Systematic screening of patients with dilated cardiomyopathy (DCM) has recently revealed that a rather broad spectrum of viruses may also chronically persist in human myocardium (Kühl et al. 2005a). It is therefore assumed that acute cardiac viral infections may not only cause acute illness, but also chronic heart disease if not definitely eliminated from the heart (Kühl et al. 2005b). The disease may progress to terminal heart failure and then require heart transplantation as a last therapeutic option.

RNA interference (RNAi) is a process of posttranscriptional gene silencing mediated by double-stranded RNA (dsRNA). The introduction of double-stranded small interfering RNAs (siRNA) or vectors expressing short hairpin RNA (shRNA) have already been successfully used to inhibit the replication of multiple viruses in vitro and in vivo including respiratory viruses (Bitko et al. 2005), hepatitis B (Carmona et al. 2006; Weinberg et al. 2007; Chen et al. 2007) and C viruses (Takigawa et al. 2004), human herpes virus-6 (Yoon et al. 2004), cytomegalovirus (Wiebusch et al. 2004), SARS coronarvirus (Lu et al. 2004), HIV-1 (Hayafune et al. 2006a,b) and CoxB3 (Merl et al. 2005; Werk et al. 2005). However, the efficiency of RNAi depends on exact homology of the siRNA to the target sequence. A single mismatch can be sufficient to abrogate silencing by RNAi (Sabariegos et al. 2006).

Many RNA viruses encode polymerase enzymes that lack proof reading abilities and as a result have a high mutations rate. Thus, there is a high probability that viruses will rapidly develop resistance to a particular siRNA during virus replication by incorporation of nucleotide mutations within the target sequence of the siRNA. Furthermore, it has been shown that hepatitis C virus becomes resistant against a particular siRNA after several cycles of replication by the incorporation of point mutations within the siRNA target sequence (Wilson and Richardson 2005). Similar data have been reported for HIV-1 (Boden et al. 2003, 2007) and poliovirus infections (Gitlin et al. 2005). Therefore, therapeutic targeting of non-variable virus-binding receptors on susceptible cells or other host cell molecules directly or indirectly involved in virus infections is an interesting alternative to targeting of the virus itself (Werk et al. 2005; Murray et al. 2005; Anderson and Akkina 2005; Arrighi et al. 2004; Gao et al. 2004; Ping et al. 2004; Kameoka et al. 2004). One approach of this type showed that downregulation of the CD4-independent attachment receptor (DC-SIGN) significantly inhibited HIV infection of dendritic cells (DC) and prevented the transfer of infectious HIV-1 particles from DC to T cells (Arrighi et al. 2004).

Adenoviruses of types 2 and 5 and CoxB3 use the Coxsackievirus–Adenovirus-Receptor (CAR) to infect their target cells, either for virus attachment (adenoviruses) or virus internalization (CoxB3) (Bergelson et al. 1998). Therefore, CAR represents an attractive target for the inhibition of cardiac CoxB3 and adenovirus infections. It has been shown recently that suppression of CAR by RNAi results in inhibition of CoxB3 infections (Werk et al. 2005; Coyne and Bergelson 2006). These studies were carried out with cells of non-cardiac origin and chemically synthesized siRNAs were used to inhibit CAR expression. However, the therapeutic use of synthetic siRNAs is significantly limited by their rapid degradation in target cells resulting in only transient gene silencing (Watanabe et al. 2004). Moreover, most cells of cardiac origin are very difficult to transfect with chemically synthesized siRNAs even in vitro. These problems are significantly aggravated in vivo. Despite intense efforts during recent years systemic and target organ-specific delivery of siRNAs remains a major hurdle for in vivo applications of RNAi.

Therefore, we investigated the antiviral potential of CAR gene silencing in cells of cardiac origin by use of newly developed AdVs generating shRNA against murine or rat CAR (shCAR). shCAR treatment of HL-1 cells and primary neonatal rat cardiomyocytes (PNCMs) led to almost complete silencing of CAR gene expression and efficient inhibition of CoxB3 infection, while efficient inhibition of adenovirus infections was only seen in PNCMs. The studies also revealed cell type-specific responses of cellular CAR protein to shCAR treatment which need to be considered as important determinants of this new antiviral approach in vivo.

3.3 *Results of an Anti-Virus Receptor Approach*

3.3.1 Selection and Specificity of Mouse CAR shRNAs

To investigate the efficacy of CAR silencing for the inhibition of adenovirus and CoxB3 infection in cardiac HL-1 cells, we generated a total of three shRNAs (shCAR2m, shCAR3m, shCAR4m) directed against extracellular domains of the

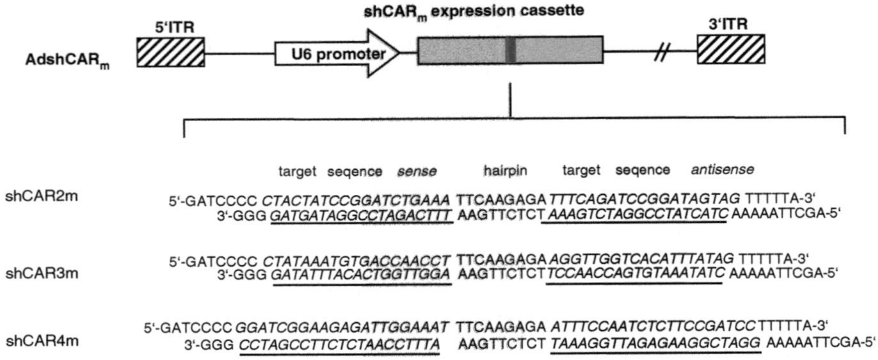

Fig. 2 Structure of adenovectors for shRNA transcription. Schematic illustration of recombinant AdshCARm and the sequence of shCAR: the *5′-ITR* represents nucleotide positions 1–342 of adenovirus type 5; the shCAR expression cassette is composed of an U6 = RNA polymerase III promoter and an shCAR sequence; the individual shCAR motifs consist of 19 (*shCAR2m, shCAR3m*) or 21 nucleotides (*shCAR4m*) corresponding to the coding regions of mCAR1 and mCAR2. The two motifs that form the sense and antisense strand of the shRNA are separated by a loop of nine nucleotides. A transcriptional termination signal of five tymidines is added at the 3′-end of the inverted repeat. *Whereas shCAR2m and shCAR3m* are directed against the IG1 extracellular domain of CAR (shCAR2m = nucleotides 246–264; shCAR3m = nucleotides 311–329), shCAR4m is directed against the nucleotides 448–468 of the IG2 extracellular domain; *3′-ITR* right inverted terminal repeat of adenovirus 5. Note: siRNA sequence in shCAR2m matches complete to mouse and rat CAR target sequences

mCAR splice variants mCAR-1 (Bergelson et al. 1998) and mCAR-2 (Tomko et al. 1997) (Fig. 2) following published siRNA selection criteria (Reynolds et al. 2004). DNA oligonucleotides encoding these shRNAs were then cloned into expression plasmids under the control of a murine RNA polymerase III U6 promotor. To test their efficacies, we co-transfected 293T cells with shCAR2m, shCAR3m, or shCAR4m-expressing plasmid, together with plasmids expressing recombinant mCAR1 and mCAR2. As a transfection control a plasmid expressing an irrelevant shRNA was used. To determine the extent of recombinant mCAR1- and mCAR2-mRNA downregulation, total cellular RNA was isolated 48 h after transfection and Northern blot hybridization was carried out. These experiments showed that two shRNAs were highly efficient: shCAR2m with 93% silencing, and shCAR4m with 97% silencing of mCAR1- and mCAR2-mRNA. The control shRNA affected neither mCAR1- nor mCAR2-mRNA expression (Fig. 3a). To analyze the specificities of shCAR2m and shCAR4m, both were also tested for downregulation of human CAR (hCAR) by co-transfection with a plasmid expressing recombinant soluble hCAR. None of the mouse CAR-specific shRNAs influenced human CAR-mRNA expression. In contrast, a human-specific shRNA efficiently silenced hCAR-mRNA expression (Fig. 3b). Considering that shCAR2m has only one mismatch and shCAR4m three mismatches as compared to hCAR-mRNA, these experiments indicate very high species-specificity of the selected mCAR-shRNAs.

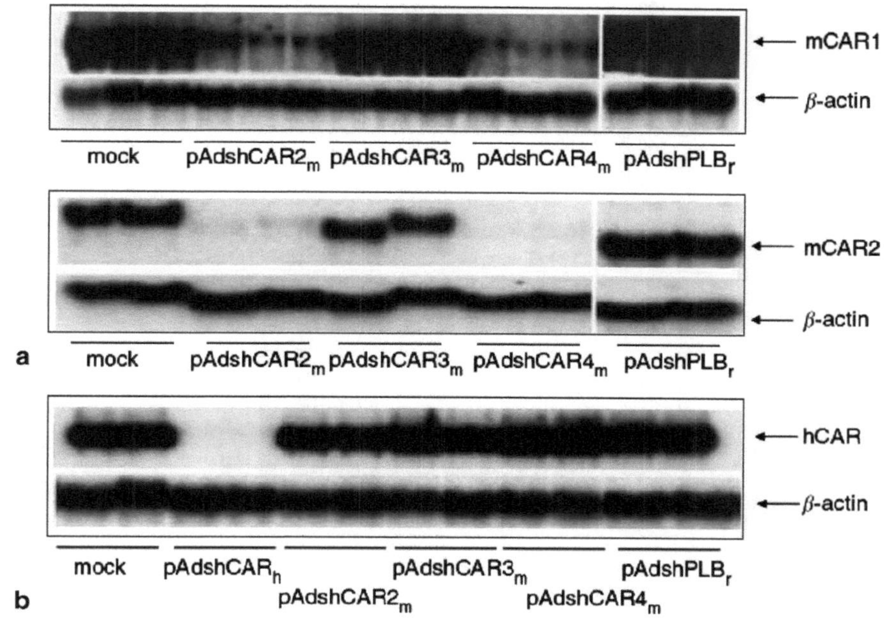

Fig. 3 Efficiency and specificity of shCARm-mediated mCAR silencing. (**a**) Selection of shCARm. 293T cells were co-transfected with mCAR1 (*upper panel*) or mCAR2 (*lower panel*) plus shCAR2m, shCAR3m, or shCAR4m expression plasmid. Northern blot analysis performed 48 h after transfection showed efficient downregulation of both mCAR1- and mCAR2-mRNA expression by shCAR2m and shCAR4m as compared to untransfected and control-shRNA (shPLBr) transfected cells. Experiments are made in triplicates (*upper panel*) and duplicates (*lower panel*). (**b**) Specificity of shCARm. 293T cells were co-transfected with human soluble CAR (hCAR) plus shCARm or human CAR-shRNA (pAdshCARh) expression plasmids. Northern blot analysis was performed as above. None of the shCARm downregulated hCAR, whereas human CAR-shRNA led to strong downregulation of hCAR-mRNA. Determination of β-actin expression was included as loading control. Experiments were made in duplicate

3.3.2 Silencing of mCAR in Mouse Cardiac HL-1 Cells

To determine the optimal AdV dose for transduction of mouse cardiac HL-1 cells, we first transduced HL-1 cells with a GFP-expressing marker AdV at an MOI from 5 to 1,000. At an MOI from 60 to 100 GFP expression was seen in 76–80% of the cells, whereas no cytotoxic effects were observed. Further dose escalation resulted in an only slight further increase of transduction rate (Fig. 4a), and at an MOI of 500 or higher cytotoxic effects became evident (increase of apoptotic cells, reduction of cell growth). HL-1 cells express mCAR1 but no mCAR2 on their cell surface (not shown). Therefore, the silencing capacity of the new AdshCAR4m (Fig. 2) which expresses the above evaluated most efficient shCAR4m was investigated with respect to mCAR1 expression in HL-1 cells only. To assess the amount of AdV-generated shCAR4m, HL-1 cells were transduced with AdshCAR4m at an MOI from 5 to 1,000. Twenty-four, 48, and 72 h

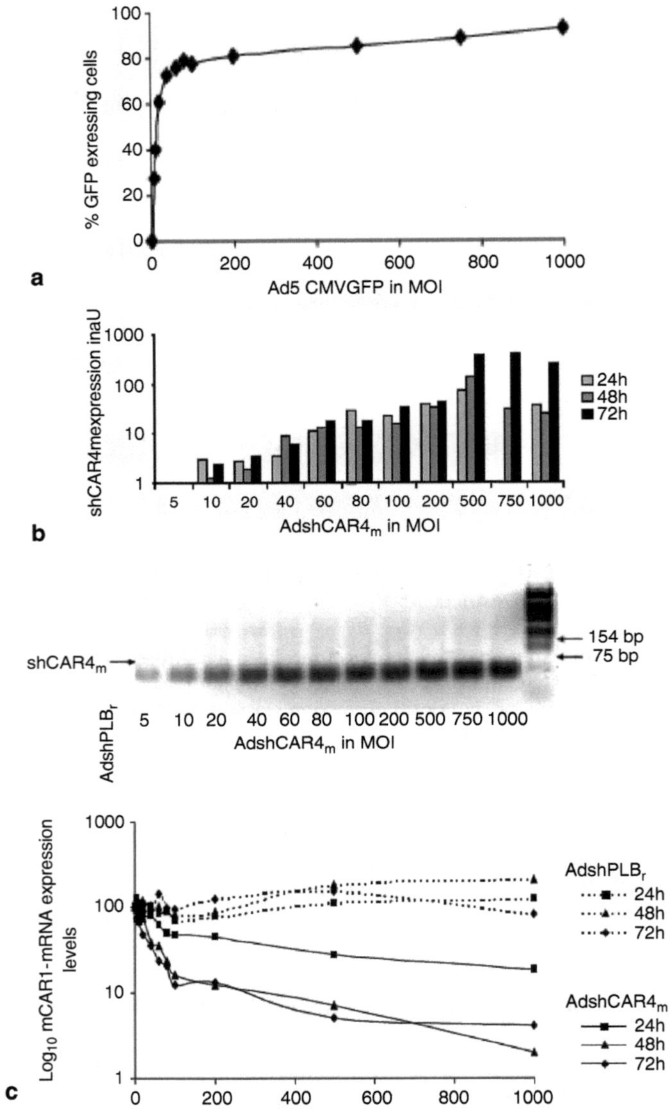

Fig. 4 Downregulation of mCAR mRNA in HL-1 cells by AdshCAR4m. (**a**) AdV-mediated gene transfer efficacy. HL-1 cells were transduced with Ad5CMVGFP (MOI from 5 to 1,000) and the fraction of cells expressing GFP was quantified by flow cytometry 48 h later. (**b**) AdV-mediated shCAR4m expression levels. HL-1 cells were transduced with AdshCAR4m (MOI from 5 to 1,000) and the shCAR4m expression level (shown as arbitrary units, aU) was then measured by real-time RT-PCR (*upper panel*) and agarose gel electorphoresis (*lower panel*, 24 h after transduction) at the indicated time points. HL-1 cells transduced with the vector AdshPLBr served as controls. (**c**) Downregulation of mCAR-mRNA expression by AdshCAR4m. HL-1 cells were transduced with AdshCAR4m or AdshPLBr (MOI from 5 to 1,000) and mCAR-mRNA expression was measured by real-time RT-PCR at the indicated time points

later the expression of shCAR4m was analyzed. A time- and dose-dependent transcription of shCAR4m was detected (Fig. 4b). As a consequence of shRNA expression, strong downregulation of mCAR1-mRNA was observed 48 and 72 h after transduction. Near maximal reduction of 85–90% was observed at an MOI of 100. Higher doses did not result in significant further downregulation of mCAR1-mRNA (Fig. 4c). These results indicate that an MOI of 100 was optimal for HL-1 transduction, since this dose allowed strong downregulation of mCAR1-mRNA in HL-1 cells, while cytotoxicity of the viral vector was still very low.

We next investigated mCAR1 protein expression after transduction of HL-1 cells with AdshCAR4m at an MOI of 100. To disrupt pre-existing membrane-associated mCAR1, HL-1 cells were trypsinized 24 h after AdshCAR4m transduction and then re-seeded. This procedure led to ablation of mCAR1 protein at the cell surface 2 days after transduction (Fig. 5a/b). In contrast, control cells rapidly re-expressed mCAR1 (Fig. 5b). Two days after AdshCAR4m treatment, a few cells still displayed some membrane-associated mCAR1 immunoreactivity (Fig. 5b), probably reflecting residual mCAR1 domains still anchored in the cell membrane after trypsinization. At days 3 and 4 after AdshCAR4m transduction mCAR1 immunoreactivity disappeared from HL-1 cells, whereas it remained permanently detectable in the control cells (Fig. 5b). Interestingly, CAR was only expressed at cell-to-cell contact sites in HL-1 cells suggesting that the observed stability of CAR in intact monolayers is determined by homophilic or heterophilic CAR interaction at cell-to-cell contact sites. Together, the results indicate that AdshCAR4m treatment efficiently inhibits de novo synthesis of mCAR1.

3.3.3 Inhibition of CoxB3 Replication in HL-1 Cells by AdshCAR4m

The antiviral effect of mCAR1 silencing was now investigated with respect to CoxB3 for which cellular CAR constitutes the internalization receptor. For this purpose, we first investigated the CoxB3 protective efficacy of AdshCAR4m as a function of dose, at an MOI of AdshCAR4m ranging from 10 to 100, using the same trypsin-involved experimental procedure as described above. HL-1 cells were infected with CoxB3 48 h after transduction with AdshCAR4m and the amount of new generated progeny virus determined six days later. Virus plaque assays clearly demonstrated that CoxB3 replicates in HL-1 cells, but at a rather low frequency, since no cytopathogenic effect was visible during the 6-day investigation period. As expected, the strongest reduction of CoxB3 progeny virus production by 97% was observed at the highest AdshCAR4m dose with an MOI of 100. Nevertheless, MOIs of 50 and 25 still led to significant inhibition of CoxB3 replication by 85 and 75%, respectively, whereas a MOI of 10 showed no significant effect (Fig. 6a). To investigate whether disruption of pre-existing CAR affects CoxB3 infection we treated HL-1 cells with the trypsin-involved experimental procedure as described above or omitted the trypsin step after AdshCAR4m transduction. Titers of CoxB3 progeny virus generated 5 and 6 days after CoxB3 infection were strongly reduced by about 97% in AdshCAR4m-transduced cells,

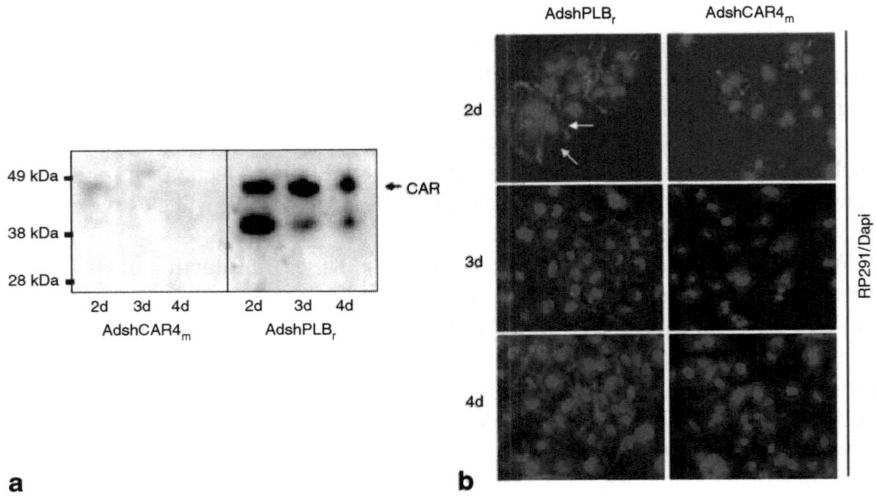

Fig. 5 Downregulation of mCAR1 protein in HL-1 cells by AdshCAR4m. (**a**) Detection of mCAR1 ablation by Western blot analysis. HL-1 cells were transduced with AdshCAR4m or the control vector AdshPLBr (MOI of 100) for 24 h. Cells were then trypsinized, re-seeded, and investigated using the polyclonal rabbit-anti-CAR antibody H-300 directed against the extracellular domain of CAR. mCAR1 protein became undetectable as early as 2 days after transduction with AdshCAR4m, whereas in the controls mCAR1 protein was abundantly expressed. (**b**) Detection of mCAR1 knockdown by indirect immunofluorescence. HL-1 cells were treated as in (**a**). mCAR1 expression was detected by indirect immunofluorescence using the polyclonal rabbit-antibody RP-291 (Sollerbrant et al. 2003) directed against the CAR variant with intracellular SIV tail (mCAR variant 1, accession no. NM_001025192). mCAR1 was abundantly expressed over the whole investigation period (4 days after transduction) in HL-1 cells transduced with control vector (*left-hand panels*). In contrast, mCAR1 was visible at low levels 2 days after transduction but became undetectable at later time points (*right-hand panels*). Cell nuclei are stained with DAPI and overlayed images are shown. Notably mCAR1 was localized only at cell–cell contact sites (*white arrows*), whereas cell membrane regions not in contact with other cells displayed no mCAR1 protein (*yellow arrows*). mCAR2 expression was not detectable in HL-1 cells by using polyclonal rabbit-antibody RP-194 [56] directed against the CAR variant with intracellular TVV tail (mCAR variant 2, accession no. NM_009988) (not shown)

irrespective of whether membrane-associated CAR was disrupted by trypsinization or not (Fig. 6b). These experiments indicate that AdshCAR4m inhibits CoxB3 replication in HL-1 cells with high efficacy. Moreover, proteolysis of membrane-associated CAR seems to be not affect CoxB3 infection.

3.3.4 Inhibition of Adenovirus Infection of HL-1 Cells by AdshCAR4m

CAR is also involved in adenovirus infection as attachment receptor. To investigate the long-term efficacy of AdshCAR4m with respect to the inhibition of

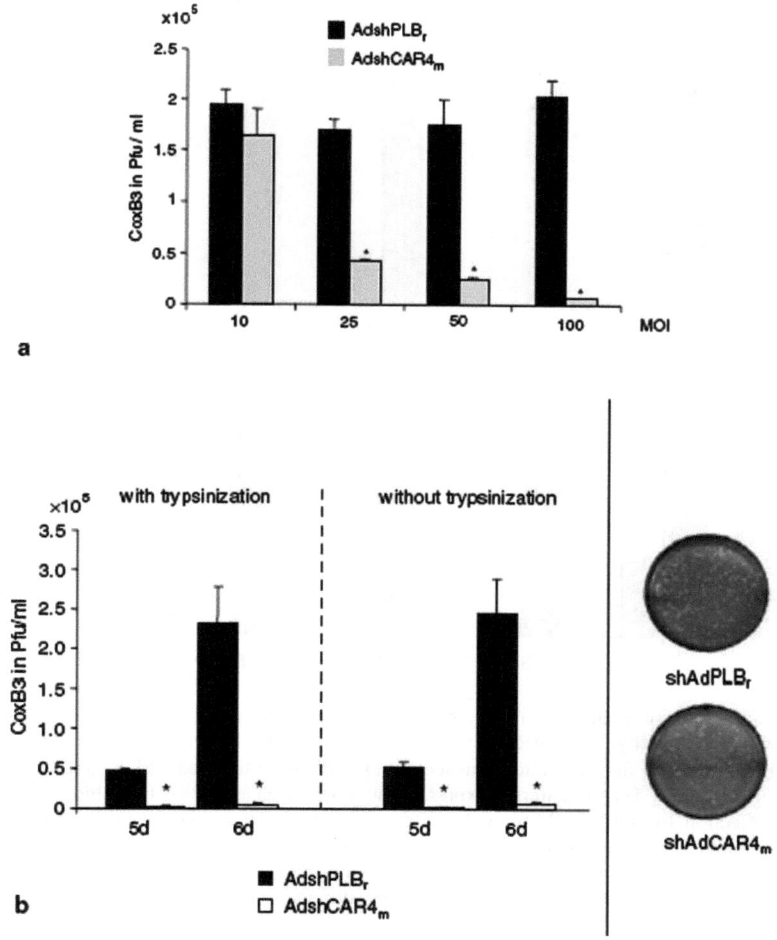

Fig. 6 Inhibition of CoxB3 infection by mCAR silencing in HL-1 cells. (**a**) Inhibition of CoxB3 replication as a function of AdshCAR4m dose. HL-1 cells were transduced with AdshCAR4m or control vector at variable MOIs ranging from 10 to 100. The cells were trypsinated 24 h later, re-seeded and infected with with CoxB3 (at an MOI of 1) 24 h later. Plaque assays were carried out 6 days post infection. *Asterisks* indicate significant differences ($p < 0.05$). (**b**) Inhibition of CoxB3 replication by AdshCAR4m as a function of trypsin treatment. HL-1 cells were transduced with AdshCAR4m or control vector (MOI of 100) as in Fig. 5. 24 h later cells were trypsinated and re-seeded (with trypsinization) or medium was changed (without trypsinization). 48 h after transduction the cells were infected with CoxB3 at an MOI of 1. Five and six days later cells were lysed by three freeze/thaw cycles and virus titers were determined by plaque assay on HeLa cells (*left-hand diagram*). *Right side* Example for plaque assay on HeLa cells using a dilution of 1:500 of HL-1 cell lysate from AdshCAR4m and control vector transduced cells

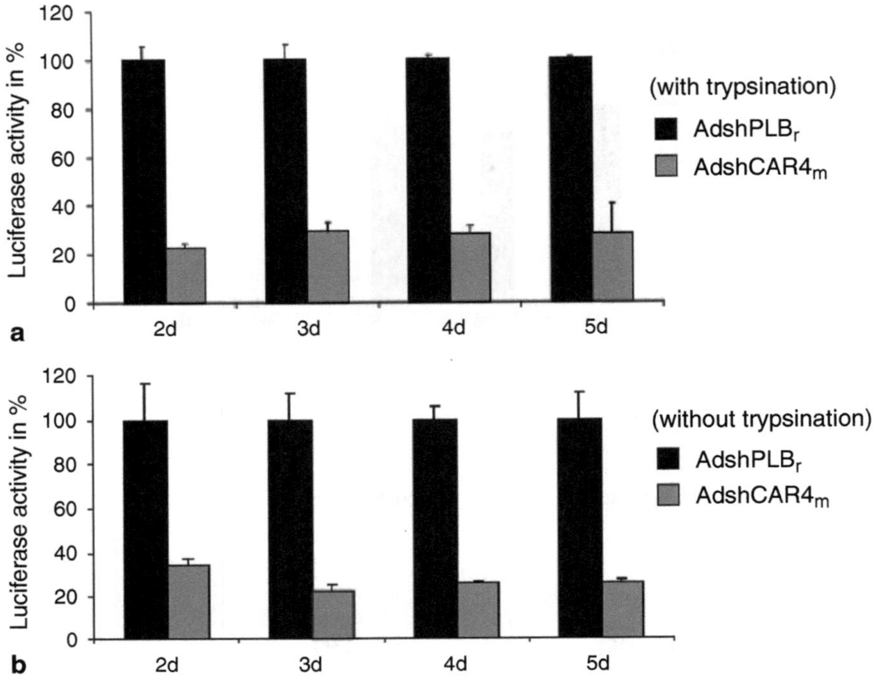

Fig. 7 Inhibition of adenovirus infection by mCAR silencing in HL-1 cells. (**a**) Inhibition of adenovirus infection by AdshCAR4m (with trypsinization). HL-1 cells were treated as in Fig. 5. Two, three, four, and five days after transduction the cells were infected with a luciferase marker adenovirus (MOI of 5) and luciferase expression was measured after 24h. (**b**) Inhibition of adenovirus infection by AdshCAR4m (without trypsiinization). HL-1 cells were transduced with AdshCAR4m or the control vector AdshPLBr (each MOI of 100). The medium was changed after 24h. Two, three, four, and five days after transduction the cells were infected with a luciferase marker adenovirus (MOI of 5) and luciferase expression was measured 24h later

adenovirus infection and to investigate whether proteolysis of pre-existing CAR affects adenovirus infection, HL-1 cells were transduced with AdshCAR4m versus the control vector at an MOI of 100. Twenty-four hours later, cells were either trypsinized and re-seeded, or the medium was changed (treatment without trypsinization). At 2, 3, 4 and 5 days after transduction, the cells were infected with a luciferase marker adenovirus (MOI of 5). The measurement of luciferase activity 24h later revealed persistent reduction of marker gene expression of about ≈75% in AdshCAR4m-treated HL-1 cells during the 5-day investigation period. This clearly demonstrates that CAR-shRNA treatment inhibits adenovirus uptake into HL-1 cells, but markedly less efficient than for CoxB3. Inhibition of adenovirus infections was closely similar between trypsinized and non-trypsinized cells (Fig. 7) indicating that proteolysis of CAR does not affect the function of CAR in adenovirus infections.

3.3.5 The Anti-Adenovirus and Anti-CoxB3 Effect of CAR Silencing in PNCMs

As HL-1 cells represent a permanent tumor cell line from heart, we were also interested in investigation of the effect of CAR silencing onto viral infection in primary cardiac cells. For this reason, PNCMs were transduced with AdshCAR2m, which shRNA sequence complete matches with mouse and rat CAR sequences or the control vector AdshGFP, both at an MOI of 100. At this dose nearly 80% of PNCMs are transduce with AdV (results not shown), which is closely similar to the HL-1 cells. Following treatment of PNCMs with AdshCAR2m CAR expression was unaffected up to day 3 after transduction, but it becomes strongly downregulated 6 days and 8 days after transduction (Fig. 8a). As a consequence of delayed CAR downregulation challenge of AdshCAR2m-treated PNCMs with CoxB3 (MOI of 1) resulted in delayed inhibition of CoxB3 replication. In fact, significant reduction of CoxB3 replication was seen if PNCMs were infected with CoxB3 at day 6 and 8 but not at day 3 after AdshCAR2m transduction. Moreover, inhibition of CoxB3 replication strongly depended on the initially used AdshCAR2m dose. An MOI of 100 of AdshCAR2m resulted in reduction of CoxB3 titers of about 66 and 90%, while an MOI of 25 resulted in reduction of CoxB3 titers of only 16 and 34% at day 6 and 8, respectively (Fig. 8b). Very similar results were obtained if AdshCAR2m (MOI 100) transduced PNCMs were infected with AdVs. Adenovirus uptake was reduced of about 85% at day 6 and about 92% at day 8 after AdshCAR2m transduction, while no effect was seen at day 3 (Fig. 8c).

3.4 Discussion of Anti-Viral Strategies

CoxB3 and adenoviruses are common agents of acute myocarditis and inflammatory cardiomyopathy (Bowles et al. 1986; Kühl et al. 2005a). Although interferon-β treatment has the potential to eliminate cardiotropic viruses and to improve heart function in patients with myocardial persistence of viral genomes (Kühl et al. 2003), no specific treatment against CoxB3 and adenovirus infections is available to date. With respect to antiviral therapy, RNAi is a promising new technology. Its therapeutic potential against viral infections has been demonstrated in multiple in vitro and in vivo studies (Bitko et al. 2005; McCaffrey et al. 2003; Carmona et al. 2006; Park et al. 2003; Merl et al. 2005; Werk et al. 2005). Despite encouraging results, however, several reports have revealed a fundamental general problem of siRNA-based antiviral therapy. Due to the requirement for strict sequence-specificity between therapeutic siRNA and target mRNA, even single point mutations in the viral target sequence may result in the rapid selection of escape mutants during persistent viral infections (Wilson and Richardson 2005; Boden et al. 2007; Gitlin et al. 2005).

In this study, we show that inhibition of CoxB3 and adenovirus infections is possible via silencing of the CAR receptor by means of vector-generated CAR-shRNA. This strategy is an attractive alternative to siRNA-based silencing of virus-encoded gene, since it drastically reduces the probability of escape mutant development (Arrighi et al.

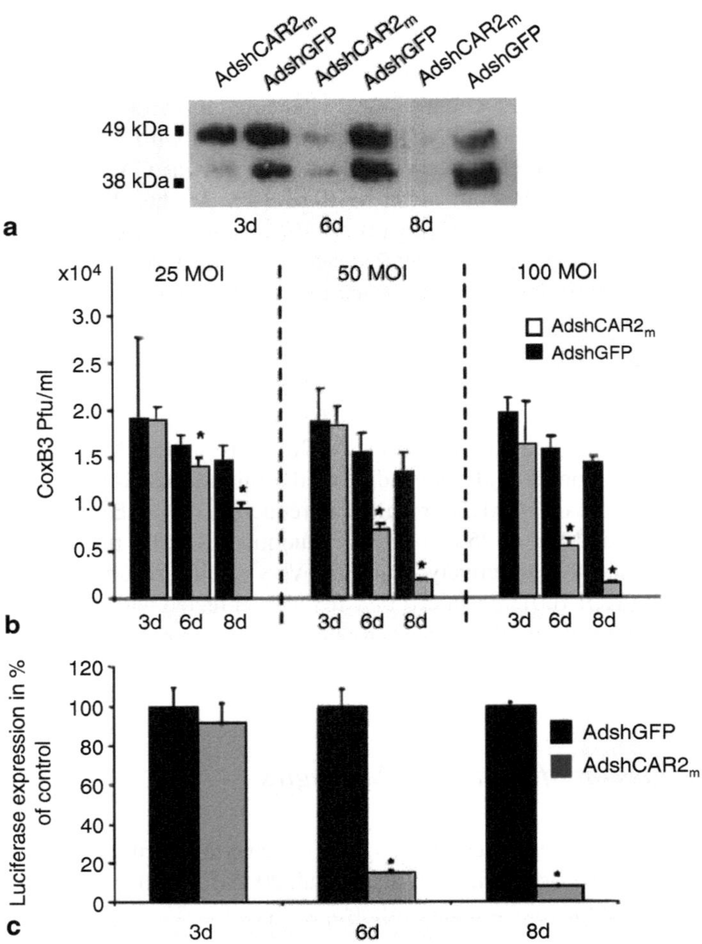

Fig. 8 Inhibition of CoxB3 and adenovirus infection in PNCMs by CAR silencing. (**a**) Detection of rat CAR downregulation by Western blot analysis. PNCM's were transduced with AdshCAR2m or the control vector AdshGFP (each MOI of 100) for 3, 6, and 8 days and than investigated for CAR expression using the polyclonal rabbit-anti-CAR antibody H-300. (**b**) PNCMs were transduced with AdshCAR2m or the control vector AdshGFP (MOI of 25, 50 and 100) for 3, 6 and 8 days and then infected with CoxB3 (MOI of 1). Cells were harvested 3 days later and CoxB3 plaque titers determined by plaque assay. *Asterisks* indicate significant differences ($p < 0.05$). (**c**) PNCMs were transduced with AdshCAR2m or the control vector AdshGFP (each at MOI of 100) for 3, 6 and 8 days and then transduced with a luciferase marker adenovirus (MOI of 5). Luciferase expression was measured 24 h later. *Asterisks* indicate significant differences ($p < 0.005$)

2004). In contrast to previous work (Werk et al. 2005), our investigations were carried out in myocardial cells representing the actual targets for CoxB3 and adenovirus infections in acute myocarditis and inflammatory cardiomyopathy (Kandolf 2004). Treatment of HL-1 cells with the shCAR vector led to efficient and stable removal of mCAR protein from the cell surface and consequently, to significant inhibition of CoxB3 and adenovirus infection of these cells. Remarkably, adenovirus internalization was reduced by only ≈75% despite near complete silencing of mCAR expression, while CoxB3 infection was inhibited by up to 97%. In PNCMs, however, CAR silencing resulted in up to 92 and 90% inhibition of adenovirus and CoxB3 infection, respectively. Adenovirus binding to cells is not exclusively mediate by CAR. Type 2 and 5 adenoviruses use CAR to attach to the cell surface (Bergelson et al. 1997), while its internalization occurs through αvβ1, αvβ3, and αvβ5 integrin (Tomko et al. 1997; (Wickham et al. 1993; Li et al. 2001). Others have shown that heparan sulfate glycosamino-glycans and MHC class I molecules may also be involved in the binding of adenovirus 2 and 5 (Hong et al. 1997; (Dechecchi et al. 2001). Moreover, expression of receptors involved in adenovirus infection is highly variable between different cell types and tissues (Fechner et al. 1999). Thus, variable receptor expression levels and interaction of adenoviruses with cell type specific expressed alternative receptors is the most likely explanation for the failure of complete CAR silencing to achieve efficient blockade of adenovirus infection in HL-1 cells. To enhance efficacy in inhibition of adenovirus infection via the anti-receptor approach, therefore, it seems to be necessary to simultaneously inhibit CAR and its co-receptors.

In contrast to the anti-adenovirus experiments, the anti-CAR approach was efficient with respect to CoxB3 in both HL-1 cells and PNCMs. The magnitude of inhibition of CoxB3 infection correlated well with the vector-induced ablation of mCAR from the cell surface and emphasizes the key role of CAR for CoxB3 infections of cardiac cells. The molecular reason for the differences of anti-CAR treatment in adenovirus versus CoxB3 infections may be due to the unique role playing CAR during CoxB3 infection as it induces conformational changes in the virus capsid that are essential for CoxB3 entry into the target cell (Coyne and Bergelson 2006). The high efficacy of the shCAR vector against this virus indicates, that CAR silencing is a promising therapeutic approach against CoxB3 infections of the myocardium. Previous studies have shown a similar degree of inhibition when using siRNAs directed against CoxB3 viral genomic RNA or its RNA intermediate (Merl et al. 2005; Werk et al. 2005; Yuan et al. 2005). Combination of direct anti-viral approaches which are prone to escape mutant development with the current anti-receptor strategy are likely to further increase the efficacy of RNAi-based antiviral therapy.

Recently, it has been shown that silencing of CAR by chemically synthetisized siRNA inhibited CoxB3 infection in human cells of non-cardiac origin (Werk et al. 2005). However, this approach decreased virus titers by only ≈60% which is markedly less than in the present study employing AdV-generating shCAR. This difference should be the result of high transduction efficacy of AdVs and the prolonged high-level expression of the shCAR. A bottleneck of chemically synthesized siRNAs is its transient silencing activity of ≈3 days in cardiac cells (Watanabe et al. 2004). Viral vectors expressing shRNAs, however, enable long term silencing of target genes in cells of cardiac origin (Fechner et al. 2007) and allow efficient delivery of transgenic sequences to the heart (Ikeda et al. 2002; Wang et al. 2005). Between them, pseudotyped AAV2/9 vectors seems to be most promising vectors for cardiac gene transfer as they enable highly efficient transduction of the heart by use of the simple intravenous application route (Pacak et al. 2006; Inagaki et al. 2006).

CAR is a cell adhesion protein mediating homotypic (Cohen et al. 2001; Honda et al. 2000) and heterotypic (Zen et al. 2005) intercellular interactions. It is a component of specialized intercellular junctions including epithelial tight junctions (Cohen et al. 2001), neuro-muscular junctions (Shaw et al. 2004), and myocardial intercalated discs (Noutsias et al. 2001). We found it to be localized exclusively at cell-to-cell contact sites between HL-1 cells, possibly as a consequence of cell membrane-associated CAR stabilization through homotypic (CAR–CAR) or heterotypic (CAR–protein) intercellular protein interactions. In contrast to previous CAR silencing attempts using synthetic siRNAs, the newly developed viral vectors AdshCAR4m and AdshCAR2m achieved near complete and stable knockout of CAR. This was an essential prerequisite to reveal significantly different kinetics of CAR protein ablation in different cell types, in particular an unexpectedly high stability of CAR in primary cells versus a stable cell line. These data may be explained by cell type-specific CAR protein kinetics. But, differences in CAR mRNA stability or processing of the CAR-shRNA through the RNA interference machinery also have to be taken into account. The delayed downregulation of CAR following vector-mediated CAR-shRNA delivery into primary cardiomyocytes is relevant for in vivo investigations, since the time lag between vector application and CAR silencing needs to be taken into consideration in antiviral therapy studies.

One important concern associated with downregulation of a cellular receptor molecule for antiviral therapy is possible side effects resulting from ablation of the normal cellular functions of the receptor. During organogenesis, CAR is highly expressed in the heart, but rapidly downregulated post partum (Fechner et al. 2003a). In a recently published genomic mouse knockout model, cardiomyocyte-specific CAR deletion resulted in severe cardiac anomalies and death between day 11.5 and 13.5 of embryonal development (Chen et al. 2006; Dorner et al. 2005). If the CAR gene became ablated late in embryonic development, however, the CAR-deficient animals survived into adulthood and had no evident cardiac anomalies (Chen et al. 2006). This finding suggests that CAR silencing by RNAi as employed in our study should be well tolerable over considerable periods of time, during which virus migration and spreading were efficiently inhibited. Furthermore, treatment of HL-1 cells with AdV-generating shCAR revealed no side effects since we found no changes in cell growth, cell morphology, or F-actin cytoskeleton protein expression (not shown). However, as CAR is an integral component of the tight junctions and may play an important role in cell methabolism in several tissues and organs, experiments have to confirm that shRNA-mediated CAR knockdown does not have undesirable side effects in vivo.

In summary, high efficacy of CAR gene silencing by shRNAs in cardiac-derived cells was observed against CoxB3, while variable results were obtained for adenoviruses. In contrast to previous silencing attempts using siRNAs or shRNA-plasmid the newly developed shRNA-vectors were able to achieve near complete and stable knockout of CAR transcription. This in turn revealed significantly different responses of cellular CAR protein to CAR-shRNA treatment which need to be considered as important determinants of this new antiviral approach in vivo.

4 RNA Interference to Improve Cardiac Function in Advanced Heart Failure

4.1 Overview

Impaired function of the phospholamban (PLB)-regulated sarcoplasmic reticulum (SR) Ca^{2+} pump (SERCA2a) contributes to cardiac dysfunction in heart failure (HF). PLB downregulation may increase SERCA2a activity and improve cardiac function. Small interfering (si)RNAs mediate efficient gene silencing by RNA interference (RNAi). However, their use for in vivo gene therapy is limited by siRNA instability in plasma and tissues, and by low siRNA transfer rates into target cells. To address these problems we developed an adenoviral vector (AdV) transcribing short hairpin (sh)RNAs against rat PLB and evaluated its potential to silence the PLB gene and to modulate SERCA2a-mediated Ca^{2+} sequestration in primary neonatal rat cardiomyocytes (PNCMs). Over a period of 13 days, vector transduction resulted in stable >99.9% ablation of PLB-mRNA at an MOI of 100. PLB protein gradually decreased until day 7 (7 ± 2% left), whereas SERCA, Na^+/Ca^{2+} exchanger (NCX1), calsequestrin (CSQ2), and troponin I (TnI) protein remained unchanged. PLB silencing was associated with a marked increase in ATP-dependent oxalate-supported Ca^{2+} uptake at 0.34 μM of free Ca^{2+}, and rapid loss of responsiveness to PKA-dependent stimulation of Ca^{2+} uptake was maintained until day 7. In summary, these results indicate that AdV-derived PLB-shRNA mediates highly efficient, specific, and stable PLB gene silencing and modulation of active Ca^{2+} sequestration in PNCMs. The availability of the new vector now enables employment of RNAi for the treatment of HF in vivo.

4.2 Regulatory RNA- and Gene-Based Therapies of Heart Failure

Heart failure (HF) remains a leading cause of mortality in the developed world. Deteriorated function of the failing heart has been partially attributed to dysfunction of the PLB-controlled SERCA2a (Piacentino et al. 2003). Reduction of both SERCA2a expression and PLB phosphorylation (Wolska et al. 2002; Schmidt et al. 1999) may contribute to this dysfunction. Nonphosphorylated PLB keeps the Ca^{2+} affinity of SERCA2a low resulting in decreased SR Ca^{2+} uptake, slowed relaxation and decreased SR Ca^{2+} load, while PLB phosphorylation in response to β-adrenergic stimulation relieves this inhibition. Germline transgenic approaches for ablation of PLB expression and function in mice (Luo et al. 1994), and somatic gene transfer for dominant negative PLB mutants (Iwanaga et al. 2004; Hoshijima et al. 2002), PLB-*antisense*-RNAs (Eizema et al. 2000; Li et al. 2005; He et al. 1999; delMonte et al. 2002), or intracellular inhibitory PLB antibodies (Dieterle et al. 2005; Meyer et al. 2004) were employed to increase cardiac SR Ca^{2+} transport activity and hence the contractile function of cardiomyocytes under physiological and diseased conditions. RNA interference (RNAi) mediated by chemically synthesized siRNAs was recently employed to silence PLB expression in cardiomyocytes (Watanabe et al. 2004). RNAi relies on post-trancriptional, sequence-specific gene silencing via small homologous double-stranded RNAs (Leung and Whittaker 2005). Its silencing efficacy is higher than that of antisense RNAs (Ogorelkova et al. 2006). Nevertheless, therapeutic use of synthetic siRNAs is significantly limited by their rapid degradation in target cells, resulting in only transient gene silencing (Watanabe et al. 2004), and by the difficulties in achieving sufficient transfer rates into multiple cells of therapeutic interest including cardiomyocytes. These limitations are significantly aggravated in vivo and despite intense efforts in recent years systemic delivery of siRNAs remains a major hurdle for in vivo applications of RNAi (Lewis et al. 2002). Since viral vector systems have been shown to be suitable to overcome these limitations in vitro and in vivo, we have developed a novel adenoviral vector (AdV) suitable for transcription of shRNA targeting PLB. Treatment with this vector resulted in highly efficient and specific PLB gene silencing in primary neonatal rat cardiac myocytes (PNCMs) which was stable over 2 weeks and associated with a marked increase in the SERCA2a-catalyzed SR Ca^{2+} sequestration.

4.3 *RNAi-Based Modulation of Cardiac Ca^{2+} Homeostasis in Heart Failure*

In order to select potential PLB-siRNA sequences for the generation of PLB-shRNAs, we initially co-transfected Cos-7 cells with an GFP-rat PLB fusion construct plus different siRNAs directed against rat PLB. Among five tested siRNAs, three were highly efficient in downregulating the fusion transcripts. The most efficient of these siRNAs was then cloned into a shRNA expression plasmid, as a DNA sequence encoding the PLB-shRNA. The co-transfection experiments with the GFP-rat PLB fusion constructs showed the very high efficacy of PLB-shRNA17 (Fig. 9). Based on these results we constructed an AdV designated AdshPLBr which generates PLB-shRNA17 (Fig. 10). To assess the efficacy of the vector AdshPLBr in downregulating endogenously expressed PLB in the actual target cells, we transduced primary neonatal rat cardiomyocytes (PNCMs) with AdshPLBr and investigated the dose- and time-dependency of its action. A control AdV (AdshCAR4m) generating an shRNA irrelevant for PLB and cardiac Ca^{2+} metabolism was employed to test for target specificity of AdshPLBr. The treatment of PNCMs with AdshPLBr resulted in strong dose-dependent downregulation of PLB-mRNA expression, by 27% at an MOI of 10 and by 87% at an MOI of 100, 2 days after transduction (Fig. 11a). Rapid, strong, and enduring PLB silencing was observed after treatment of PNCMs with AdshPLBr (MOI = 100) over a prolonged 13-day investigation period. PLB-mRNA abundance was extremely low <0.1% of baseline as quantitated by real-time-RT-PCR (not shown) as early as day 1 and remaining at this level until day 13 (end of experiment). By

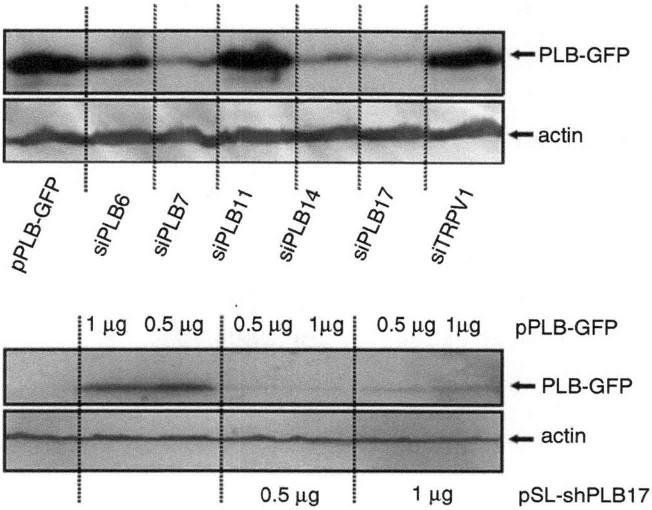

Fig. 9 Evaluation of phospholamban (PLB) siRNAs. *Upper* Selection of PLB siRNAs. Cos-7 cells were co-transfected with the PLB-GFP fusions construct pPLB-GFP and synthetic PLB-siRNAs. Cells were harvested 20 h later and western-blot carried out. PLB-GFP fusion protein was detected by a rabbit antiserum against GFP. To confirm equal loading of the samples, membranes were stripped and reprobed with a monoclonal antibody against actin. Note: siTRPV1 is a synthetic siRNA directed against the vanilloid receptor 1 (TRPV1). *Lower* Dose-dependency of pSL-shPLB17 action. Cos-7 cells were co-transfected with the pSL-shPLB17 and pPLB-GFP as indicated. Immunoblots were carried out 24 h after transfection as described above

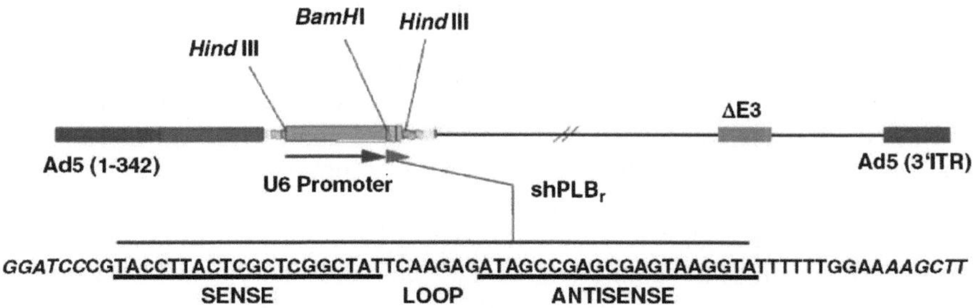

Fig. 10 Map of AdshPLBr vector and PLB-shRNA sequence

contrast, PLB-mRNA expression stayed at baseline level both in non-transduced controls and AdshCAR4m-treated PNCMs (Fig. 11b). Similar but delayed changes occurred at the PLB protein level. A decrease in PLB protein became first visible on day 3 (51 ± 4% remaining) as compared to controls. PLB protein then further decreased continuously until day 7 (7 ± 2% remaining). By contrast, the protein levels of SERCA2, NCX1, TnI, and CSQ2 remained unchanged indicating the absence of unspecific side effects on these cardiac proteins of both the

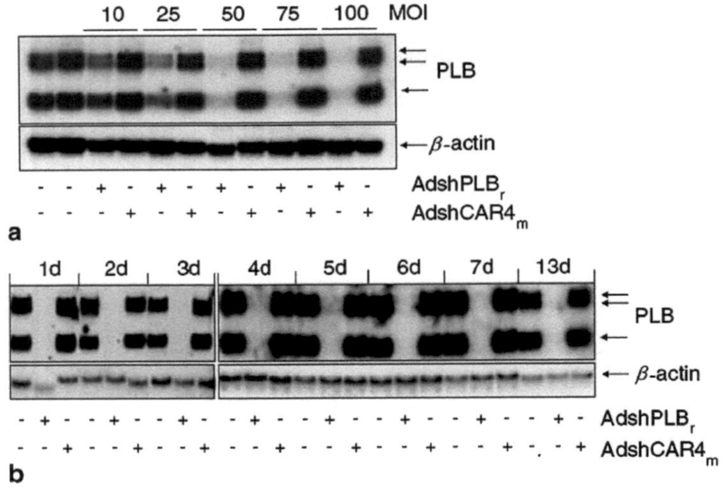

Fig. 11 Silencing of PLB-mRNA expression by AdshPLBr-generated PLB-shRNA. (**a**) Dose-dependency of AdshPLBr-mediated PLB-mRNA downregulation. PNCMs were transduced with AdshPLBr or a control vector as indicated. Cells were harvested 48 h later and Northern-blot carried out using a rat PLB specific probe. To confirm equal RNA, loading blots were striped and rehybridized with an α-actin specific probe. (**b**) Time-dependency of AdshPLBr-mediated PLB-mRNA downregulation. PNCMs were transduced with AdshPLBr or a control vector at an MOI of 100. Cells were harvested at indicated time points after transduction and Northern-blot carried out as described above. Note: PLB-mRNA was undetectable during the complete 13 days investigation period in PNCMs transduced with AdshPLBr while its expression stayed unaffected in cells transduced with the control vector

vector itself and of the shRNA that it generates (Fig. 12a/b). To examine the functional consequences of the PLB-shRNA-induced decline in the PLB protein level, PNCMs were transduced with AdshPLBr or control vector AdshCAR4m (each at an MOI of 100). Cell homogenate SR Ca^{2+} uptake rates were determined at submicromolar (0.34 μM) and saturating (3.68 μM) free Ca^{2+} concentrations. At the latter, the SR Ca^{2+}-ATPase is known to be insensitive to non-phosphorylated PLB, whereas at submicromolar Ca^{2+} non-phosphorylated PLB has been shown to decrease the Ca^{2+} affinity of this enzyme 17. The rate of Ca^{2+} uptake determined at 0.34 μM free Ca^{2+} and normalized to the maximum uptake rate (Vmax) values at saturating Ca^{2+} (relative Ca^{2+} uptake rate) did not differ between non-transduced and AdshCAR4m- or AdshPLBr-treated PNCMs one day after transduction day 1 (approximately 45% of Vmax each). The relative Ca^{2+} uptake rate remained at this level both in non-transduced and AdshCAR4m-transduced PNCMs until day 7 after transduction. In contrast, it increased steadily in AdshPLBr-transduced PNCMs reaching 100% of Vmax on day 5 after transduction and remaining at this high level until the end of the experiment on day 7 (Fig. 13a). This indicates an increase in the Ca^{2+} affinity of the SR Ca^{2+}-ATPase in the PLB-deficient PNCMs.

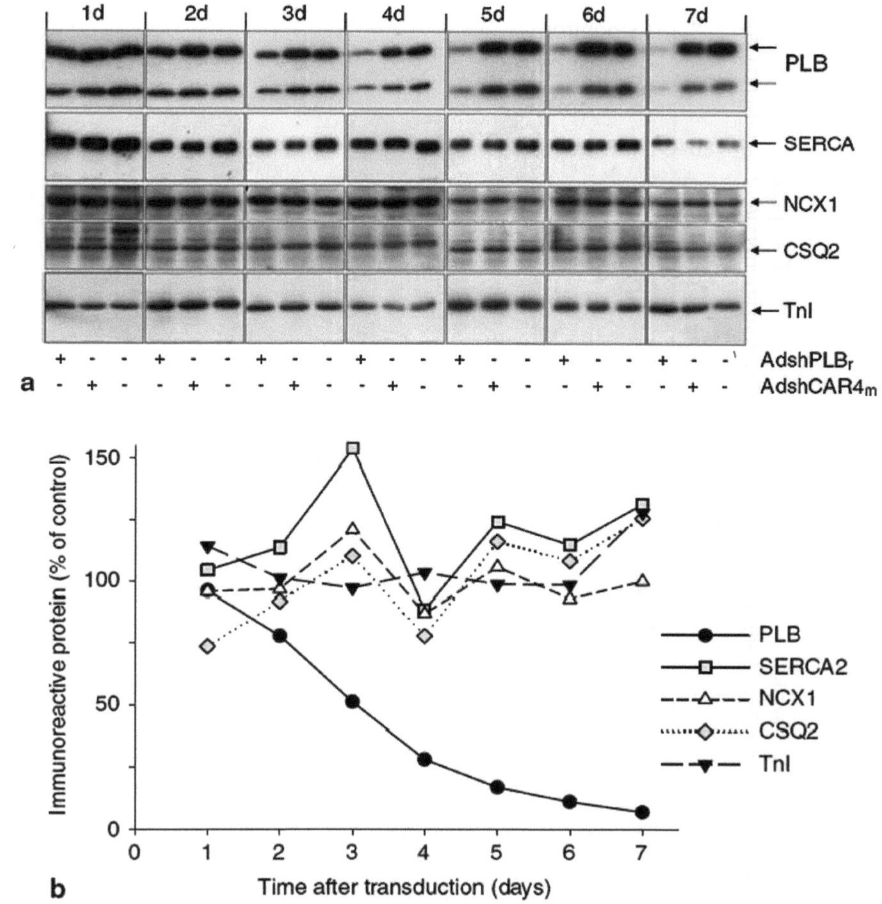

Fig. 12 Time-dependency of AdshPLBr-mediated PLB protein downregulation. (**a**) Western blot analysis of PLB protein expression. PNCMs were transduced with AdshPLBr or a control vector at an MOI of 100. Cells were harvested at indicated time points after transduction and Western-blots carried out. Significant PLB downregulation became visible 3 days post-transduction while the expression levels of SERCA, NCX1, TnI, and CSQ2 were unaffected during the 7-day investigation period (**b**) Relative expression levels of PLB, SERCA, NCX1, TnI, and CSQ2 protein normalized to the respective Western blot signals at the corresponding time points in control vector-transduced PNCMs

As shown in Fig. 13b, a linear relationship between the relative rates of Ca^{2+} uptake and the respective relative PLB protein levels was observed in AdshPLBr-transduced PNCMs. In addition we investigated the degree of stimulation of Ca^{2+} uptake by protein kinase A (PKA)-dependent in vitro phosphorylation following transduction of PNCMs with AdshPLBr at an MOI of 100. On day 1 after transduction the PKA-induced increase in the rate of Ca^{2+} uptake was 82% both in

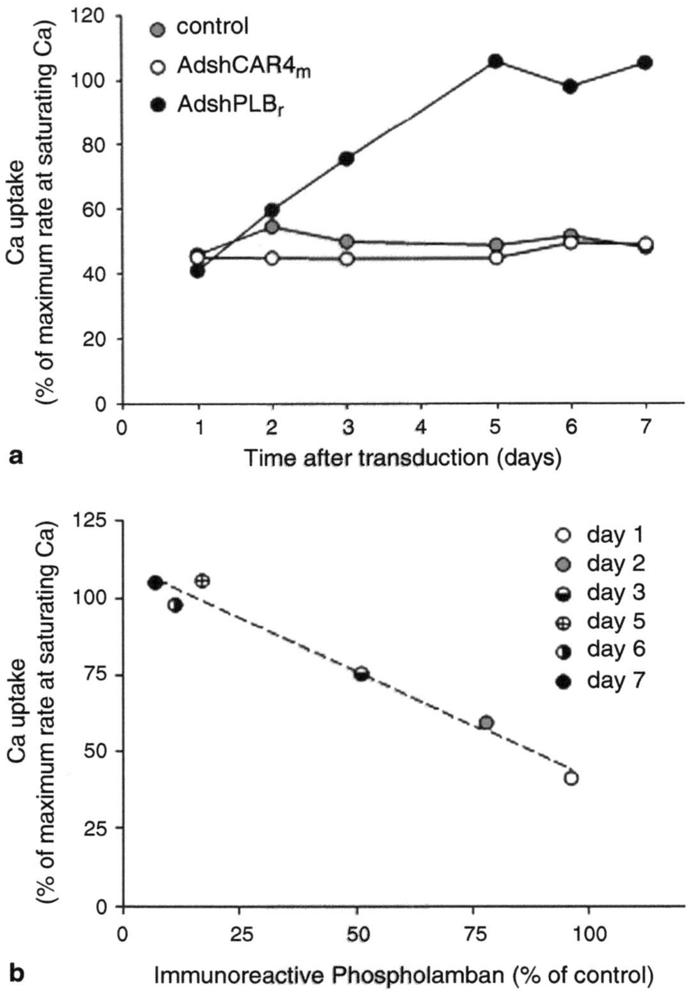

Fig. 13 SR Ca^{2+} uptake in PNCM homogenates after AdshPLBr-mediated PLB silencing. (**a**) Relative Ca^{2+} uptake at 0.34 μM free Ca^{2+} in homogenates of PNCMs. PNCMs were transduced with AdshPLBr or a control vector at an MOI of 100. Cells were harvested at indicated time points and SR Ca^{2+} uptake rates were determined at submicromolar (0.34 μM) and saturating (3.68 μM) free Ca^{2+} concentrations. Values after 3 min of uptake were normalized to the maximum uptake at a saturating Ca^{2+} of 3.68 μM. *Control* Non-transduced cells. (**b**) Relative Ca^{2+} uptake versus percent change in PLB protein in PNCMs homogenates on days 1–7 post-transduction. Data were calculated from the results obtained from Ca^{2+} uptake experiments (**a**) and PLB protein expression (Fig. 12b)

non-transduced and in AdshPLBr-transduced PNCMs. In homogenates of AdshPLBr-transduced PNCMs this PKA stimulation became completely lost around day 4/5 after transduction, whereas approximately 1.5-fold stimulation of Ca^{2+} uptake was observed on days 4–7 in non-transduced PNCMs (Fig. 14).

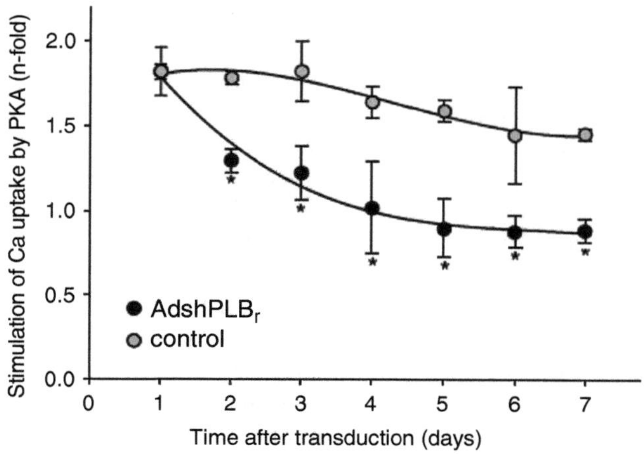

Fig. 14 Stimulation of Ca2+ uptake in PNCMs homogenates by the catalytic subunit of PKA. Values in the presence of added PKA normalized to those measured in the presence of 2 μM synthetic PKA peptide inhibitor. Shown are means ± SD for four separate transductions. *Control* Non-transduced cells

4.4 Discussion of an RNAi-Based Approach to Heart Failure Therapy

We demonstrated highly efficient and specific ablation of endogenous PLB expression in PNCMs by PLB-shRNA expressed from an AdV. shRNA transcription resulted in downregulation of endogenous PLB-mRNA below 0.1% of baseline, persisting for 13 days. No changes of the expression of other cardiac proteins including Ca^{2+} handling proteins occurred, indicating high target specificity of the PLB-shRNA vector. A control vector had no effect on PLB indicating absence of unspecific effects of shRNA per se on PNCMs. At the functional level, the SERCA2a Ca^{2+} affinity was markedly increased after PLB silencing accompanied by a loss of responsiveness to PKA-dependent stimulation of Ca^{2+} uptake. SERCA2a and PLB form a functional complex regulating Ca^{2+} uptake. Altered levels of either protein may therefore have profound effects by changing the PLB/SERCA2a ratio and thereby altering intracardiac Ca^{2+} homeostasis. This was demonstrated in normal and diseased hearts by use of transgenic animal models or gene transfer approches targeting PLB or SERCA2a (Iwanaga et al. 2004; del-Monte et al. 2001; Vetter et al. 2002; Miyamoto et al. 2000). Reduction of the PLB/SERCA2a ratio was achieved by classical overexpression of cDNAs encoding SERCA2a, a dominant negative PL mutant, or PLB-targeted antibodies. We and others have employed antisense RNAs directed against PLB-mRNA (Eizema et al. 2000; He et al. 1999).

In contrast, RNAi-based therapeutic strategies are not yet widely employed in the cardiovascular field. Synthetic siRNAs were recently employed to downregulate

PLB in cardiomyocytes (Watanabe et al. 2004). Since siRNAs enable only transient gene silencing, we have developed an AdV generating shRNAs mediating efficient, specific, and stable PLB silencing. With respect to the duration of silencing a striking difference between previous work using synthetic siRNAs (Watanabe et al. 2004) and the current AdV-based approach was observed. siRNAs resulted in downregulation of PLB-mRNA to 5% of baseline within 12 h, but this effect was almost completely lost after 4 days. In contrast, PLB-mRNA levels remained below 0.1% of baseline over a 2-week period after PLB-shRNA-AdV treatment. Vector-based shRNA generation was clearly superior to synthetic siRNAs with respect to silencing stability. With respect to the efficacy of silencing the PLB-shRNA-AdV resulted in >99.9% ablation of PLB-mRNA, whereas a former study using a PLB-antisense-RNA-AdV achieved maximal PLB-mRNA ablation ≈75% (Eizema et al. 2000). In line with previous reports, we found PLB ablation to improve SERCA2a dependent intracellular Ca^{2+} handling. Significant upregulation of SERCA2a Ca^{2+} affinity was first observed when PLB protein was down to 50% of baseline. Interestingly, PLB downregulation to 20% was sufficient to mediate maximal increase in the Ca^{2+} affinity of SERCA2a, further reduction had no additional effect. There was a steady loss of responsiveness to PKA-dependent Ca^{2+} uptake stimulation by PLB-shRNA, and complete loss of responsiveness was observed on day 4–7. There was no loss of responsiveness to PKA-dependent Ca^{2+} uptake stimulation in the control groups. This indicates that the RNAi-mediated PLB silencing was linked to the anticipated loss of responsiveness of the SR Ca^{2+} transport system to PKA-dependent phosphorylation normally mediated through PLB.

Although RNAi is a powerful method for gene silencing, its implementation for therapeutic purposes in humans requires that two technical problems are solved. First, instability of chemically synthesized siRNAs in plasma and cells requires their repetitive administration in vivo. The loss of initially efficient target gene silencing in cardiomyocytes within 3 days allows an estimate of the required siRNA application frequency for this specific target tissue. The problem may be overcome by viral vector systems producing shRNAs over long time periods in vivo. For intermediate stability, AdV-shRNA systems as used here may suffice, and even provide advantages over long-term stable AAV vectors (Wang et al. 2005; Minamisawa et al. 1999a,b) if the RNAi effect is needed only temporarily in an acute and potentially reversible condition (e.g., heart failure due to viral or autoimmune myocarditis). Second, cardiac targeting of synthetic siRNAs is currently only possible by experimental methods unsuitable for possible transfer to any clinical setting (Lewis et al. 2002). In contrast, recent developments in vector technology (e.g., pseudotyped AAVs with cardiotropic properties) are likely to allow cardiac targeting of transgene and shRNA expression cassettes by simple intravenous injection (Inagaki et al. 2006; Pacak et al. 2006). Importantly, the small expression cassette and specific PLB-shRNA sequence used in this study can easily be incorporated in different AAV and other vectors selected for purpose. Within the framework of our proof-of-concept study which introduces a novel PL-shRNA tool we have not performed in vivo work, since the efficacy of PLB ablation for heart failure therapy in animal models has already been demonstrated (Iwanaga et al. 2004; delMonte et al. 2001; Vetter et al. 2002; Miyamoto et al. 2000). In future cardiac gene therapy studies, we will use pseudotyped AAV8 and AAV9 vectors, since two recent studies (Inagaki et al. 2006; Pacak et al. 2006) have demonstrated important advantages of these pseudotypes over lentiviruses, adenoviruses, and previously used AAV vectors. One remaining challenge is the modification of the cassette in such a way as to allow exogenously regulatable shRNA expression and adjustment of the degree of PLB modulation to changing physiological conditions. This appears to be particularly important in humans since chronic PLB deficiency due to genomic mutations was associated with cardiomyopathies (Schmitt et al. 2003; Zhao et al. 2006; Haghighi et al. 2003, 2006). Whereas in mice complete knockout of PLB (as may also be achieved by RNAi) was able to rescue the severe cardiomyopathic phenotype of MLP knockout mice (Minamisawa 1999a,b), suggesting that unregulated PLB silencing is appropriate in this species, application in humans most probably requires regulatable RNAi. Recent studies by our group have extended the repertoire of RNAi vector for the suppression of PLB by a series of AAV9-based vectors. A long-term in vivo study of both the original AdV described above and analogous AAV9 vectors have demonstrate that over periods of 1 month (for the AdV vector) and 3 months (for the AAV9 vector) these RNAi vector significantly improve systolic and diastolic cardiac function, reduce cardiac hypertrophy and dilation, and improve survival (in the case of the AAV vectors only) (unpublished data). Furthermore, the catecholamine-induced deregulations of several functionally relevant miRNA in cardiomyocytes were restore to normal by RNAi vector treatment (unpublished data).

5 Clinical Perspectives and Unanswered Questions

5.1 Three Levels of Cardiac Targeting

During the past decade, the molecular and structural foundations and the key problems associated with cardiac gene therapy – and thereby also many of those encountered in regulatory RNA-based approaches – have been defined in considerable detail. It has become obvious that efficient and selective transport of any regulatory RNA or cDNA vector to the myocardium (Targeting) is a first key issue and major problem. Future clinical cardiac RNA and gene therapy protocols will probably involve direct intracoronary vector infusion to achieve target selectivity (level I of targeting). This may be enhanced by the use cardiac-specific promotors resulting in further confinement of regulatory RNA or cDNA expression (level II). Importantly, for regulatory RNA, e.g., shRNA transcription common tissue-specific type II polymerase promoters cannot be used since proper shRNA transcription has until recently only been achieved with type III promoters none of which was tissue-specific. This problem may be overcome, however, by the introduction of microRNA-based shRNA expression systems which also allow the use of cardiac-specific and drug-regulatable type II polymerase promotors. This makes level II of targeting also available for regulatory RNA-based therapies. Another major breakthrough in the field of cardiac RNA and gene therapy was the discovery that certain pseudotyped AAV vectors (in particular

AAV9 and derivates thereof) are cardiotropic even after simple intravenous injection (level III of targeting). Although AAV9 is not yet absolutely cardiac-specific but also reaches the liver by the intravenous route, further modification of the AAV capsid may solve or alleviate this issue. Currently, the use of AAV9 in combination with level I and II targeting already enables highly cardiac-specific RNA and gene therapy protocols one of which is evaluated in a first clinical trial.

5.2 Stability and Control of Regulatory RNA and Gene Therapy

Depending on the target disease, the therapeutic goal may be achieved by regulatory RNA or transgene expression for only few days or weeks or require long-term of even life-long stability. Currently available adenovectors are capable of achieving the short-term goal which may also suffice in bridge-to-recovery or bridge-to-transplant situations. Nevertheless, even for short-term therapies the generally better safety profile of AAV vectors will lead to the use of this vector type in any clinical RNA or gene therapy protocol. For all long-term therapies AAV vectors are a priori the system of choice due to their impressive long-term cardiac stability and function in several experimental studies. Whereas control of transgene expression is available for some years via drug-regulatable system, this goal has only recently been achieved for regulatory RNA therapies. As for conventional gene therapy, optimized clinical RNA therapy protocols are also likely to involve drug-regulatable systems since under certain conditions shRNAs (and likely also other regulatory RNA, e.g., miRNAs) have shown serious adverse effects. The option to shut off their expression will therefore significantly add to therapeutic safety in the clinical setting.

5.3 Practical Issues of Clinical Translation

Even when the above basic issues are solved satisfactorily, the intensity of efforts to establish a regulatory RNA or gene therapeutic regime in the clinic will be critically influenced by four determinants. First, that the specific therapeutic target is not accessible by other means, e.g., conventional pharmacotherapy. Second, that the target disease is sufficiently severe to justify the efforts and risks of any experimental therapeutic approach. Third, that the number of patients which may possibly benefit from the experimental therapy to be established is relatively high since this will greatly facilitate fundraising and incorporation of industrial partners. Fourth, any therapeutic protocol which addresses a pathomechanism (e.g., disturbed cardiac Ca^{2+} homeostasis) common to cardiac diseases of diverse etiologies will have a broader range of application than strategies which target specific pathogenic steps, and need to be tailor-made for every different type of pathogenesis. Without doubt, in cardiovascular medicine in general and specifically in the fields of cardio-myopathies and heart failure, a large number of patients are in urgent need of improved treatment, and the first regulatory RNA and gene therapeutic protocols adequately fulfill the above criteria. Thus, actual clinical translation of first therapies of this type is likely to occur within the next decade.

References

Afione S, Conrad C et al (1996) In vivo model of adeno-associated virus vector persistence and rescue. J Virol 70(5):3235–3241

Anderson J, Akkina R (2005) HIV-1 resistance conferred by siRNA cosuppression of CXCR4 and CCR5 coreceptors by a bispecific lentiviral vector. AIDS Res Ther 2(1):1

Arrighi JF, Pion M et al (2004) Lentivirus-mediated RNA interference of DC-SIGN expression inhibits human immunodeficiency virus transmission from dendritic cells to T cells. J Virol 78(20):10848–10855

Auricchio A, Acton P et al (2003) In vivo quantitative noninvasiive imaging of gene transfer by single-photon emission computerized tomography. Human Gene Ther 14:255–261

Beeri R, Guerrero L et al (2002) New efficient catheter-based system for myoccardial gene delivery. Circulation 106:1756–1759

Bergelson J, Cunningham J et al (1997) Isolation of a common receptor for coxsackie B viruses and adenoviruses 2 and 5. Science 275:1320–1323

Bergelson J, Krithivas A et al (1998) The murine CAR homolog is a receptor for coxsackie B viruses and adenoviruses. J Virol 72(1):415–419

Bitko V, Musiyenko A et al (2005) Inhibition of respiratory viruses by nasally administered siRNA. Nat Med 11(1):50–55

Bliznakov E (2002) Therapeutic angiogenesis: hope or hype. Circulation 106(25):e220–e221

Boden D, Pusch O et al (2003) Human immunodeficiency virus type 1 escape from RNA interference. J Virol 77(21):11531–11535

Boden D, Pusch O et al (2007) Overcoming HIV-1 resistance to RNA interference. Front Biosci 12:3104–3116

Boekstegers P, vonDegenfeld G et al (2000) Myocardial gene transfer by selective pressure-regulated retroinfusion of coronary veins. Gene Ther 7:232–240

Bowles NE, Richardson PJ et al (1986) Detection of coxsackie-B-virus-specific RNA sequences in myocardial biopsy samples from patients with myocarditis and dilated cardiomyopathy. Lancet 1(8490):1120–1123

Bowles NE, Ni J et al (2003) Detection of viruses in myocardial tissues by polymerase chain reaction. evidence of adenovirus as a common cause of myocarditis in children and adults. J Am Coll Cardiol 42(3):466–472

Büning H, Ried M et al (2003) Receptor targeting of adeno-associated virus vectors. Gene Ther 10:1142–1151

Carmona S, Ely A et al (2006) Effective inhibition of HBV replication in vivo by anti-HBx short hairpin RNAs. Mol Ther 13(2):411–421

Check E (2003) Cancer risk prompts US to curb gene therapy. Nature 422:7

Chen JW, Zhou B et al (2006) Cardiomyocyte-specific deletion of the coxsackievirus and adenovirus receptor results in hyperplasia of the embryonic left ventricle and abnormalities of sinuatrial valves. Circ Res 98:923–930

Chen CC, Ko TM et al (2007) Long-term inhibition of hepatitis B virus in transgenic mice by double-stranded adeno-associated virus 8-delivered short hairpin RNA. Gene Ther 14(1):11–19

Chtarto A, Bender H et al (2003) Tetracycline-inducible transgene expression mediated by a single AAV vector. Gene Ther 10:84–94

Cohen C, Shieh J et al (2001) The coxsackievirus and adenovirus receptor is a transmembrane component of the tight junction. Proceedings of the National Academy of Science USA 98(26):15191–15196

Communal C, Huq F et al (2003) Decreased efficiency of adenovirus-mediated gene transfer in aging cardiomyocytes. Circulation 107(8):1170–1175

Coyne CB, Bergelson JM (2006) Virus-induced Abl and Fyn kinase signals permit coxsackievirus entry through epithelial tight junctions. Cell 124(1):119–131
Curiel D (1999) Considerations and challenges for the achievement of targeted gene delivery (Perspective) Gene Ther 6(9):1497–1498
Davidson M, Jones J et al (2001) Cardiac gene delivery with cardiopulmonary bypass. Circulation 104:131–133
Dechecchi MC, Melotti P et al (2001) Heparan sulfate glycosaminoglycans are receptors sufficient to mediate the initial binding of adenovirus types 2 and 5. J Virol 75(18):8772–8780
delMonte F, Williams E et al (2001) Improvement in survival and cardiac metabolism after gene transfer of sarcoplasmic reticulum Ca^{2+} ATPase in a rat model of heart failure. Circulation 104:1424–1429
delMonte F, Harding S et al (2002) Targeting phospholamban by gene transfer in human heart failure. Circulation 105:904–907
Dieterle T, Meyer M et al (2005) Gene transfer of a phospholamban-targeted antibody improves calcium handling and cardiac function in heart failure. Cardiovasc Res 67(4):678–688
Donahue J, Kikkawa K et al (1997) Ultrarapid, highly efficient viral gene transfer to the heart. Proc Natl Acad Sci USA 94:4664–4668
Dorner AA, Wegmann F et al (2005) Coxsackievirus–adenovirus receptor (CAR) is essential for early embryonic cardiac development. J Cell Sci 118(Pt 15):3509–3521
Eizema K, Fechner H et al (2000) Adenovirus-based phospholamban-antisense-mRNA expression as a novel approach to improve cardiac contractile dysfunction – comparison of a constitutive viral versus an endothelin-1-responsive cardiac promoter. Circulation 101:2193–2199
Fechner H, Haack A et al (1999) Expression of coxsackie–adenovirus-receptor and α_v-integrin does not correlate with adenovector targeting in vivo indicating anatomical vector barriers. Gene Ther 6(9):1520–1535
Fechner H, Noutsias M et al (2003a) Induction of coxsackievirus–adenovirus-receptor expression during myocardial tissue formation and remodeling: identification of a cell-to-cell contact-dependent regulatory mechanism. Circulation 107(6):876–882
Fechner H, Wang X et al (2003b) A novel tetracycline-controlled transactivator-transrepressor system enables external control of oncolytic adenovirus replication. Gene Ther 10:1680–1690
Fechner H, Suckau L et al (2007) Highly efficient and specific modulation of cardiac calcium homeostasis by adenovector-derived short hairpin RNA targeting phospholamban. Gene Ther 14:211–218
Franz W, Rothmann T et al (1997) Analysis of tissue-specific gene delivery by recombinant adenoviruses containing cardiac-specific promoters. Cardiovascular Research 35:560–566
Gao X, Wang H et al (2004) Inhibition of Epstein-Barr virus (EBV) reactivation by short interfering RNAs targeting p38 mitogen-activated protein kinase or c-myc in EBV-positive epithelial cells. J Virol 78(21):11798–11806
Giordano F, Ping P et al (1996) Intracoronary gene transfer of fibroblast growth factor-5 increases blood flow and contractile function in an ischemic region of the heart. Nature Medicine 2(5):534–539
Gitlin L, Stone JK et al (2005) Poliovirus escape from RNA interference: short interfering RNA-target recognition and implications for therapeutic approaches. J Virol 79(2):1027–1035
Grines C, Watkins M et al (2002) Angiogenic gene therapy (AGENT) trial in patients with stable angina pectoris. Circulation 105:1291–1297
Hacein-Bey-Abina S, LeDeist F et al (2002) Sustained correction of X-linked severe combined immunodeficiency by ex vivo gene therapy. N Engl J Med 346(16):1185–1193
Hacein-Bey-Abina S, vonKalle C et al (2003) A serious adverse event after successful gene therapy for X-linked severe combined immunodeficiency. N Engl J Med 348(3):255–256
Haghighi K, Kolokathis F et al (2003) Human phospholamban null results in lethal dilated cardiomyopathy revealing a critical difference between mouse and human. J Clin Invest 111(6):869–876
Haghighi K, Kolokathis F et al (2006) A mutation in the human phospholamban gene, deleting arginine 14, results in lethal, hereditary cardiomyopathy. Proc Natl Acad Sci USA 103(5):1388–1393
Hajjar R, Schmidt U et al (1998) Modulation of ventricular function through gene transfer in vivo. Proc Natl Acad Sci USA 95:5251–5256

Hayafune M, Miyano-Kurosaki N et al (2006a) HIV gene therapy using RNA virus systems. Nucleic Acids Symp Ser (Oxf) (50):79–80
Hayafune M, Miyano-Kurosaki N et al (2006b) Silencing of HIV-1 gene expression by two types of siRNA expression systems. Antivir Chem Chemother 17(5):241–249
He H, Meyer M et al (1999) Effects of mutant and antisense RNA of phospholamban on SR Ca^{2+}-ATPase activity and cardiac myocyte contractility. Circulation 100:974–980
Hedman M, Hartikainen J et al (2003) Safety and feasibility of catheter-based local intracoronary vascular endothelial growth factor gene transfer in the prevention of postangioplasty and in-stent restenosis and in the treatment of chronic myocardial ischemia – phase II results of the Kuopio Angiogenesis Trial (KAT). Circulation 107:2677–2683
Henderson S, Spencer M et al (1989) Structure, organization, and expression of the rat cardiac myosin light chain-2 gene. Identification of a 250-base pair fragment which confers cardiac-specific expression. J Biol Chem 264(30):18142–18148
Herzog R, Fields P et al (2002) Influence of vector dose on factor IX-specific T and B cell responses in muscle-directed gene therapy. Hum Gene Ther 13:1281–1291
Honda T, Saitoh H et al (2000) The coxsackievirus–adenovirus receptor protein as a cell adhesion molecule in the developing mouse brain. Brain Res Mol Brain Res 77:19–28
Hong SS, Karayan L et al (1997) Adenovirus type 5 fiber knob binds to MHC class I alpha2 domain at the surface of human epithelial and B lymphoblastoid cells. EMBO J 16(9):2294–2306
Hoshijima M, Ikeda Y et al (2002) Chronic suppression of heart-failure progression by a pseudophosphorylated mutant of phospholamban via in vivo cardiac rAAV gene delivery. Nat Med 8(8):864–871
Ikeda Y, Gu Y et al (2002) Restoration of deficient membrane proteins in the cardiomyopathic hamster by in vivo cardiac gene transfer. Circulation 105:502–508
Inagaki K, Fuess S et al (2006) Robust systemic transduction with AAV9 vectors in mice: efficient global cardiac gene transfer superior to that of AAV8. Mol Ther 14(1):45–53
Inubushi M, Wu J et al (2003) Positron-emission tomography reporter gene expression imaging in rat myocardium. Circulation 107:326–332
Iwanaga Y, Hoshijima M et al (2004) Chronic phospholamban inhibition prevents progressive cardiac dysfunction and pathological remodeling after infarction in rats. J Clin Invest 113(5):727–736
Kameoka M, Nukuzuma S et al (2004) RNA interference directed against Poly(ADP-Ribose) polymerase 1 efficiently suppresses human immunodeficiency virus type 1 replication in human cells. J Virol 78(16):8931–8934
Kandolf R (2004) [Virus etiology of inflammatory cardiomyopathy]. Dtsch Med Wochenschr 129(41):2187–2192
Kawada T, Nakazawa M et al (2002) Rescue of hereditary form of dilated cardiomyopathy by rAAV-mediated somatic gene therapy: Amelioration of morphological findings, sarcolemmal permeability, cardiac performances, and the prognosis of TO-2 hamsters. Proc Natl Acad Sci USA 99:901–906
Kay M, Manno C et al (2000) Evidence for gene transfer and expression of factor IX in haemophilia B patients treated with an AAV vector. Nat Genet 24:257–261
Koch W, Lefkowitz R et al (2000) Functional consequences of altering myocardial adrenergic receptor signaling. Annu Rev Physiol 62:237–260
Kohn D, Sadelain M et al (2003) Occurrence of leukaemia following gene therapy of X-linked SCID. Nat Rev Cancer 3:477–488
Kühl U, Pauschinger M et al (2003) Interferon-β treatment eliminates cardiotropic viruses and improves left ventricular function in patients with myocardial persistence of viral genomes and left ventricular dysfunction. Circulation 107(22):2793–2798
Kühl U, Pauschinger M et al (2005a) High prevalence of viral genomes and multiple viral infections in the myocardium of adults with idiopathic left ventricular dysfunction. Circulation 111:887–893
Kühl U, Pauschinger M et al (2005b) Virus persistence in the myocardium is associated with progressive cardiac dysfunction. Circulation 112(13):1965–1970
Kypson A, Hendrickson S et al (1999) Adenovirus-mediated gene transfer of the b2-adrenergic receptor to donor hearts enhances cardiac function. Gene Ther 6:1298–1304
Lehrman S (1999) Virus treatment questioned after gene therapy death. Nature 401:517–518

Leung RK, Whittaker PA (2005) RNA interference: from gene silencing to gene-specific therapeutics. Pharmacol Ther 107(2):222–239

Lewis D, Hagstrom J et al (2002) Efficient delivery of siRNA for inhibition of gene expression in postnatal mice. Nat Genet 32:107–108

Li E, SL Brown et al (2001) Integrin alpha(v)beta1 is an adenovirus coreceptor. J Virol 75(11):5405–5409

Li J, Hu SJ et al (2005) Construction of phospholamban antisense RNA recombinant adeno-associated virus vector and its effects in rat cardiomyocytes. Acta Pharmacol Sin 26(1):51–55

Lu A, Zhang H et al (2004) Attenuation of SARS coronavirus by a short hairpin RNA expression plasmid targeting RNA-dependent RNA polymerase. Virology 324(1):84–89

Luo W, Grupp IL et al (1994) Targeted ablation of the phospholamban gene is associated with markedly enhanced myocardial contractility and loss of beta-agonist stimulation. Circ Res 75(3):401–409

Marshall E (2000) Gene therapy on trial. Science 288:951–957

Maurice J, Hata J et al (1999) Enhancement of cardiac function after adenoviral-mediated in vivo intracoronary β2-adrenergic receptor gene delivery. J Clin Invest 104:21–29

McCaffrey AP, Nakai H et al (2003) Inhibition of hepatitis B virus in mice by RNA interference. Nat Biotechnol 21(6):639–644

Merl S, Michaelis C et al (2005) Targeting 2A protease by RNA interference attenuates coxsackieviral cytopathogenicity and promotes survival in highly susceptible mice. Circulation 111(13):1583–1592

Meyer M, Belke DD et al (2004) A recombinant antibody increases cardiac contractility by mimicking phospholamban phosphorylation. FASEB J 18(11):1312–1314

Miller D, Rutledge E et al (2002) Chromosomal effects of adeno-associated virus vector integration. Nat Genet 30:147–148

Miller DG, Petek LM et al (2004) Adeno-associated virus vectors integrate at chromosome breakage sites. Nat Genet 36(7):767–773

Minamisawa S, Hoshijima M et al (1999a) Genetic complementation identifies chronic phospholamban-sarcoplasmic reticulum calcium ATPas interaction as a critical calcium cycling defect in the progression of dilated cardiomyopathy. Cell 99(3):313–322

Minamisawa S, Hoshijima M et al (1999b) Chronic phospholamban-sarcoplasmic reticulum calcium ATPase interaction is the critical calcium cycling defect in dilated cardiomyopathy. Cell 99(3):313–322

Miyamoto M, delMonte F et al (2000) Adenoviral gene transfer of SERCA2a improves left-ventricular function in aortic-banded rats in transition to heart failure. Proc Natl Acad Sci USA 97:793–798

Monahan P, Samulski R et al (1998) Direct intramuscular injection with recombinant AAV vectors results in sustained expression in a dog model of hemophilia. Gene Ther 5(1):40–49

Murray JL, Mavrakis M et al (2005) Rab9 GTPase is required for replication of human immunodeficiency virus type 1, filoviruses, and measles virus. J Virol 79(18):11742–11751

Nakai H, Iwaki Y et al (1999) Isolation of recombinant adeno-associated virus vector-cellular DNA junctions from mouse liver. J Virol 73(7):5438–5447

Nakai H, Yant S et al (2001) Extrachromosomal recombinant adeno-associated virus vector genomes are primarily responsible for stable liver transduction in vivo. J Virol 75(15):6969–6976

Nakai H, Storm T et al (2003) Pathways of removal of free DNA vector ends in normal and DNA-PKcs–deficient SCID mouse hepatocytes transduced with rAAV vectors. Hum Gene Ther 14:871–881

Nicklin S, Baker A (2002) Tropism-modified adenoviral and adeno-associated viral vectors for gene therapy. Curr Gene Ther 2(3):273–293

Noutsias M, Fechner H et al (2001) Human coxsackie–adenovirus-receptor is co-localized with integrins $\alpha_v\beta_3$ and $\alpha_v\beta_5$ on the cardiomyocyte sarcolemma and upregulated in dilated cardiomyopathy – implications for cardiotropic viral infections. Circulation 104:275–280

Ogorelkova M, Zwaagstra J et al (2006) Adenovirus-delivered antisense RNA and shRNA exhibit different silencing efficiencies for the endogenous transforming growth factor-beta (TGF-beta) type II receptor. Oligonucleotides 16(1):2–14

Pacak CA, Mah CS et al (2006) Recombinant adeno-associated virus serotype 9 leads to preferential cardiac transduction in vivo. Circ Res 99(4):e3–e9

Park WS, Hayafune M et al (2003) Specific HIV-1 env gene silencing by small interfering RNAs in human peripheral blood mononuclear cells. Gene Ther 10(24):2046–2050
Perabo L, Büning H et al (2003) In vitro selection of viral vectors with modified tropism: the adeno-associated virus display. Mol Ther 8(1):151–157
Piacentino V 3rd, Weber CR et al (2003) Cellular basis of abnormal calcium transients of failing human ventricular myocytes. Circ Res 92(6):651–658
Ping YH, Chu CY et al (2004) Modulating HIV-1 replication by RNA interference directed against human transcription elongation factor SPT5. Retrovirology 1(1):46
Poller W, Fechner H et al (2002a) The molecular basis of cardiotropic viral infections. Eur Heart J 4(Suppl I):I18–I30
Poller W, Fechner H et al (2002b) Highly variable expression of virus receptors in the human cardiovascular system. Implications for cardiotropic viral infections and gene therapy. Z Kardiol 91(12):978–991
Ponnazhagan S, Hoover F (2004) Delivery of DNA to tumor cells in vivo using adeno-associated virus. Methods Mol Biol 246:237–243
Ponnazhagan S, Mahendra G et al (2002) Conjugate-based targeting of recombinant adeno-associated virus type 2 vectors by using avidin-linked ligands. J Virol 76(24):12900–12907
Price R, Skyba D et al (1998) Delivery of colloidal particles and red blood cells to tissue through microvessel ruptures created by targeted microbbubble destruction with ultrasound. Circulation 98:1264–1267
Rabinowitz J, Rolling F et al (2002) Cross-packaging of a single adeno-associated virus (AAV) type 2 vector genome into multiple AAV serotypes enables transduction with broad specificity. J Virol 76(2):791–801
Raper S, Yudkoff M et al (2002) A pilot study of in vivo liver-directed gene transfer with an adenoviral vector in partial ornithine transcarbamylase deficiency. Hum Gene Ther 13:163–175
Reid T, Warren R et al (2002) Intravascular adenoviral agents in cancer patients: lessons from clinical trials. Cancer Gene Ther 9(12):979–986
Reynolds P, Nicklin S et al (2001) Combined transductional and transcriptional targeting. Nat Biotechnol 19:838–842
Reynolds A, Leake D et al (2004) Rational siRNA design for RNA interference. Nat Biotechnol 22(3):326–330
Rockman H, Chien K et al (1998) Expression of a β-adrenergic receptor kinase 1 inhibitor prevents the development of myocardial failure in gene-targeted mice. Proc Natl Acad Sci USA 95:7000–7005
Rutledge E, Russell D (1997) Adeno-associated virus vector integration junctions. J Virol 71(11):8429–8436
Rutledge EA, Halbert CL et al (1998) Infectious clones and vectors derived from adeno-associated virus (AAV) serotypes other than AAV type 2. J Virol 72(1):309–319
Sabariegos R, Gimenez-Barcons M et al (2006) Sequence homology required by human immunodeficiency virus type 1 to escape from short interfering RNAs. J Virol 80(2):571–577
Schmidt U, Hajjar RJ et al (1999) Human heart failure: cAMP stimulation of SR Ca^{2+}-ATPase activity and phosphorylation level of phospholamban. Am J Physiol 277(2 Pt 2):H474–H480
Schmitt J, Kamisago M et al (2003) Dilated cardiomyopathy and heart failure caused by a mutation in phospholamban. Science 299:1410–1413
Shaw CA, Holland PC et al (2004) Isoform-specific expression of the coxsackie and adenovirus receptor (CAR) in neuromuscular junction and cardiac intercalated discs. BMC Cell Biol 5(1):42
Shi W, Bartlett J (2003) RGD inclusion in VP3 provides adeno-associated virus type 2 (AAV2)-based vectors with a heparan sulfate-independent cell entry mechanism. Mol Ther 7(4):515–525
Simons M, Bonow R et al (2000) Clinical trials in coronary angiogenesis: Issues, problems, consensus. An expert panel summary. Circulation 102:e73–e86
Sollerbrant K, Raschperger E et al (2003) The coxsackievirus and adenovirus receptor (CAR) forms a complex with the PDZ domain-containing protein ligand-of-numb protein-X (LNX) J Biol Chem 278(9):7439–7444
Srour M, Fechner H et al (2003) Regulation of human factor IX expression using a doxycycline-inducible gene expression system. Thromb Haemost 90:398–405

Sun J, Chatterjee S et al (2002) Immunogenic issues concerning recombinant adeno-associated virus vectors for gene therapy. Curr Gene Ther 2:485–500

Sun J, Anand-Jawa V et al (2003) Immune responses to adeno-associated virus and its recombinant vectors. Gene Ther 10:964–976

Takahashi H, Hirai Y et al (2002) Long-term systemic therapy of Fabry disease in a knockout mouse by adeno-associated virus-mediated muscle-directed gene transfer. Proc Natl Acad Sci USA 99(21):13777–13782

Takigawa Y, Nagano-Fujii M et al (2004) Suppression of hepatitis C virus replicon by RNA interference directed against the NS3 and NS5B regions of the viral genome. Microbiol Immunol 48(8):591–598

Tomko R, Xu R et al (1997) HCAR and MCAR: the human and mouse cellular receptors for subgroup C adenoviruses and group B coxsackieviruses. Proc Natl Acad Sci USA 94:3352–3356

Vale P, Losordo D et al (1999) Catheter-based myocardial gene transfer using nonfluoroscopic electromechanical left ventricular mapping. J Am Coll Cardiol 34(1):246–254

Vetter R, Rehfeld U et al (2002) Transgenic overexpression of the sarcoplasmic reticulum Ca^{2+} ATPase improves reticular Ca^{2+} handling in normal and diabetic rat hearts. FASEB J 16:1657–1659

Wang Z, Zhu T et al (2005) Adeno-associated virus serotype 8 efficiently delivers genes to muscle and heart. Nat Biotechnol 23(3):321–328

Watanabe A, Arai M et al (2004) Phospholamban ablation by RNA interference increases Ca2+ uptake into rat cardiac myocyte sarcoplasmic reticulum. J Mol Cell Cardiol 37(3):691–698

Weinberg MS, Ely A et al (2007) Specific inhibition of HBV replication in vitro and in vivo with expressed long hairpin RNA. Mol Ther 15(3):534–541

Werk D, Schubert S et al (2005) Developing an effective RNA interference strategy against a plus-strand RNA virus: silencing of coxsackievirus B3 and its cognate coxsackievirus–adenovirus receptor. Biol Chem 386(9):857–863

White D, Hata J et al (2000) Preservation of myocardial b-adrenergic receptor signaling delays the development of heart failure after myocardial infarction. Proc Natl Acad Sci USA 97(10):5428–5433

Wickham T (2000) Targeting adenovirus. Gene Ther 7:110–114

Wickham T, Mathias P et al (1993) Integrins $\alpha_v\beta_3$ and $\alpha_v\beta_5$ promote adenovirus internalization but not virus attachment. Cell 73(2):309–319

Wickham T, Tzeng E et al (1997) Increased in vitro and in vivo gene transfer by adenovirus vectors containing chimeric fiber proteins. J Virol 71:8221–8229

Wiebusch L, Truss M et al (2004) Inhibition of human cytomegalovirus replication by small interfering RNAs. J Gen Virol 85(Pt 1):179–184

Wilson JA, Richardson CD (2005) Hepatitis C virus replicons escape RNA interference induced by a short interfering RNA directed against the NS5b coding region. J Virol 79(11):7050–7058

Wolska BM, Arteaga GM et al (2002) Expression of slow skeletal troponin I in hearts of phospholamban knockout mice alters the relaxant effect of beta-adrenergic stimulation. Circ Res 90(8):882–888

Wu J, Inubushi M et al (2002) Positron emission tomography imaging of cardiac reporter gene expression in living rats. Circulation 106:180–183

Ylä-Herttuala S, Alitalo K (2003) Gene transfer as a tool to induce therapeutic vascular growth. Nat Med 9(6):694–701

Yoon JS, Kim SH et al (2004) Inhibition of herpesvirus-6B RNA replication by short interference RNAs. J Biochem Mol Biol 37(3):383–385

Yuan J, Cheung P et al (2005) Inhibition of coxsackievirus B3 replication by small interfering RNAs requires perfect sequence match in the central region of the viral positive strand. J Virol 79(4):2151–2159

Zen K, Liu Y et al (2005) Neutrophil migration across tight junctions is mediated by adhesive interactions between epithelial coxsackie and adenovirus receptor and a junctional adhesion molecule-like protein on neutrophils. Mol Biol Cell 16(6):2694–2703

Zhao W, Yuan Q et al (2006) The presence of Lys27 instead of Asn27 in human phospholamban promotes sarcoplasmic reticulum Ca^{2+}-ATPase superinhibition and cardiac remodeling. Circulation 113(7):995–1004

Cardiac Delivery of Nucleic Acids by Transcriptional and Transductional Targeting of Adeno-Associated Viral Vectors

O.J. Müller(✉) and H.A. Katus

Contents

Abstract RNAi-based approaches show promising effects in preclinical models of cardiac diseases. However, a potential clinical use may be limited by the low efficiency of cardiac transfer. Efficient and sustained cardiac delivery in large animal models and finally clinical trials requires transfer of vectorized shRNA with suitable application systems. Packaging shRNA constructs in targeted adeno-associated viral (AAV) vectors may be versatile for systemic delivery into rodent hearts. Expressing microRNA under control of cardiac-specific promoters may further increase efficiency and specificity of delivery. For large animals and finally clinical studies, careful selection of the vector and application system will be necessary to obtain valid results.

1 Introduction

RNA interference (RNAi) has been identified as promising modality for sequence-specific knockdown of gene expression. Small interfering RNAs (siRNAs) are chemically synthesized sequences which trigger RNAi. In cardiology, RNAi-based therapeutic approaches have been developed for heart failure (El-Armouche et al. 2007; Fechner et al. 2007; Watanabe et al. 2004), postinfarction myocardial protection

O.J. Müller
Innere Medizin III, Universitatsklinikum Heidelberg, Im Neuenheimer Feld 410, 69120 Heidelberg, Germany
e-mail: oliver.mueller@med.uni-heidelberg.de

V.A. Erdmann et al. (eds.) *RNA Technologies in Cardiovascular Medicine and Research*,
doi: 10.1007/978-3-540-78709-9,

(Sugano et al. 2005) or virus myocarditis (Merl et al. 2005). Heart failure is associated with reduced phospholamban phosphorylation resulting in decreased activity of the sarcoplasmic reticulum calcium phosphatase and increased enddiastolic calcium levels (Schmidt et al. 1999). Therefore, a promising future therapeutic modality might be knockdown of phospolamban expression in cardiomyocytes. In vitro studies using siRNA against phospholamban enabled increased calcium uptake into the sarcoplasmic reticulum in cultivated rat cardiomyocytes (Watanabe et al. 2004). An approach to treat virus myocarditis showed an therapeutic effect of synthesized siRNA against a highly conserved region encoding a protease within the coxsackie-virus B3 genome (Merl et al. 2005). Hydrodynamic tail vein injections of this siRNA in a murine model of coxsackie-virus infection resulted in transfer of the siRNA in about 50% of cardiomyocytes and significant reduction of viral titers, reduced tissue damage and prolonged survival. However, hydrodynamic injections of a large volume are not considered applicable for nucleic acid delivery in patients (Snove and Rossi 2006). In contrast to local delivery of synthesized siRNA into the eye or oropharynx, which is already studied in clinical trials (de Fougerolles et al. 2007), cardiac delivery is limited by the accessibility on the one hand and rapid degradation of chemically synthesized siRNAs resulting in a only transient effect on the other hand. An alternative to synthesized siRNAs are short hairpin *RNAs* (shRNA), which can be expressed from promoters and delivered either by shRNA-expressing plasmids or viral vectors such as adenoviral or adeno-associated viral (AAV) vectors (Snove and Rossi 2006). Using vector-expressed shRNA, efficiency, specificity, and duration of expression depends on the characteristics of the vector system itself. Improving nucleic acid delivery in order to achieve efficient, specific, and homogeneous delivery into the (human) heart remains a challenge for the development of cardiac vectors and application modes. For safety and efficiency reasons, targeting of gene transfer is preferable. Targeting can be generally achieved at four levels:

1. Targeting of tissue-specific cellular pathways
2. Transductional targeting using modifications of the vector surface
3. Transcriptionally targeting using tissue-specific promoters
4. Targeting by a tissue-specific application mode

In contrast to conventional gene transfer approaches, expression of shRNA allows a certain level of targeting by silencing genes which are expressed only in a distinct tissue like the heart. At the vector level, there are two approaches for targeting cardiac gene expression: transductional targeting by modification of the vector surface itself and transcriptional targeting using a cardiac-specific promoter. Transductional targeting utilizes the ability of distinct modifications of the vector surface to redirect vector transduction to a certain cell type. Transcriptional targeting is an additional level of targeting using tissue-specific regulatory sequences. Transcriptional targeting cannot prevent transduction of non-target tissue such as antigen presenting cells, but is able to prevent expression of the encoded nucleic acid which could otherwise result in side effects or an immune response. The suitability of shRNAs for use under tissue-specific promoters depends on the design of the hairpin structure. Conventional shRNAs based on complementary hairpins require polIII promoters such as the polIII U6 promoter which enable an unspecific, but highly abundant expression (Snove and Rossi 2006). Since uncontrolled high intracellular shRNA levels may lead to toxicity (Grimm et al. 2006), the use of a tissue-specific promoter could improve safety. Tissue-specific polII promoters can be used for expression of shRNAs with structural features of micro-RNA precursors (Lee et al. 2004). Finally, a suitable application mode may further increase efficiency and specificity of transduction in small animal models and will be a prerequisite for pre-clinical studies in large animal models and finally clinical trials.

The aim of this chapter is to give an overview about current strategies to target nucleic acid transfer into the myocardium with a focus on AAV vectors – a vector system which enables an efficient and long-term gene expression in the myocardium.

2 Vector Systems for Cardiac Nucleic Acid Transfer

Vectors are gene delivery vehicles that are used to carry the genetic material. They may be based on plasmid DNA or viral particles. Irrespective of the vector system, gene transfer vectors can be applied to the myocardium either by local injection or a transvascular route (Müller et al. 2007). Intravascular delivery requires a vector which escapes neutralizing antibodies and is able to cross the vascular barrier, ideally specifically in the target region. Once the vector reaches the myocardium, either after passage through the vascular wall or by direct injection, it has to efficiently spread through the extracellular matrix. Even successful uptake in the target cell is not sufficient for efficient gene transfer, since the vector genome needs to be delivered into the nucleus. For high level expression of a therapeutic gene, an efficient regulatory sequence is required. Finally, the duration of expression depends on the ability of the vector to prevent clearance of vector genomes in the cell on the one hand, and evasion from an immune response either against vector epitopes or the therapeutic factor itself on the other.

Advantages of non-viral vectors are low immunogenic properties, low costs of production, low toxicity and high organ specificity when injected locally. However, important disadvantages remain. Low transduction efficiency and short duration of gene expression (Li and Huang 2006) still limit their applicability in conditions requiring long-term expression like heart failure or myocarditis. However, transient expression by plasmid DNA could be sufficient for certain applications, such as cardioprotection after acute myocardial infarction. Intramyocardial injections of plasmid-DNA encoding shRNA against the Src homology domain 2-containing tyrosine phosphatase-1, which is involved in triggering proapoptototic pathways, resulted in an impressive effect on infarct size reduction in a rat model of acute myocardial infarction (Sugano et al. 2005).

The major advantage of delivery with viral vectors such as adenoviral or adeno-associated viral vectors is the high transduction rate. Virus-mediated gene transfer resulted in 30–360-fold higher levels of cardiac transduction with adeno-associated viral or adenoviral vectors after direct intramyocardial injections in rabbits compared to plasmid approaches (Wright et al. 2001). Transduction efficiencies were up to 75% of cardiomyocytes around the needle track after direct injection of adenoviral vectors in adult pigs (French et al. 1994). The higher efficiency of viral approaches is explained by the efficient cellular uptake and intracellular transport of packaged DNA to the nucleus since viral vectors are taken up by specific surface receptors and escape lysosomal degradation (Ding et al. 2005; Greber et al. 1993; Roelvink et al. 1999; Summerford and Samulski 1998).

Adenoviral vectors are frequently used in experimental and clinical gene transfer studies targeting the heart since they enable highly efficient cardiac gene delivery and can be produced in sufficient amounts. Adenoviral vectors have been successfully used to transfer shRNA against phospholamban modulating calcium sequestration in primary neonatal rat cardiomyocytes (Fechner et al. 2007). Another study showed that adenoviral transfer of shRNA against proteinkinase Cα, a negative regulator of contractility, resulted in a significant downregulation of protein levels and improved contractility in neonatal rat cardiomyocytes and engineered heart tissues (El-Armouche et al. 2007). However, adenoviral vectors are limited by a transient gene expression in vivo caused by an immune response against viral gene products resulting in rapid clearance of transduced cells (Chirmule et al. 1999; Jooss et al. 1998; Yang et al. 1994). The immunogenicity of adenoviral vectors could be reduced in third generation (so called "gutless" or "high capacity") adenoviral vectors that lack the complete viral genome except for the packaging sequence (Kochanek et al. 1996). Nevertheless, an immune response may still occur due to transduction of and gene expression in dendritic cells that play a key role in triggering a cytotoxic immune response (Jooss et al. 1998).

Adeno-associated virus (AAV) is increasingly used as alternative to adenoviral vectors because of its safety profile: AAV is a non-pathogenic parvovirus that cannot be amplified without co-infection with a helper virus. AAV vectors transduce the myocardium as efficiently as adenoviral vectors and – in contrast to shorter expression with adenoviral vectors – allow stable expression of transgenes over several months (Chu et al. 2003; Vassalli et al. 2003). It has been shown that AAV vectors are able to integrate into active genes (Nakai et al. 2003). Although integration appears to occur rarely with wildtype AAV as well as recombinant AAV vectors (McCarty et al. 2004; Schnepp et al. 2005), there may be a potential oncogenic risk (Donsante et al. 2007) which is controversially discussed (Kay 2007). AAV vectors have been successfully used in several experimental therapeutic approaches such as protection from ischemia/reperfusion injury in a rat model (Melo et al. 2002) or positive inotropic therapy in cardiomyopathy hamsters (Hoshijima et al. 2002; Kawada et al. 2002). AAV-approaches revealed beneficial effects on neoangiogenesis, infarct-size, and cardiac function in a murine model of myocardial ischemia (Su et al. 2004). To achieve a highly selective transduction of myocardial tissue, AAV vectors were administered via intramyocardial injection or perfusions of coronary arteries (Hoshijima et al. 2002; Kawada et al. 2002; Melo et al. 2002; Su et al. 2004).

3 Transductional Targeting of AAV Vectors for Cardiac Nucleic Acid Transfer

Transductional targeting is an approach used to restrict vector uptake to the target tissue. In addition to specificity, efficiency should also be increased due to reduced transduction of non-target tissue. Transductional targeting can be achieved either by distinct modifications of the AAV capsid surface or, in a broader sense, by identifying vectors from different naturally occurring serotypes with a distinct transduction pattern.

Comparison of the transduction efficiency and specificity of different AAV serotype vector by intravenous injection in rodents showed that AAV-6, -8 and -9 serotype vectors resulted in uniform and extensive cardiac transfer in adult mice (Gregorevic et al. 2004; Inagaki et al. 2006; Müller et al. 2006; Nakai et al. 2005; Pacak et al. 2006). Highest cardiac efficiency and specificity was observed with AAV-9 vectors (Inagaki et al. 2006; Pacak et al. 2006). The efficient gene transfer via a transvascular route is explained by the ability of certain AAV serotypes to efficiently cross the blood vessel barrier in rodents. Furthermore, efficient spreading through the extracellular matrix as previously shown for AAV-6 and -8 may additionally contribute to the increased transduction (Du et al. 2004; Palomeque et al. 2007). The efficiency of systemic delivery is also reflected by impressive therapeutic effects in animal models of cardiac diseases. Intravenous injection of AAV-8 vectors enabled reconstitution of δ-sarcoglycan in δ-sarcoglycan-deficient TO-2 cardiomyopathy hamsters (Zhu et al. 2005) and extended their lifespan by preventing heart failure. Systemic transfer of AAV-7 pseudotyped vectors overexpressing acid α-glucosidase has successfully reduced cardiac glycogen content in a murine model of glycogen storage disease type II (Pompe disease) that causes death in infancy from cardiorespiratory failure (Sun et al. 2005). Although AAV serotype 9 revealed the highest cardiac gene transfer efficiency, there is still significant transduction of extracardiac tissue. Elucidation of the superior transduction mechanisms of distinct AAV serotypes may help to develop even more efficient and specific vectors.

Modification of the natural tropism of AAVs is a promising approach to increase transduction efficiency and decrease potential side effects due to extracardiac gene transfer. Undesired liver transduction could be dramatically reduced with mutant AAV-2(R484E; R585E) vectors, which lack binding to the AAV-2 primary receptor heparan sulfate proteoglycane (Kern et al. 2003), resulting in a 100-fold increased ratio of cardiac to hepatic reporter activity (Müller et al. 2006). Interestingly, the same region is also involved in uptake of AAV-2 in human dendritic cells and elucidation of a specific T-cell response against AAV capsids (Vandenberghe et al. 2006). Thus, deletion of binding to the AAV-2 primary receptor may not only improve safety by reducing hepatic transduction, but also by reducing immunogenicity. Such elaborated guess-driven approaches are limited by several unknown variables determining successful transduction such as penetration of endothelium within the vascular wall, physical barriers of the target tissue such as the extracellular matrix, or the

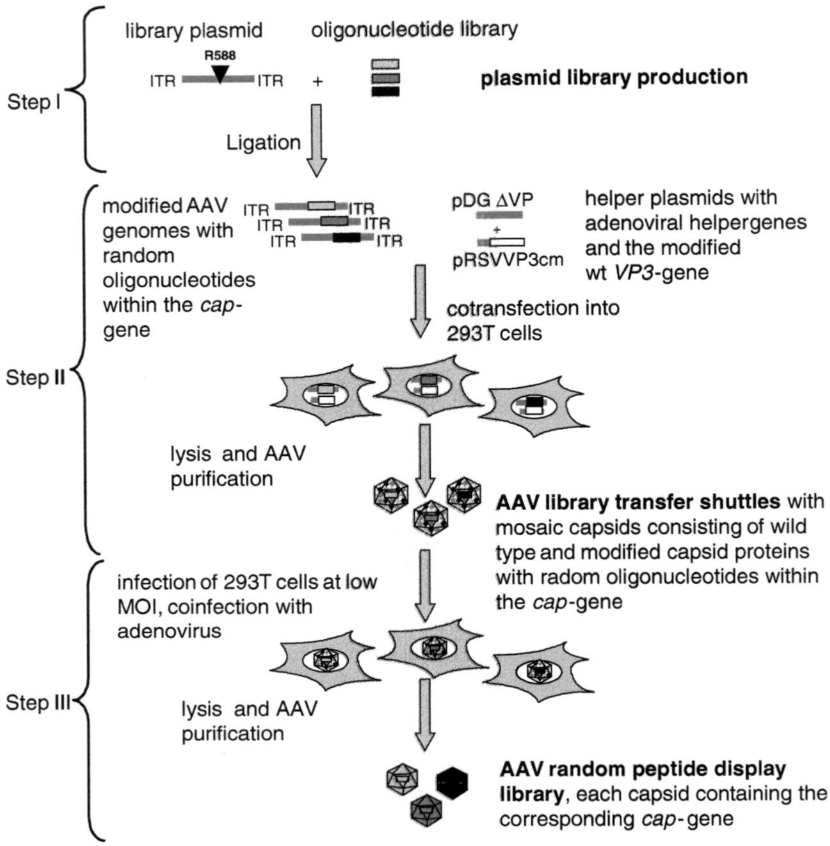

Fig. 1 Three step procedure to produce random AAV display peptide libraries. First, an AAV plasmid library is produced based on the library plasmid containing the ITR flanked AAV-2 rep and cap gene. A random oligonucleotide library is cloned adjacent to nucleotide position 3967 (corresponding to arginine 588) of the AAV-2 cap gene (Step I). The transfer shuttle library is generated by cotransfection of the library plasmids flanked by ITRs, the pDGΔVP adenoviral helper plasmid and an ITR-less sequence encoding wild type capsids (pRSVVP3 cm), leading to chimeric AAV capsids with packaged library genomes (Step II). The final virus peptide display library is generated by infecting 293T cells with transfer shuttles at a low MOI (1 replicative unit/cell) to achieve the uptake and propagation of one library genome per cell (Step III) (Müller et al. 2003; Waterkamp et al. 2006)

intracellular fate of vectors. Combinatorial approaches displaying a library of randomized peptide motifs on the AAV surface (Müller et al. 2003; Perabo et al. 2003) have resulted in a vector targeted to H9C2 cardiomyoblasts in vitro (Waterkamp et al. 2006). Figure 1 shows the elaborated production process of an AAV peptide library. Figure 2 gives an overview about an in vitro AAV-library selection process. In vivo selection of randomized AAV peptide libraries on murine hearts may allow identification of novel AAV-vectors targeted to myocardium in vivo.

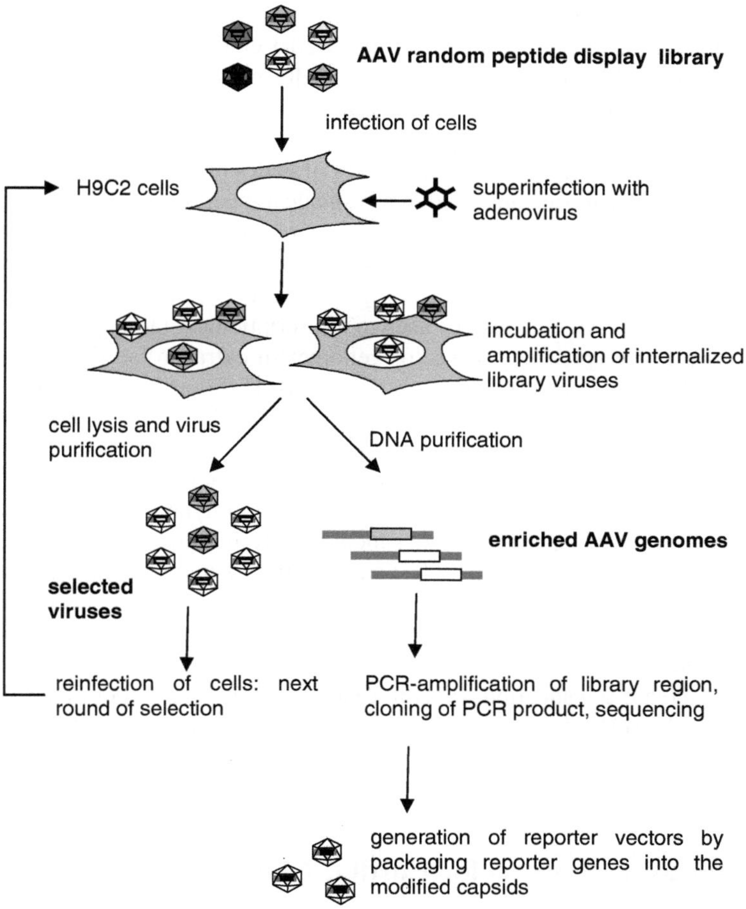

Fig. 2 Selection of an AAV random peptide display library in vitro. Upon transduction with the AAV library, H9C2 are superinfected with adenovirus in order to amplify successful transducing vectors. Vectors are recovered and used for further selections rounds. In parallel, DNA is extracted from a small aliquot and used for PCR amplification of the library region. PCR products are sequenced. Based on the sequences, reporter vectors are generated for validating an increased transduction efficiency

4 Transcriptional Targeting of AAV Vectors for Cardiac Nucleic Acid Transfer

Transcriptional targeting uses promoters that regulate expression of myocardial proteins. This approach will not prevent transduction of extracardiac tissue, but may restrict gene expression to the myocardium. Promoter elements not only determine tissue-specificity and expression levels, but also duration of gene expression.

In order to obtain a high expression efficiency, the strong and ubiquitous active CMV-promoter has been used traditionally (Müller et al. 2006). The use of tissue-specific regulatory sequences results in markedly prolonged gene expression by prevention of promoter downregulation (Brooks et al. 2004; Qin et al. 1997) and transgene expression in antigen presenting cells (Cordier et al. 2001; Schiedner et al. 2002). Originally, regulatory sequences of genes predominantly expressed in the myocardium have been analyzed in transgenic animal models resulting in identification of suitable promoter sequences that could induce a cardiac phenotype in transgenic animal models (Franz et al. 1997a). Further studies using viral vectors showed that the regulatory sequences of genes encoding α-myosin heavy chain or myosin light chain-2v could also be used to transcriptionally target gene expression to the heart despite presence of vector genomes in extracardiac tissues (Aikawa et al. 2002; Boecker et al. 2004; Champion et al. 2003; Franz et al. 1997b; Griscelli et al. 1998; Müller et al. 2006; Phillips et al. 2002; Su et al. 2004) (Table 1). Another novel regulatory sequence enabling efficient gene expression in the heart is the α-cardiac actin enhancer/elongation factor 1α- promoter (Pleger et al. 2007). A potent, but rather unspecific promoter is the β-actin hybrid promoter that leads to early and widespread cardiac transduction after intramyocardial injection of AAV-6 vectors (Kawamoto et al. 2005) and a hybrid α-myosin heavy-chain/creatine kinase promoter (Salva et al. 2007). In order to overcome the low expression levels of tissue-specific regulatory sequences, fusion with strong viral enhancers or hypoxia-regulatory elements, as shown for the (MLC) 2v-promoter, has been suggested to increase transduction levels in rodents (Müller et al. 2006; Pachori et al. 2004; Su et al. 2004). Table 1 lists regulatory sequences having enabled successful cardiac gene expression using adenoviral and AAV vectors.

5 Application Systems for Cardiac Gene Transfer

Although several AAV serotypes are suitable for systemic cardiac gene transfer in rodents (Champion et al. 2003; Gregorevic et al. 2004; Inagaki et al. 2006; Müller et al. 2006; Pacak et al. 2006; Sun et al. 2005; Vandendriessche et al. 2007; Wang et al. 2005), nucleic acid transfer in larger animal models or even patients in clinical studies require targeted delivery approaches. Although direct intramyocardial injection of vectors are possible and result in a specific gene transfer, this approach is limited by a patchy pattern of gene transfer and surgery related risks (Chu et al. 2003; French et al. 1994; Svensson et al. 1999; Wright et al. 2001). Further developments of percutaneous needle catheters in combination with cardiac mapping systems enabled vector injections in the myocardium in large animal studies without the risk associated with open chest surgery (Kornowski et al. 2000; Rutanen et al. 2004). This approach has been successfully transferred into a clinical study. Percutaneous intramyocardial injections of plasmid-DNA encoding VEGF with a needle-catheter under guidance of an intramyocardial mapping system (NOGA) have shown beneficial effects on local wall motions although endpoints of this

Table 1 Regulatory sequences used for cardiac gene expression in animal models

Regulatory sequence	Vector system	Reporter gene	Route of gene transfer	Species	References
Hypoxia-response elements + mSV40 promoter	AAV-2	lacZ	Intramyocardial	Adult rats	Pachori et al. (2004)
Hypoxia-response elements + 250 bp myosin light chain-2v promoter	AAV-2	VEGF	Intramyocardial	Adult mice	Su et al. (2004)
α-myosin heavy chain promoter	AAV-2	human growth hormone	Intramyocardial	Adult mice	Aikawa et al. (2002)
	Adenovirus	phospholam-ban	Intracoronary	Adult mice	Champion et al. (2003)
α-myosin heavy chain promoter; 0.8 kb myosin light chain-2v promoter	Adenovirus	Luciferase	Intramyocardial	Neonat rats	Franz et al. 1997b
1.7 kb myosin light chain-2v promoter	AAV-2	GFP	"Intracardiac"	Neonat rats	Phillips et al. (2002)
2.1 kb and 260 bp myosin light chain-2v promoter	Adenovirus	lacZ	Intramyocardial	Adult rats	Griscelli et al. (1998)
250 bp myosin light chain-2v promoter		Luciferase	Intramyocardial	Adult rats	Boecker et al. (2004)
CMV-enhanced 1.5 kb myosin light chain-2v promoter	AAV-2 and -6	Luciferase	Intravenous	Adult mice	Müller et al. (2006)
CMV-enhanced β-actin promoter[a]	AAV-2 and -6	Alkaline phosphatise	Intramyocardial	Adult rats	Kawamoto et al. (2005)
	AAV-2, -5, -6, -8	GFP		Adult mice	Wang et al. (2005)
α-myosin heavy-chain/creatine kinase promoter	AAV-6	Micro-dystrophin	Intravenous	Mice	Salva et al. (2007)
Muscle-specific creatine kinase	AAV-7	Acid α-glucosidase	Intravenous	Adult mice	Sun et al. (2005)
Human brain natriuretic peptide promoter	Adenovirus	Luciferase	Intramyocardial	Adult mice	LaPointe et al. (2002)
Cardiac actin enhancer/elongation factor 1α-promoter (Pleger et al. 2007)	AAV-6	EGFP, S100A1	Coronary perfusion	Adult rats	Pleger et al. (2007)

Ref., reference; neonat, neonatal
[a]Not specific for muscle tissue

study were not significantly improved (Kastrup et al. 2005). Injection of viral vectors into the pericardial sack has been described as an alternative application mode. However, this technique did not demonstrate efficient transmural gene expression (Lamping et al. 1997; March et al. 1999) and has not been further investigated.

A less invasive alternative are transvascular approaches. Infusion of an adenoviral vector into coronary arteries using a percutaneous transluminal approach resulted in low transfer efficiency (Muhlhauser et al. 1996). This was explained by the short exposure time of the vector within the coronary endothelium and loss of the vector into the systemic circulation following intracoronary injection. Since the KAT study (Hedmann et al. 2003) and AGENT 2 study (Grines et al. 2003) suggested a potential application of intracoronary delicery in clinical studies, this approach has been analyzed in the multicenter AGENT 3/4 study which failed to reach a significant endpoint except for subgroup analyzes (http://www.cardiumthx.com).

In order to improve vector delivery via the coronary circulation, several critical parameters have been identified such as coronary flow, vector concentration, and endothelial permeability (Donahue et al. 1997, 1998). The main barrier appears to be the vascular wall, especially the microvascular endothelium. Thus, several targets have been identified in order to increase endothelial passage of the vector such as endothelial permeability, endothelial exposure time and intravascular pressure. Endothelial permeability and transduction efficiency could be increased by pharmacological agents such as histamin and serotonin (Donahue et al. 1998; Logeart et al. 2001), VEGF (Gregorevic et al. 2004; Raake et al. 2008; Logeart et al. 2001), phosphodiesterase-5 inhibitors (Nagata et al. 2001), and nitroglycerin (Sasano et al. 2007). Endothelial permeability can also be increased by application of ultrasound targeted destruction of microbubbles loaded with vector particles. (Beeri et al. 2002; Shohet et al. 2000; Bekeredjian et al. 2003).

Endothelial exposure time of the vector can be increased with different application modes. Simultaneous clamping of the aorta and pulmonary artery during vector application in rats resulted in repeated recirculation of the vector in the coronary arteries and an efficient cardiac gene transfer (Hajjar et al. 1998). Crossclamping time could be increased by hypothermia or cardiac arrest in small animals (Ding et al. 2004; Ikeda et al. 2002; Iwanaga et al. 2004; Iwatate et al. 2003) and cardiopulmonary bypass in large animals (Davidson et al. 2001). Another study reported a novel percutaneous closed-loop recirculatory system for homogeneous cardiac gene transfer in a large animal model (Kaye et al. 2007). Using a membrane oxygenator, this system allows perfusion of an AAV-1 vector through the coronary circulation for 10 min.

Intravascular pressure can be elevated by crossclamping the aorta (Champion et al. 2003; Hajjar et al. 1998; Ikeda et al. 2002) or percutaneous occlusion with a balloon catheter (Beeri et al. 2002; Ding et al. 2004). Vector delivery by selective retroinfusion through the coronary veins enables not only an increased intravascular pressure, but also prolonged exposure of the vector to the endothelium as well as allowing application of factors permeabilizing the endothelium. Retroinfusion of adenoviral or AAV vectors resulted in an efficient gene transfer into the target area with low spillover in control organs in a porcine model (Boekstegers et al. 2000;

Raake et al. 2004, 2008). Combination of retroinfusion with antegrade delivery through the left anterior descending artery resulted in adenoviral gene transfer in up to 78% of cardiomyocytes in a recent study (Sasano et al. 2007). The role of delaying venous drainage was also shown by retroinfusion of saline during intracoronary delivery which resulted in increased gene transfer compared to intracoronary transfer alone (Logeart et al. 2001).

6 Perspective

RNAi based approaches show promising effects in preclinical models of cardiac diseases. A potential clinical use for cardiac delivery in large animal models and finally clinical trials requires transfer of vectorized shRNA with suitable application systems. Future studies will show whether shRNA constructs packaged in targeted AAV vectors may be versatile for knockdown of distinct cardiac gene products after systemic delivery into rodent hearts. Furthermore, it remains to be elucidated whether transcriptional targeting of microRNA with cardiac promoters allows efficient and specific silencing of genes expressed in the heart. Once those vectors have been established in rodents, suitable application systems such as retroinfusion into coronary veins are necessary in order to analyze a vectorized shRNA-approach in large animal models and finally clinical trials.

References

Aikawa R, Huggins GS, Snyder RO (2002) Cardiomyocyte-specific gene expression following recombinant adeno-associated viral vector transduction. J Biol Chem 277:18979–18985

Beeri R, Guerrero JL, Supple G et al (2002) New efficient catheter-based system for myocardial gene delivery. Circulation 106:1756–1759

Bekeredjian R, Chen S, Frenkel PA et al (2003) Ultrasound-targeted microbubble destruction can repeatedly direct highly specific plasmid expression to the heart. Circulation 108:1022–1026

Boecker W, Bernecker OY, Wu JC et al (2004) Cardiac-specific gene expression facilitated by an enhanced myosin light chain promoter. Mol Imaging 3:69–75

Boekstegers P, von Degenfeld G, Giehrl W et al (2000) Myocardial gene transfer by selective pressure-regulated retroinfusion of coronary veins. Gene Ther 7:232–240

Brooks AR, Harkins RN, Wang P et al (2004) Transcriptional silencing is associated with extensive methylation of the CMV promoter following adenoviral gene delivery to muscle. J Gene Med 6:395–404

Champion HC, Georgakopoulos D, Haldar S et al (2003) Robust adenoviral and adeno-associated viral gene transfer to the in vivo murine heart: application to study of phospholamban physiology. Circulation 108:2790–2797

Chirmule N, Moscioni AD, Qian Y et al (1999) Fas–Fas ligand interactions play a major role in effector functions of cytotoxic T lymphocytes after adenovirus vector-mediated gene transfer. Hum Gene Ther 10:259–269

Chu D, Sullivan CC, Weitzman MD, Du L et al (2003) Direct comparison of efficiency and stability of gene transfer into the mammalian heart using adeno-associated virus versus adenovirus vectors. J Thorac Cardiovasc Surg 126:671–679

Cordier L, Gao GP, Hack AA et al (2001) Muscle-specific promoters may be necessary for adeno-associated virus-mediated gene transfer in the treatment of muscular dystrophies. Hum Gene Ther 12:205–215
Davidson MJ, Jones JM, Emani SM et al (2001) Cardiac gene delivery with cardiopulmonary bypass. Circulation 104:131–133
de Fougerolles A, Vornlocher HP, Maraganore J et al (2007) Interfering with disease: a progress report on siRNA-based therapeutics. Nat Rev Drug Discov 6:443–453
Ding Z, Fach C, Sasse A et al (2004) A minimally invasive approach for efficient gene delivery to rodent hearts. Gene Ther 11:260–265
Ding W, Zhang L, Yan Z et al (2005) Intracellular trafficking of adeno-associated viral vectors. Gene Ther 12:873–880
Donahue JK, Kikkawa K, Johns DC et al (1997) Ultrarapid, highly efficient viral gene transfer to the heart. Proc Natl Acad Sci USA 94:4664–4668
Donahue JK, Kikkawa K, Thomas AD et al (1998) Acceleration of widespread adenoviral gene transfer to intact rabbit hearts by coronary perfusion with low calcium and serotonin. Gene Ther 5:630–634
Donsante A, Miller DG, Li Y et al (2007) AAV vector integration sites in mouse hepatocellular carcinoma. Science 317:477
Du L, Kido M, Lee DV et al (2004) Differential myocardial gene delivery by recombinant serotype-specific adeno-associated viral vectors. Mol Ther 10:604–608
El-Armouche A, Singh J, Naito H et al (2007) Adenovirus-delivered short hairpin RNA targeting PKCalpha improves contractile function in reconstituted heart tissue. J Mol Cell Cardiol 43:371–376
Fechner H, Suckau L, Kurreck J et al (2007) Highly efficient and specific modulation of cardiac calcium homeostasis by adenovector-derived short hairpin RNA targeting phospholamban. Gene Ther 14:211–218
French BA, Mazur W, Geske RS et al (1994) Direct in vivo gene transfer into porcine myocardium using replication-deficient adenoviral vectors. Circulation 90:2414–2424
Franz WM, Mueller OJ, Hartong R et al (1997a) Transgenic animal models: new avenues in cardiovascular physiology. J Mol Med 75:115–129
Franz WM, Rothmann T, Frey N et al (1997b) Analysis of tissue-specific gene delivery by recombinant adenoviruses containing cardiac-specific promoters. Cardiovasc Res 35:560–566
Greber UF, Willetts M, Webster P et al (1993) Stepwise dismantling of adenovirus 2 during entry into cells. Cell 75:477–486
Gregorevic P, Blankinship MJ, Allen JM et al (2004) Systemic delivery of genes to striated muscles using adeno-associated viral vectors. Nat Med 10:828–834
Grimm D, Streetz KL, Jopling CL et al (2006) Fatality in mice due to oversaturation of cellular microRNA/short hairpin RNA pathways. Nature 441:537–541
Grines C, Rubanyi GM, Kleiman NS et al (2003) Angiogenic gene therapy with adenovirus 5 fibroblast growth factor-4 (Ad5FGF-4): a new option for the treatment of coronary artery disease. Am J Cardiol 92:24N–31N
Griscelli F, Gilardi-Hebenstreit P, Hanania N et al (1998) Heart-specific targeting of beta-galactosidase by the ventricle-specific cardiac myosin light chain 2 promoter using adenovirus vectors. Hum Gene Ther 9:1919–1928
Hajjar RJ, Schmidt U, Matsui T et al (1998) Modulation of ventricular function through gene transfer in vivo. Proc Natl Acad Sci USA 95:5251–5256
Hedman M, Hartikainen J, Syvänne M, Stjernvall J, Hedman A, Kivelä A, Vanninen E, Mussalo H, Kauppila E, Simula S, Närvänen O, Rantala A, Peuhkurinen K, Nieminen MS, Laakso M, Ylä-Herttuala S (2003) Safety and feasibility of catheter-based local intracoronary vascular endothelial growth factor gene transfer in the prevention of postangioplasty and in-stent restenosis and in the treatment of chronic myocardial ischemia: phase II results of the Kuopio Angiogenesis Trial (KAT). Circulation 107(21):2677–2683
Hoshijima M, Ikeda Y, Iwanaga Y et al (2002) Chronic suppression of heart-failure progression by a pseudophosphorylated mutant of phospholamban via in vivo cardiac rAAV gene delivery. Nat Med 8:864–871

Ikeda Y, Gu Y, Iwanaga Y et al (2002) Restoration of deficient membrane proteins in the cardiomyopathic hamster by in vivo cardiac gene transfer. Circulation 105:502–508

Inagaki K, Fuess S, Storm TA et al (2006) Robust systemic transduction with AAV9 vectors in mice: efficient global cardiac gene transfer superior to that of AAV8. Mol Ther 14:45–53

Iwanaga Y, Hoshijima M, Gu Y et al (2004) Chronic phospholamban inhibition prevents progressive cardiac dysfunction and pathological remodeling after infarction in rats. J Clin Invest 113:727–736

Iwatate M, Gu Y, Dieterle T et al (2003) In vivo high-efficiency transcoronary gene delivery and Cre-LoxP gene switching in the adult mouse heart. Gene Ther 10:1814–1820

Jooss K, Yang Y, Fisher KJ et al (1998) Transduction of dendritic cells by DNA viral vectors directs the immune response to transgene products in muscle fibers. J Virol 72: 4212–4223

Kastrup J, Jorgensen E, Ruck A et al (2005) Direct intramyocardial plasmid vascular endothelial growth factor-A165 gene therapy in patients with stable severe angina pectoris A randomized double-blind placebo-controlled study: the Euroinject One trial. J Am Coll Cardiol 45:982–988

Kawada T, Nakazawa M, Nakauchi S et al (2002) Rescue of hereditary form of dilated cardiomyopathy by rAAV-mediated somatic gene therapy: amelioration of morphological findings, sarcolemmal permeability, cardiac performances, and the prognosis of TO-2 hamsters. Proc Natl Acad Sci USA 99:901–906

Kawamoto S, Shi Q, Nitta Y et al (2005) Widespread and early myocardial gene expression by adeno-associated virus vector type 6 with a beta-actin hybrid promoter. Mol Ther 11:980–985

Kay MA (2007) AAV vectors and tumorigenicity. Nat Biotechnol 25:1111–1113

Kaye DM, Preovolos A, Marshall T et al (2007) Percutaneous cardiac recirculation-mediated gene transfer of an inhibitory phospholamban peptide reverses advanced heart failure in large animals. J Am Coll Cardiol 50:253–260

Kern A, Schmidt K, Leder C et al (2003) Identification of a heparin-binding motif on adeno-associated virus type 2 capsids. J Virol 77:11072–11081

Kochanek S, Clemens PR, Mitani K et al (1996) A new adenoviral vector: replacement of all viral coding sequences with 28 kb of DNA independently expressing both full-length dystrophin and beta-galactosidase. Proc Natl Acad Sci USA 93:5731–5736

Kornowski R, Leon MB, Fuchs S et al (2000) Electromagnetic guidance for catheter-based transendocardial injection: a platform for intramyocardial angiogenesis therapy. Results in normal and ischemic porcine models. J Am Coll Cardiol 35:1031–1039

Lamping KG, Rios CD, Chun JA et al (1997) Intrapericardial administration of adenovirus for gene transfer. Am J Physiol 272:H310–H317

LaPointe MC, Yang XP, Carretero OA et al (2002) Left ventricular targeting of reporter gene expression in vivo by human BNP promoter in an adenoviral vector. Am J Physiol Heart Circ Physiol 283:H1439–H1445

Lee Y, Kim M, Han J et al (2004) MicroRNA genes are transcribed by RNA polymerase II. EMBO J 23:4051–4060

Li SD, Huang L (2006) Gene therapy progress and prospects: non-viral gene therapy by systemic delivery. Gene Ther 13:1313–1319

Logeart D, Hatem SN, Heimburger M et al (2001) How to optimize in vivo gene transfer to cardiac myocytes: mechanical or pharmacological procedures? Hum Gene Ther 12:1601–1610

March KL, Woody M, Mehdi K et al (1999) Efficient in vivo catheter-based pericardial gene transfer mediated by adenoviral vectors. Clin Cardiol 22:I23–I29

McCarty DM, Young SM Jr, Samulski RJ (2004) Integration of adeno-associated virus (AAV) and recombinant AAV vectors. Annu Rev Genet 38:819–845

Melo LG, Agrawal R, Zhang L et al (2002) Gene therapy strategy for long-term myocardial protection using adeno-associated virus-mediated delivery of heme oxygenase gene. Circulation 105:602–607

Merl S, Michaelis C, Jaschke B et al (2005) Targeting 2A protease by RNA interference attenuates coxsackieviral cytopathogenicity and promotes survival in highly susceptible mice. Circulation 111:1583–1592

Muhlhauser J, Jones M, Yamada I et al (1996) Safety and efficacy of in vivo gene transfer into the porcine heart with replication-deficient, recombinant adenovirus vectors. Gene Ther 3:145–153
Müller OJ, Kaul F, Weitzman MD et al (2003) Random peptide libraries displayed on adeno-associated virus to select for targeted gene therapy vectors. Nat Biotechnol 21:1040–1046
Müller OJ, Leuchs B, Pleger ST et al (2006) Improved cardiac gene transfer by transcriptional and transductional targeting of adeno-associated viral vectors. Cardiovasc Res 70:70–78
Müller OJ, Katus HA, Bekeredjian R (2007) Targeting the heart with gene therapy-optimized gene delivery methods. Cardiovasc Res 73:453–462
Nagata K, Marban E, Lawrence JH et al (2001) Phosphodiesterase inhibitor-mediated potentiation of adenovirus delivery to myocardium. J Mol Cell Cardiol 33:575–580
Nakai H, Montini E, Fuess S et al (2003) AAV serotype 2 vectors preferentially integrate into active genes in mice. Nat Genet 34:297–302
Nakai H, Fuess S, Storm TA et al (2005) Unrestricted hepatocyte transduction with adeno-associated virus serotype 8 vectors in mice. J Virol 79:214–224
Pacak CA, Mah CS, Thattaliyath BD et al (2006) Recombinant adeno-associated virus serotype 9 leads to preferential cardiac transduction in vivo. Circ Res 99:e3–e9
Pachori AS, Melo LG, Hart ML et al (2004) Hypoxia-regulated therapeutic gene as a preemptive treatment strategy against ischemia/reperfusion tissue injury. Proc Natl Acad Sci USA 101:12282–12287
Palomeque J, Chemaly ER, Colosi P et al (2007) Efficiency of eight different AAV serotypes in transducing rat myocardium in vivo. Gene Ther 14:989–997
Perabo L, Buning H, Kofler DM et al (2003) In vitro selection of viral vectors with modified tropism: the adeno-associated virus display. Mol Ther 8:151–157
Phillips MI, Tang Y, Schmidt-Ott K et al (2002) Vigilant vector: heart-specific promoter in an adeno-associated virus vector for cardioprotection. Hypertension 39:651–655
Pleger ST, Boucher M, Most P et al (2007) Targeting myocardial beta-adrenergic receptor signaling and calcium cycling for heart failure gene therapy. J Card Fail 13:401–414
Qin L, Ding Y, Pahud DR et al (1997) Promoter attenuation in gene therapy: interferon-gamma and tumor necrosis factor-alpha inhibit transgene expression. Hum Gene Ther 8:2019–2029
Raake P, von Degenfeld G, Hinkel R et al (2004) Myocardial gene transfer by selective pressure-regulated retroinfusion of coronary veins: comparison with surgical and percutaneous intramyocardial gene delivery. J Am Coll Cardiol 44:1124–1129
Raake P, Hinkel R, Müller S et al (2008) Cardio-specific long-term gene expression in a porcine model after selective pressure-regulated retroinfusion of adeno-associated viral (AAV) vectors. Gene Ther 15:12–17
Roelvink PW, Mi Lee G, Einfeld DA et al (1999) Identification of a conserved receptor-binding site on the fiber proteins of CAR-recognizing adenoviridae. Science 286:1568–1571
Rutanen J, Rissanen TT, Markkanen JE et al (2004) Adenoviral catheter-mediated intramyocardial gene transfer using the mature form of vascular endothelial growth factor-D induces transmural angiogenesis in porcine heart. Circulation 109:1029–1035
Salva MZ, Himeda CL, Tai PW et al (2007) Design of tissue-specific regulatory cassettes for high-level rAAV-mediated expression in skeletal and cardiac muscle. Mol Ther 15:320–329
Sasano T, Kikuchi K, McDonald AD et al (2007) Targeted high-efficiency, homogeneous myocardial gene transfer. J Mol Cell Cardiol 42:954–961
Schiedner G, Hertel S, Johnston M et al (2002) Variables affecting in vivo performance of high-capacity adenovirus vectors. J Virol 76:1600–1609
Schmidt U, Hajjar RJ, Kim CS et al (1999) Human heart failure: cAMP stimulation of SR Ca^{2+}-ATPase activity and phosphorylation level of phospholamban. Am J Physiol 277: H474–H480
Schnepp BC, Jensen RL, Chen CL et al (2005) Characterization of adeno-associated virus genomes isolated from human tissues. J Virol 79:14793–14803
Shohet RV, Chen S, Zhou YT et al (2000) Echocardiographic destruction of albumin microbubbles directs gene delivery to the myocardium. Circulation 101:2554–2556

Snove O Jr, Rossi JJ (2006) Expressing short hairpin RNAs in vivo. Nat Methods 3:689–695
Su H, Joho S, Huang Y et al (2004) Adeno-associated viral vector delivers cardiac-specific and hypoxia-inducible VEGF expression in ischemic mouse hearts. Proc Natl Acad Sci USA 101:16280–16285
Sugano M, Tsuchida K, Hata T et al (2005) RNA interference targeting SHP-1 attenuates myocardial infarction in rats. FASEB J 19:2054–2056
Summerford C, Samulski RJ (1998) Membrane-associated heparan sulfate proteoglycan is a receptor for adeno-associated virus type 2 virions. J Virol 72:1438–1445
Sun B, Zhang H, Franco LM et al (2005) Correction of glycogen storage disease type II by an adeno-associated virus vector containing a muscle-specific promoter. Mol Ther 11:889–898
Svensson EC, Marshall DJ, Woodard K et al (1999) Efficient and stable transduction of cardiomyocytes after intramyocardial injection or intracoronary perfusion with recombinant adeno-associated virus vectors. Circulation 99:201–205
Vandenberghe LH, Wang L, Somanathan S et al (2006) Heparin binding directs activation of T cells against adeno-associated virus serotype 2 capsid. Nat Med 12:967–971
Vandendriessche T, Thorrez L, Acosta-Sanchez A et al (2007) Efficacy and safety of adeno-associated viral vectors based on serotype 8 and 9 vs. lentiviral vectors for hemophilia B gene therapy. J Thromb Haemost 5:16–24
Vassalli G, Bueler H, Dudler J et al (2003) Adeno-associated virus (AAV) vectors achieve prolonged transgene expression in mouse myocardium and arteries in vivo: a comparative study with adenovirus vectors. Int J Cardiol 90:229–238
Wang Z, Zhu T, Qiao C et al (2005) Adeno-associated virus serotype 8 efficiently delivers genes to muscle and heart. Nat Biotechnol 23:321–328
Watanabe A, Arai M, Yamazaki M et al (2004) Phospholamban ablation by RNA interference increases Ca2+ uptake into rat cardiac myocyte sarcoplasmic reticulum. J Mol Cell Cardiol 37:691–698
Waterkamp DA, Müller OJ, Ying Y et al (2006) Isolation of targeted AAV2 vectors from novel virus display libraries. J Gene Med 8:1307–1319
Wright MJ, Wightman LM, Lilley C et al (2001) In vivo myocardial gene transfer: optimization, evaluation and direct comparison of gene transfer vectors. Basic Res Cardiol 96:227–236
Yang Y, Nunes FA, Berencsi K et al (1994) Cellular immunity to viral antigens limits E1-deleted adenoviruses for gene therapy. Proc Natl Acad Sci USA 91:4407–4411
Zhu T, Zhou L, Mori S et al (2005) Sustained whole-body functional rescue in congestive heart failure and muscular dystrophy hamsters by systemic gene transfer. Circulation 112:2650–2659

Part III
Ribozymes

Characterization of Hammerhead Ribozymes Potentially Suitable for the Treatment of Hyper-Proliferative Vascular Diseases

G. Grassi(✉) and M. Grassi

Contents

Abstract Nucleic acid based drugs have emerged as attractive and novel alternatives to commonly used drugs for the treatment of different human diseases. Hammerhead ribozymes were the first type of nucleic acid based drugs to be extensively studied and predicted to be of potential practical utility. These are catalytic RNA molecules capable of inducing the site-specific cleavage of a phosphodiester bond within an RNA molecule. Thus, they can be used to reduce the intracellular level of a specific mRNA coding for a protein which affects cellular metabolism or environment, causing disease. Here we present a description of hammerhead ribozyme cleavage kinetic properties, some of the problematic related to their delivery from viral vectors and the description of their potential application in the cardiovascular field with a particular accent on the pathological condition known as artery-restenosis.

G. Grassi
Department of Internal Medicine, University Hospital of Trieste, Cattinara 34149, Trieste, Italy and Department of Molecular Pathology, University Hospital of Tübingen, Liebermeisterstr. 8, D-72076, Tübingen, Germany
e-mail: ggrassi@units.it

V.A. Erdmann et al. (eds.) *RNA Technologies in Cardiovascular Medicine and Research*,
doi: 10.1007/978-3-540-78709-9,

1 Introduction

In the last 20 years, considerable progress has been made in the identification of RNA molecules (Scherr et al. 2003) with potential therapeutic values. Among these, ribozymes were the first to be discovered (Kruger et al. 1982). The name ribozymes derives from the fact that these molecules are made of ribonucleic acid stretches (ribo-) and behave like true enzymes (-zymes) which are able to specifically catalyze the cleavage of an RNA phosphodiester back-bone. This feature implies that ribozymes can be used to cut and destroy deleterious RNAs (Puerta-Fernandez et al. 2003). Their potential therapeutic value has been demonstrated in a number of pathological conditions including virally-induced diseases, cancer, cardiovascular, neurological and genetic diseases (Grassi et al. 2004; Grassi and Marini 1996; Puerta-Fernandez et al. 2003), against which conventional therapies have little success. Among the huge scientific production in this field, we will focus here on the most commonly used form of ribozyme, namely hammerhead ribozyme (HRz), describing its therapeutic potential in the treatment of a particular human pathological condition, i.e., artery restenosis, the prototype of hyper-proliferative vascular diseases. A complete description of the potential use of HRz in the treatment other human hyper-proliferative diseases, such as glomerulonephritis, rheumatoid arthritis, ocular pathologies and tumors, has been reported elsewhere (Grassi et al. 2004). To provide a comprehensive picture relative to the potential use of HRz as therapeutics, an overview about their selection, kinetic characterization and delivery precedes the description of their potential use in the treatment of artery restenosis.

2 Hammerhead Ribozyme

HRzs represent a class of ribozyme (Grassi and Marini 1996) which were originally isolated from viroid RNA (Uhlenbeck 1987). They are capable of inducing the site-specific cleavage of a phosphodiester bond within an RNA molecule (Birikh et al. 1997b). Despite the fact that naturally occurring HRzs are *cis*-acting molecules, meaning that the ribozyme and the substrate are on the same molecule, it is possible to design *trans*-acting HRzs able to cut a separate RNA molecule (Symons 1992). The *trans*-acting HRz has three essential features (Fig. 1): (1) three double helices numbered I, II, and III with helices I and III also called binding arms; (2) a triplet cleavage site within the target RNA which is composed of the tri-nucleotide triplet NUH where N represents any nucleotide and H stands for A, C, or U (but HRzs which can cleave triplets ending with G have been also described; Vaish et al. 1998); and (3) two highly conserved sequences representing the catalytically active core (Haseloff and Gerlach 1988). HRzs specifically bind to their RNA targets by the binding arms which recognize complementary nucleotide regions on the target. As the binding arms are not conserved (Symons 1992), it is in principle possible to generate HRzs with flanking regions complementary to a substrate of any sequence.

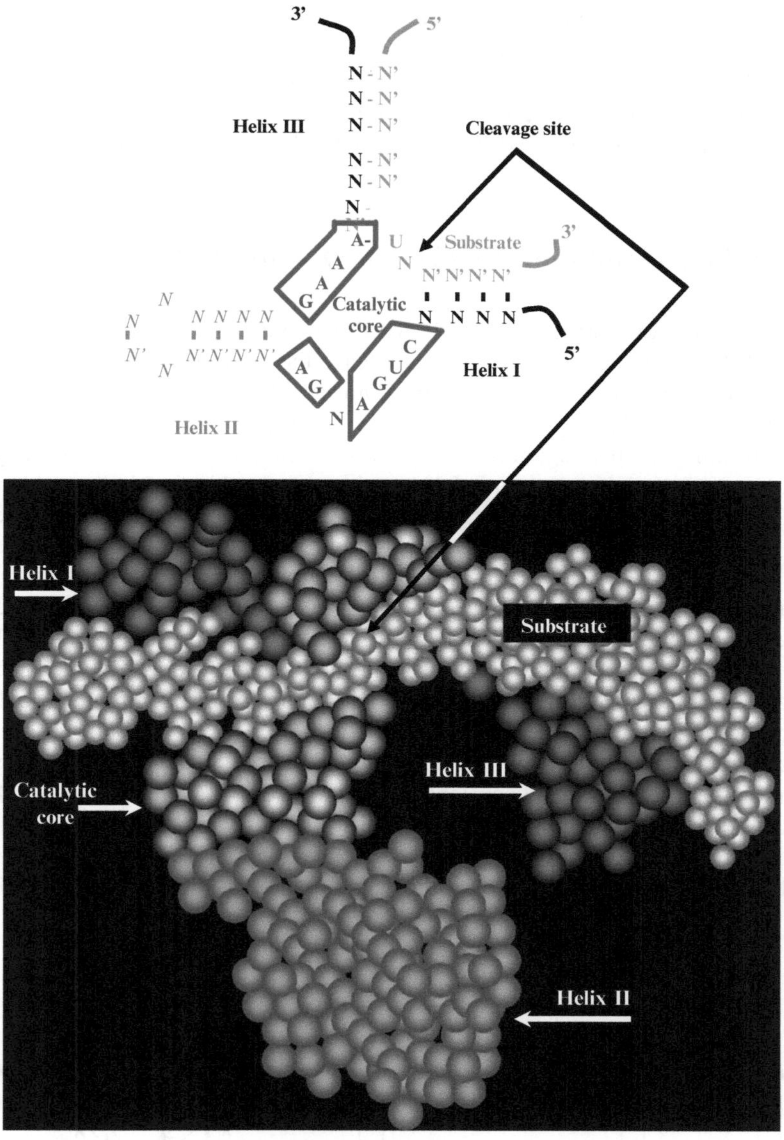

Adapted from P HeinzW.; Nature (1994), 372: 68-74

Fig. 1 Structure of hammerhead ribozymes. Two (*top*) and three (*bottom*) dimensional structure of a hammerhead ribozyme in the complex with its target. Hammerhead ribozyme structure consists of three double helices marked *I, II*, and *III*, and a triplet cleavage site within the target RNA composed of the tri-nucleotide NUH where N represents any nucleotide and H represents A, C, or U. Two stretches of highly conserved sequences (*boxed* in the two-dimensional structure) represent the catalytically active core

Following binding, a Mg^{+2}-dependent *trans*-esterification reaction takes place breaking a covalent bond 3′ of the triplet cleavage site (Dahm and Uhlenbeck 1991). The HRz kinetic pathway terminates with the release of the products (cut target RNA) from the HRzs, free to bind another target molecule (Stage-Zimmermann and Uhlenbeck 1998) as detailed in Sect. 2.3 (see also Fig. 2).

2.1 Selection and Characterization of Active HRzs

The design of HRzs for the selective destruction of a target RNA is not an easy task as not all HRzs can efficiently reach and cleave the target RNA. Due to the RNA folding (secondary and tertiary structure), not all of the triplets are available for HRz cleavage. In this regard, it is assumed that triplets surrounded by an RNA stretch forming an open region (loop), can be efficiently reached and cleaved by the HRz. In contrast, triplets present in RNA regions involved in the binding with other RNA stretches of the same molecules are considered to be far less accessible for HRz cleavage. Computer algorithms (Zuker and Jacobson 1998; Mercatanti et al. 2002) were firstly developed to predict the folding of the target RNA. The limits of this approach suggested that experimental strategies should have been also developed. We and others (Grassi et al. 2001; Birikh et al. 1997a; Cairns et al. 1999) have used a completely randomized DNA oligonucleotide library together with RNaseH to map HRz accessible cleavage sites. Others have followed even more articulated approaches (Warashina et al. 2001).

Following the identification of the accessible cleavage sites, an initial in vitro determination of HRz kinetic constants k_{cat}, representing the chemical step to form products, and K_m, which reflects the affinity of the HRz for the substrate, are required (see Sect. 2.3 for details). For in vivo applications, it is usually suggested to select HRzs with low K_m values as they can function efficiently at comparatively

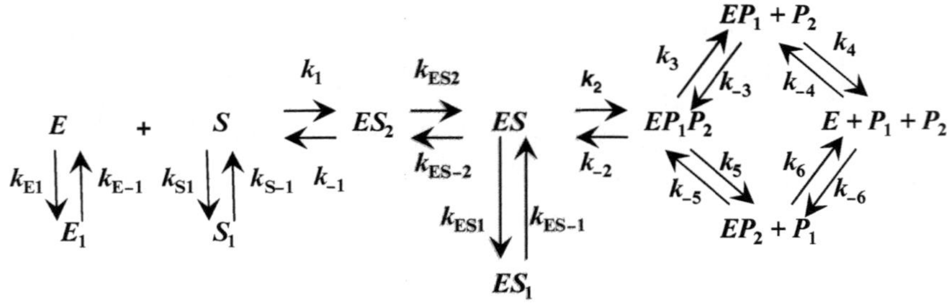

Fig. 2 Ribozyme cleavage kinetic pathway. In the hammerhead ribozyme cleavage kinetic pathway different ribozymes (*E, E1*) substrates (*S, S1*) and activated complex (*ES, ES1, ES2*) conformations can occur; moreover, different mechanisms can rule the conversion of the complex EP1P2 (hammerhead ribozyme and the cleavage products) to E and P1–P2. All these variables eventually determine hammerhead ribozyme cleavage kinetics

low concentrations (Birikh et al. 1997b). To correctly calculate kinetic constants the appropriate experimental conditions and mathematical models are required. Whereas experimental conditions were defined (Birikh et al. 1997b), less effort has been put into the development of an optimal mathematical model. Although currently used mathematical systems (Birikh et al. 1997b) show an acceptable accuracy in kinetic constant calculation, in the presence of extra ribozyme sequences (Platz et al. 2007) (promoter-derived sequences appended for delivery requirements; see Sect. 2.2 for details) and/or when HRz are targeted against long and structured targets, they may not be reliable. For these specific cases, the calculation procedure, detailed below, and involving equations (10) and (12), (12′), is strongly recommended (Grassi et al. 2002).

2.2 Delivery Systems: Exogenous and Endogenous Approaches

If administered as naked ribonucleic molecules, only a minor fraction of HRzs would reach the target resulting in negligible biological effects. The barrier which HRzs have to pass is firstly represented by the extra-cellular matrix compartment and its fluids where there are substances able to rapidly induce HRz degradation. Subsequently, HRzs have to cross the cellular membrane which, due to the negatively charged phosphate groups present in the HRz structure, cannot be crossed by passive diffusion but by endocytosis. Finally, once in the cellular environment, HRzs are susceptible to further degradation by cellular nucleases. To try to overcome these obstacles, two major strategies have been followed so far: the delivery of pre-synthesized HRz conjugated with a carrier (exogenous method), and the delivery of HRzs embedded in viral vectors (endogenous method).

The exogenous approach implies the use of chemically synthesized HRzs to be applied to the target cell/tissue complexed with different molecules generally termed “tranfection agents.” Among these, liposomes have been the most extensively used. These are minute hollow spheres composed of a lipid membrane surrounding an aqueous sphere (Balicki and Beutler 2002) which confer a certain protection against HRz degradation and allow a relatively efficiently crossing of the cellular membrane. Other substances used to trasnfer HRz to the target cells include polyethylenimine, chitosan and even polymeric nano-particles (Grassi et al. 2004). We are in the process of studying the delivery of liposome-complexed HRz from a polymeric matrix for endovascular application (Grassi et al. 2006a).

A limitation of the exogenous delivery systems is the relatively reduced release time. Viral systems (endogenous systems) can in part overcome this problem providing an extended expression of the HRz. Among the several viral vectors tested so far (Balicki and Beutler 2002), baculoviral vectors can be considered a novel and attractive tool for cardiovascular delivery. Baculovirval vectors are derived from the baculovirus Autographa californica nuclear polyhedrosis virus (AcNPV), an insect virus characterized by a large double-stranded circular DNA genome packaged in a rod-shaped capsid enveloped by a membrane (Ghosh et al.

2002). These vectors bear great potential in the field of gene therapy as they do not replicate in mammalian cells, have large insert capacity, and give rise to only minor cytopathic effects observable microscopically (Kost and Condreay 2002). Interestingly, baculoviral vectors have the ability to infect a broad range of mammalian cells (Sarkis et al. 2000) and our data (Grassi et al. 2006b) indicate that this vector is also particularly suited to transfer HRzs into human cardiomyocytes and VSMCs.

In addition to the viral vector choice, the successful administration of HRzs also depends on the promoter system employed to express the HRz. The promoter can influence HRz activity in the cell environment in variable ways which includes promoter transcription rate and intracellular localization of the transcript (Rossi 1995; Bertrand et al. 1997). Moreover, any HRz transcribed from promoters will invariably carry various promoter-derived sequences which can unpredictably affect HRz-cleavage. This last fact implies that, for the proper characterization and choice of HRz to be expressed from promoters, an evaluation of the effects of promoter derived sequences on HRz cleavage efficacy is necessary through an appropriate mathematical approach (see Sects. 2.3 and 2.4).

2.3 Enzymatic Mechanisms of HRz Action

The creation of mathematical models able to describe HRz cleavage reaction is of utmost importance to calculate kinetic constants thus allowing the selection of the most appropriate molecules. In the minimal kinetic pathway of HRz cleavage reaction four main species are present: the HRz (E, molar concentration), the substrate (S, molar concentration), the HRz-substrate complex (ES, molar concentration) and the HRz-product complex (EP_1P_2; P_1 and P_2 represent the two parts in which the HRz splits the substrate; molar concentration). Accordingly, E reacts with S (direct reaction kinetic constant k_1, inverse reaction kinetic constant k_{-1}) to form ES that, in turn, converts into EP_1P_2 (direct reaction kinetic constant k_2, inverse reaction kinetic constant k_{-2}) to finally give E plus P_1 and P_2 passing through two different EP_1P_2 decomposition paths (direct reactions kinetic constants k_3, k_4, k_5 and k_6; inverse reactions kinetic constants k_{-3}, k_{-4}, k_{-5} and k_{-6}) (Fig. 2), (Grassi et al. 2002). Unfortunately, this minimal kinetic pathway does not always reflects the real situation as alternate structures of E, S and ES (named, respectively, E_1, S_1, ES_1 and ES_2) (Fig. 2) can occur in solution (Stage-Zimmermann and Uhlenbeck 1998). In order to take into account the presence of these alternate structures, the simple reaction scheme needs to comprehend other reversible reactions accounting for E, S and ES conversion into possible alternate forms (E_1, S_1, ES_1 and ES_2, respectively) and characterized by proper kinetics constants (k_{E1}, k_{E-1}, k_{S1}, k_{S-1}, k_{ES1}, k_{ES-1}, k_{ES2} and k_{ES-2}). In this regard, we have developed a general model (Grassi et al. 2002) to describe HRz kinetic reaction. The time variation of E, S, S_1, ES, ES_1 and ES_2 can be theoretically determined by means of the following set of differential equations:

$$\frac{dE}{dt} = -k_{E_1} E + k_{E-1} E_1 - k_1 ES + k_{-1} ES_2 + k_3 EP_1P_2 - k_{-3} EP, \tag{1}$$

$$\frac{dS}{dt} = -k_{s_1} S + k_{s-1} S_1 - k_1 ES + k_{-1} ES_2, \tag{2}$$

$$\frac{dE_1}{dt} = k_{E_1} E - k_{E-1} E_1, \tag{3}$$

$$\frac{dS_1}{dt} = k_{s_1} S - k_{s-1} S_1, \tag{4}$$

$$\frac{dES}{dt} = k_{ES_2} ES_2 - (k_{ES_{-2}} + k_{ES_1} + k_2) ES + k_{ES-1} ES_1 + k_{-2} EP_1P_2, \tag{5}$$

$$\frac{dES_1}{dt} = k_{ES_1} ES - k_{ES-1} ES_1, \tag{6}$$

$$\frac{dES_2}{dt} = k_1 ES - (k_{-1} + k_{ES_2}) ES_2 + k_{ES_{-2}} ES, \tag{7}$$

$$P = S_0 - (S + S_1) - (E_0 - (E + E_1)), (P = P_1 + P_2), \tag{8}$$

$$EP_1P_2 = E_0 - E - E_1 - ES - ES_1 - ES_2, \tag{9}$$

where t is time while E_0 and S_0 are, respectively, the initial HRz and substrate molar concentrations. While (1)–(7) represent kinetics equations, (8), (9) are expressions of species mass balances, made up on the whole reaction volume and regard P (sum of P_1 and P_2) and EP_1P_2. This mathematical model assumes, for the sake of simplicity, that the EP_1P_2 conversion into E, P_1 and P_2 depends only kinetic constants k_3 and k_{-3} (this means that $k_5 = k_6 = k_{-5} = k_{-6} = k_{-4} = 0$ and k_3, k_{-3} are $\ll k_4$). Due to the non-linear character of this system of differential equations, a fifth order adaptive step size Runge Kutta method is used to get the numerical solution (Press et al. 1992). In the absence of alternate conformations, assuming $E_0 = 10\,\text{nM}$, $S_0 = 1\,\text{nM}$ and typical values for the kinetics constants (Grassi et al. 2002) [$k_1 = 0.05\ (\text{nM min})^{-1}$, $k_{-1} = 0.06\,\text{min}^{-1}$, $k_2 = 0.01\,\text{min}^{-1}$, $k_{-2} = 0.0001\,\text{min}^{-1}$, $k_3 = 55\,\text{min}^{-1}$, $k_{-3} = 0.05\ (\text{nM min})^{-1}$] product molar concentration ($P = P_1 + P_2$) follows, approximately, an exponential increase (see Fig. 3a, solid thick line). When, on the contrary, alternate conformations occur, P trend can be highly modified. For example, in the case of an irreversible conversion into alternate conformations occurring "*off*" the kinetic pathway (E_1, S_1 and ES_1), a complete S conversion can not take place as shown in Fig. 3a,c (dashed line) in the case of ES conversion into ES_1 ($k_{\text{ES1}} = 0.02\,\text{min}^{-1}$ and $k_{\text{ES}-1} = 0$; E_1 and S_1 absent) and S conversion into S_1 ($k_{\text{S1}} = 0.2\,\text{min}^{-1}$ and $k_{\text{S}-1} = 0$; ES_1 and E_1 absent). Interestingly, if an ES fast conversion into ES_1 ($k_{\text{ES1}} = 0.2\,\text{min}^{-1}$ ($k_{\text{ES1}} > S_0k_1$), $k_{\text{ES}-1} = 0.05\,\text{min}^{-1}$) slows down P increase (see Fig. 3a, solid thick line), an ES slow conversion into ES_1 [$k_{\text{ES1}} = 0.02\,\text{min}^{-1}$ ($k_{\text{ES1}} < S_0k_1$), $k_{\text{ES}-1} = 0.005\,\text{min}^{-1}$]

leads to a biphasic behavior characterized by an initial fast P increase followed by a slower P increase (see Fig. 3a, dotted line). The presence of ES_2 confers to P increase a sigmoidal character that is more pronounced in the case of slow conversion (see Fig. 3b: slow conversion, dotted line, $k_{ES2} = 0.02\,\text{min}^{-1}$, $k_{ES-2} = 0.005\,\text{min}^{-1}$; fast conversion, dotted line, $k_{ES2} = 0.2\,\text{min}^{-1}$, $k_{ES-2} = 0.05\,\text{min}^{-1}$). Our general model (1)–(9) predicts that the smaller the ES_2 conversion to ES, the lower the product concentration increases. Finally, the presence of S_1 reflects into a reduction of P formation kinetics as witnessed by Fig. 3c. Interestingly, in this case, the difference between fast (thin solid line, Fig. 2c; $k_{S1} = 2\,\text{min}^{-1}$, $k_{S-1} = 0.5\,\text{min}^{-1}$) and slow (dotted line, Fig. 3c; $k_{S1} = 0.2\,\text{min}^{-1}$, $k_{S-1} = 0.05\,\text{min}^{-1}$) conversion is neatly less evident than in previous cases (ES_1, ES_2).

These considerations lead to the conclusion that if a biphasic behavior is usually due to the presence of ES_1, a simple P lowering may be due to the presence of S_1, E_1 and ES_2. Obviously, the situation becomes more difficult if more than one alternate conformations exist contemporarily.

Despite the complexity depicted above, in the absence of alternate conformations, HRz kinetics path can be considerably simplified assuming that the kinetic constant k_{-2} is negligible in comparison with k_2 (Birikh et al. 1997a,b). This implies an ES conversion to EP_1P_2 ruled only by k_2 and that EP_1P_2 conversion to E, P_1 and P_2 develops instantaneously (*multiple turnover*). Accordingly, kinetics path reduces to the well known Michaelis and Menten mechanism (Michaelis and Menten 1913):

$$E + S \underset{K_{-1}}{\overset{K_1}{\rightleftarrows}} ES \xrightarrow{k_2\,(=k_{cat})} P + E \quad \text{multiple turnover,}$$

where k_2 is usually called k_{cat} and P represents the sum of P_1 and P_2. This reaction scheme is acceptable only for HRz carrying the minimal structural motif and targeted against short RNA substrates (in this case E_0 can be small in comparison to S_0) (Birikh et al. 1997b). On the contrary, when long and structured RNA substrates and/or HRz carrying additional RNA stretches are present, the *multiple turnover* scheme no longer holds. Indeed, in this case, substrate and/or HRz dimensions can yield to a very slow HRz-product (EP_1P_2) conversion to free HRz (E) and product ($P = P_1 + P_2$) in comparison to the ES conversion to EP_1P_2. Accordingly, the reaction product is represented by the complex EP_1P_2 and, consequently, HRz regeneration does not occur. This is the reason why this condition is termed *single turnover*:

$$E + S \underset{K_{-1}}{\overset{K_1}{\rightleftarrows}} ES \xrightarrow{k_2\,(=k_{cat})} P(= EP_1P_2), \text{ single turnover,}$$

In this case, the reaction pathway needs an excess of HRz over the substrate (Birikh et al. 1997b). We (Grassi et al. 2002) demonstrated that when $E_0 > 10S_0$, *single turnover* condition (that is, ultimately, a simplification of the whole reaction scheme shown in Fig. 2 and mathematically represented by equations (1)–(9)) leads to the following product time dependence:

$$P = \frac{S_0}{\alpha_1 - \alpha_2}[\alpha_1(1 - e^{\alpha_2 t}) - \alpha_2(1 - e^{\alpha_1 t})], \tag{10}$$

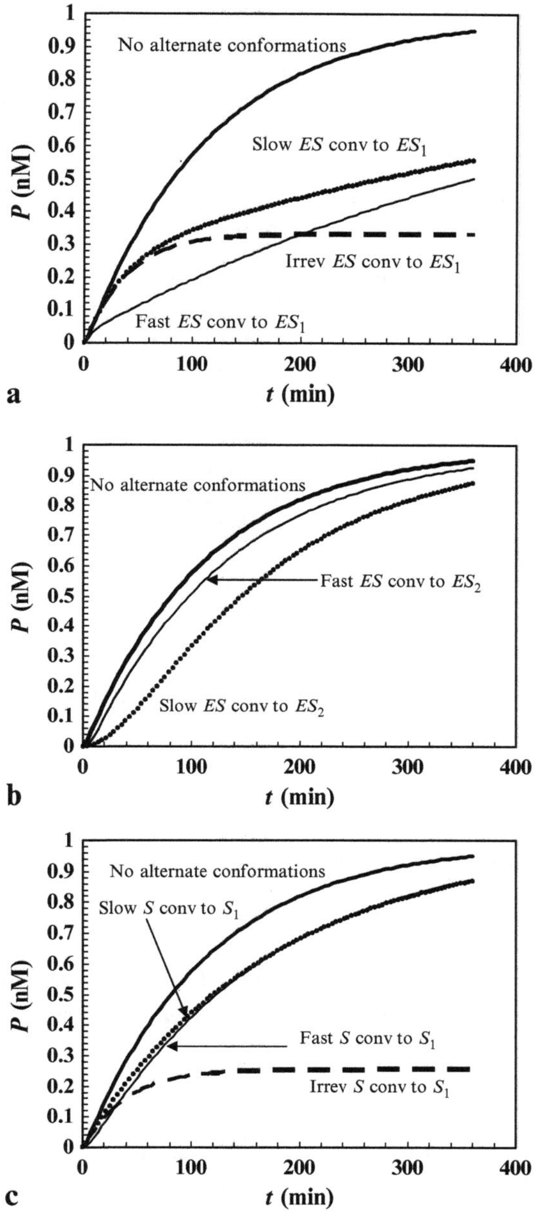

Fig. 3 Product (*P*) formation vs time (*t*) in presence of alternate conformations. **(a)** ES_1, **(b)** ES_2, **(c)** S_1, according to the proposed model (1)–(9). *Solid lines* indicate the absence of alternate conformation, *dotted lines* indicate slow conversion into the alternate conformation, *solid thin lines* indicate fast conversion into the alternate conformation and *dashed lines* indicate irreversible conversion into the alternate conformation

where α_1 and α_2 are constants depending on k_1, k_{-1}, k_{cat}, E_0 and S_0. If, additionally, S conversion to ES is very fast (in presence of long HRz/substrate this in not always the case), equation (10) becomes the usual equation considered for kinetic constants calculation (Birikh et al. 1997a; Heidenreich et al. 1994; Heidenreich and Eckstein 1992)

$$P = S_0\left(1-\exp\left(-\frac{k_{cat}E_0}{K_m+E_0}t\right)\right),\ K_m = \frac{k_{cat}+k_{-1}}{k_1} \tag{11}$$

Due to the particular mathematical nature of equation (10), the determination of the kinetics constants (k_1, k_{-1}, k_{cat}) resorting to a data fitting performed on the time course curve (the experimental P vs t data) is not possible. However, what is biologically relevant is just the determination of k_{cat} (representing the chemical step to form products) and K_m (defined in (11) and representing the affinity of the HRz for the substrate). This can be accomplished according to a double fitting procedure. Equation (10) has to be firstly fitted on each time course curve (P vs t) corresponding to different E_0 concentrations (S_0 constant and $\leq 10E_0$) in order to get α_1 and α_2 dependence on E_0. Consequently, it is possible to build the pseudo-experimental curve R vs E_0, where R is defined by:

$$R = -\frac{\alpha_1\alpha_2}{\alpha_1+\alpha_2}. \tag{12}$$

As it can be demonstrated that the following relation holds (Grassi et al. 2002):

$$R = \frac{k_{cat}(E_0-S_0)}{K_m+(E_0-S_0)}, \tag{12'}$$

k_{cat} and K_m determination comes from equation (12′) fitting on pseudo-experimental data R vs E_0 calculated according to equation (12).

2.4 Calculation of Cleavage Kinetic Constants in the Presence of Promoter-Derived Sequences: An Example

In this paragraph, we show the application of our developed mathematical model (10), (12), (12′) to calculate HRz kinetic constants under single turnover conditions. In particular, we focus on the kinetic constant determination for a HRz directed against cyclin E1. This is a pivotal cell cycle gene whose non-regulated expression is present in many human pathologies including artery restenosis (Grassi et al. 2005). To achieve prolonged biological effects, the cyclin E1-HRz, as all HRzs, needs to be introduced into target cells by specific constructs (such as viral vectors) able to drive their expression from a defined expression cassette (promoter). This strategy implies the presence of promoter-derived sequences bound to the HRz-structure which can variably and unpredictably affect HRz-cleavage kinetic constants. To select the most appropriate expression cassette, sequences derived from different promoters were added to the minimal HRz. The constructs, described in details elsewhere (Platz et al. 2007), are derived from the RNA bacteriophage T7 promoter (T7-L), RNA polymerase I promoter (PoL1 and PoL1-2SL) and RNA polymerase III (PoL3-pGval) (Fig. 4a). T7-L contains 34 and 86 nucleotides before and after the minimal HRz core, respectively. In the 86-nucleotide stretch, a stem loop region, required for polymerase termination, is contained. In the PoL1 construct, the minimal HRz core is preceded and followed by 12 and 18 nucleotides, respectively. PoL1-2SL is preceded and flanked by longer stretches of 39 and 36 nucleotides, respectively. In both stretches, a stem lop region, known to increase stability in the cellular environment, was added. Finally, in the pGval construct, the HRz is placed in a stem-loop region preceded and followed by 138 and 160 nucleotides, respectively. The stem-loop region is thought to confer higher HRz stability in the cellular environment.

HRz cleavage assays were performed under *single turn-over* conditions using a target-RNA concentration (S_0) of 1 nM in the presence of increasing molar ratios of HRz-constructs/target-RNA (E_0) as described in (Grassi et al. 2002). For each different molar ratio (E_0/S_0), the percentage of cleaved target-RNA (P/S_0) was evaluated at different time intervals, after resolving the undigested substrate and products on a 40-cm denaturing gel (3.5% polyacrilamide 7M urea). Substrate- and product-radioactivity was quantified by a Fuji PhosphorImager (Tokyo, Japan). As discussed in paragraph 2.3, K_m and k_{cat} were evaluated on the basis of a double fitting procedure. Figure 4b, showing equation (12′) best fitting (solid line) on pseudo-experimental data (symbols) R vs E_0 for the minimal HRz structure and for the other four variants considered, represents the second step of this procedure. It is evident that HRz cleavage efficiency reduces when different promoter sequences are added to the minimal core. Indeed, the R vs E_0 curve competing to the four variants is always lower than that competing to the minimal core (open circles). This effect becomes more and more evident passing from PoL1 (filled circles) to PoL1-2SL (open triangles), T7-L (filled triangles) and PoL3-pGvall (filled diamonds). This behavior translates in the following k_{cat} and K_m values: minimal core (k_{cat} = (8.8 ± 0.8)10^{-3} min^{-1}, K_m = 25 ± 7 nM), PoL1 (k_{cat} = (7.9 ± 0.6)10^{-3} min^{-1}, K_m = 67 ± 11 nM) PoL1-2SL (k_{cat} = (3.9 ± 0.4) 10^{-3} min^{-1}, K_m = 41 ± 16 nM), T7-L (k_{cat} = (3.1 ± 0.2)10^{-3} min^{-1}, K_m = 104 ± 22 nM) and PoL3-pGvall (k_{cat} = (2.3 ± 0.8) 10^{-3} min^{-1}, K_m = 595 ± 100 nM).

These data show the general negative effects exerted by promoter derived sequences on HRz cleavage reaction, stressing the importance of the proper promoter choice. Moreover, these data indicate the general tendency of an inverse relation between the length of the promoter derived sequences and the cleavage efficiency of HRz-cyclinE1. The phenomenon may be explained with the formation of alternate conformations of the different reaction molecules which eventually affect cleavage kinetic constants. The complex cleavage reaction pathway clearly requires an appropriate mathematical tool to calculate kinetic consents. Our model allows the proper calculation graphically displaying, through the relation between R and E_0, the efficacy of HRzs.

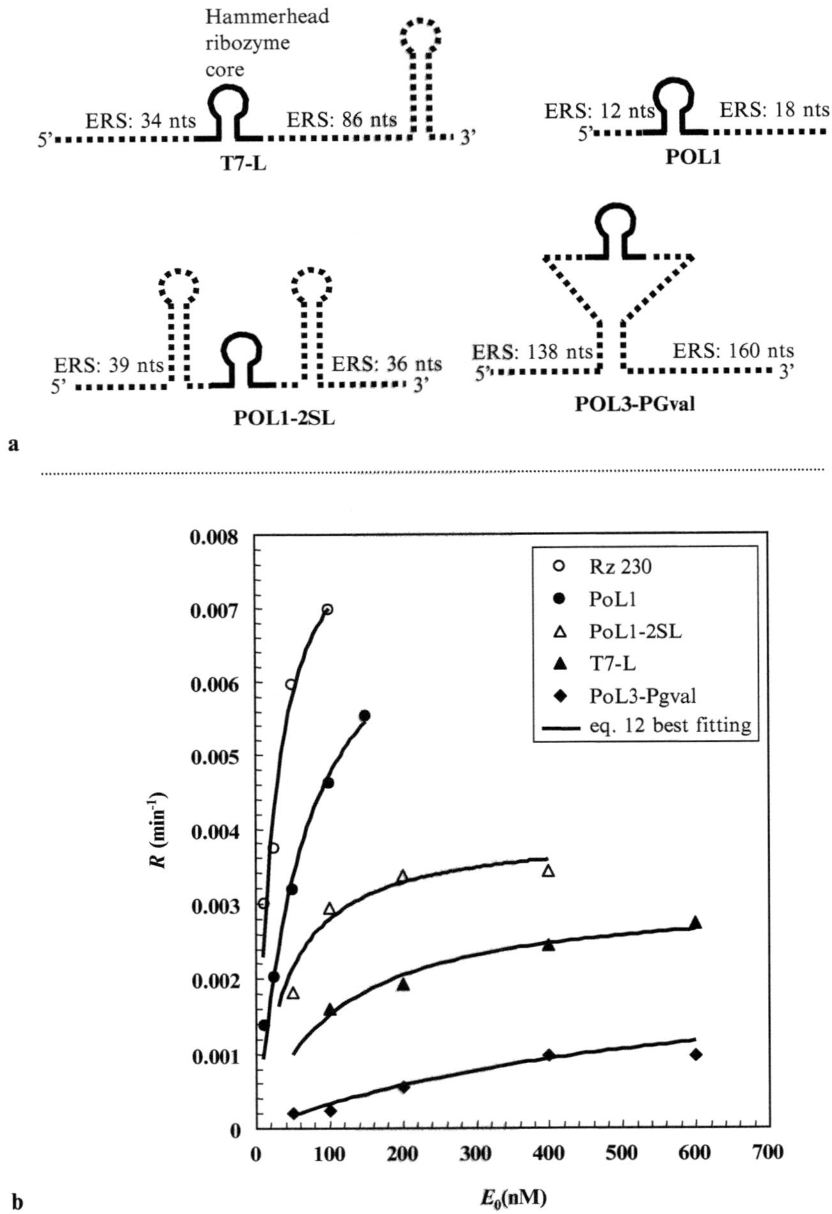

Fig. 4 Structure and cleavage efficiencies of a hammerhead ribozyme bearing different promoter-derived sequences. (**a**) A minimal hammerhead ribozyme core targeted against the mRNA of cyclin E1 has been embedded into sequences derived from the RNA bacteriophage T7 promoter (*T7-L*), the RNA polymerase I promoter (*PolI* and *PolI-2SL*) and RNA polymerase III (*pGval*) promoter; (**b**) the effects on the cleavage efficiency of the different promoter derived sequences are compared to that of the minimal hammerhead ribozyme core in terms of *R* (see equation (12)) vs E_0 (initial ribozyme concentration)

3 Potential Use of HRzs as Therapeutics for Artery Restenosis

The most common cause of small and large artery occlusions (stenosis) is the progressive development of atherosclerosis. When this pathologic process involves coronary arteries, clinical symptoms ranging from angina pectoris to heart attack can occur. In order to revascularize stenotic coronary arteries, since 1979 the so-called percutaneous transluminal coronary angioplasty (PTCA) has been introduced (Gruntzig et al. 1979). This non-surgical method, shown to be safe and effective (Pocock et al. 1995), involves: (1) advancing a balloon catheter to an area of coronary narrowing, (2) inflating the balloon, and (3) retrieving the catheter following balloon deflation (Fig. 5a). More than 500,000 percutaneous coronary intervention procedures are performed yearly in the USA, and about 1 million procedures worldwide (American Heart Association 2001). However, PTCA has been shown to induce (Fig. 5b) the development of symptomatic re-occlusion (restenosis) caused by early elastic recoil, intimal thickening, late constricting remodeling of the vessel (Ruygrok et al. 2003) and formation of mural thrombus in about 30–50% of treated patients (Califf 1995).

To try to overcome the PTCA related problems, the expansion of the balloon during angioplasty has been associated with the deployment of a stent. This is an expandable metal tubular mesh (Fig. 5c) (Sigwart et al. 1987) which has been shown to significantly reduce restenosis rate down to 20–30% (Serruys et al. 1994; Fischman et al. 1994). The partial success of the stents is due to the induction of the intimal thickening (in-stent restenosis, ISR), a phenomenon particularly evident in small caliber vessels (Ruygrok et al. 2003; Moreno et al. 2004). This pathological event is characterized by an exuberant proliferation of vascular smooth muscle cells (VSMCs) which migrate from the tunica media of the artery, where they normally reside in a quiescent state, to the intima artery layer where they synthesize extracellular matrix and give origin to the intimal thickening, also called neointima (Edelman and Rogers 1998; Ferns and Avades 2000). To try to overcome this problem, devices able to locally deliver anti-proliferative drugs such as Sirolimus and Paclitaxel (drug-eluting stent, DES) have been developed. DES have significantly reduced ISR compared to bare metal stents (Moses et al. 2003; Stone et al. 2004), in patients with discrete, de novo lesions in native vessels. However, their revascularization benefit is attenuated in high-risk patients (diabetes; the acute coronary syndromes, including ST-segment elevation myocardial infarction; smaller-diameter lesions and longer lesions; several stents or overlapping stents), compared to low risk patients displaying, for example, 26% ISR rate in coronary bifurcation lesions (Colombo et al. 2004). Moreover, some other concerns such as stent thrombosis (Tung et al. 2006; Serruys and Daemen 2007) are now emerging with regard to the use of DES. These limitations suggest that alternative approaches and/or the selection of novel antiproliferative drugs, such as HRz, may be beneficial for ISR treatment.

Platelet-derived growth factor A (PDGF-A), a known stimulator of VSMC proliferation (Hart et al. 1988), has been directly implicated in the pathogenesis of arterial proliferative diseases (Nilsson et al. 1985) which includes atherosclerosis,

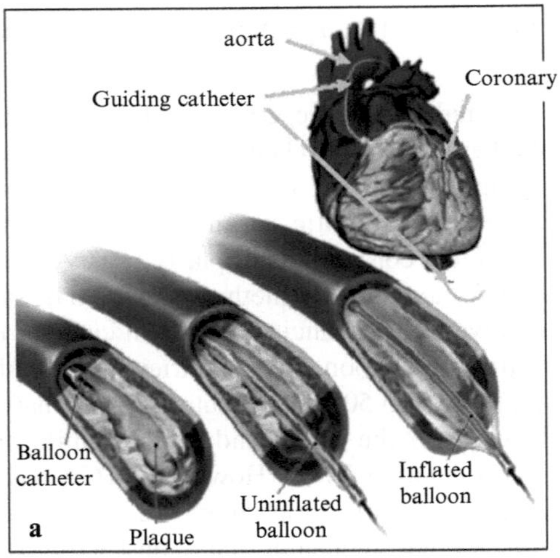

From
http://www.medicinet.com
/images/illustrations/
balloon_angio.jpg

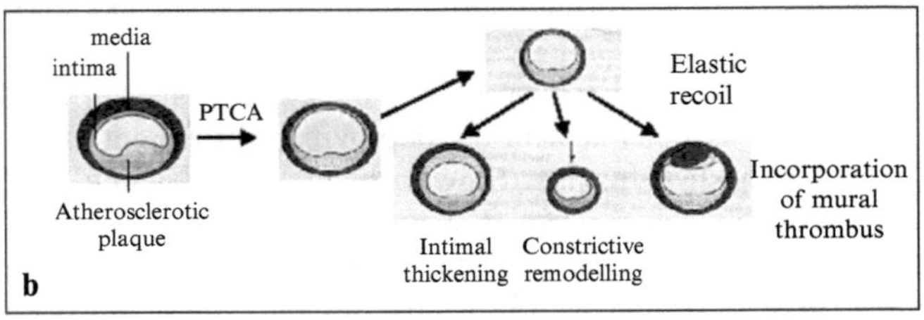

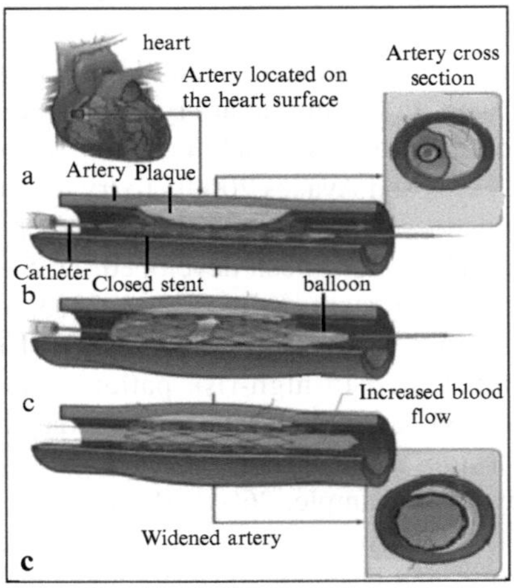

Fig. 5 Coronary revascularization procedure and its complications. (**a**) Percutaneous transluminal coronary angioplasty (*PTCA*) dilates the area of vascular narrowing by means of a balloon catheter inflated in the artery lumen; (**b**) *PTCA* can be followed by early artery elastic recoil, intimal thickening, incorporation of mural thrombus and late constrictive remodeling, all phenomenon which determines the re-occlusion of the treated vessel; (**c**) the implantation of endovascular implants (*stent*) following PTCA efficiently prevent the early artery elastic recoil and late constrictive remodeling. However, stents trigger a complex series of patho-biological events, among which the exuberant proliferation of vascular smooth muscle cells, which lead to the formation of a neo-intima with consequent reduction of the artery lumen

hypertension and artery restenosis. Thus, HRz-based approaches against PDGF-A have been explored. A PDGF-A targeted HRz has been shown to be successful in inhibiting human VSMC proliferation in vitro (Hu et al. 2001b). Notably, the authors have used for their experiments VSMC isolated from a spontaneously hypertensive rat strain (SHR) which are characterized by an higher growth rate compared to cells isolated from normotensive rats. This indicates that the HRz approach followed might also be effective in the presence of increased VSMC proliferation rate, an un-favorable condition typically found in hypertensive and diabetic patients. Finally, the authors could show a good correlation between the in vitro (cell-free environment) HRz cleavage efficacy and that observed in cultured rat VSMCs (ex vivo). This observation shows the good predictive power of HRz cell free analysis for ex vivo/in vivo applications.

A chimeric DNA-RNA HRz targeted against PDGF-A mRNA induced, in a rat model, about 45% reduction in the ratio between neointima/media (Kotani et al. 2003), a measurement commonly chosen to evaluate the efficacy of anti restenotic treatment. Interestingly, in this work, the authors also investigated, by microarray analysis, the gene expression pattern in vessel treated by the anti-PDGF-A HRz. As a result of this analysis, they could show that only the expression of PDGF-A was completely inhibited, thus indicating a high specific action of the selected HRz. The authors also observed a substantial reduction in the expression of some cell cycle related genes such as Cdc-2-related protein kinase, cyclin D3, cyclin D1 and cyclin B. This reduction was interpreted as a consequence of the inhibition of PDGF-A and might have contributed to the reduced intima hyperplasia. However, some other cell cycle genes such as cyclin D2 and cdk4 kinase were upregulated, possibly suggesting a compensative reaction of VSMCs to the anti-proliferative action of the HRz. This last observation may contribute to explain why the inhibition of intimal growth did not reach 100%.

Significant but partial success in inhibiting the neo-intimal formation was also achieved by expressing an anti-PDGF-A HRz from an adenoviral vector (Lin et al. 2004). In this case, the intima/media ratio decreased down to 30% of control in a rat model of restenosis. Comparable degree of reduction was achieved in a rat model (Yamamoto et al. 2000) by an HRz targeted against the mRNA of transforming growth factor β 1 (TGFβ-1). TGFβ-1 has been chosen as target due to its involvement in the pathogenesis of artery restenosis (Nikol et al. 1992). The expression of this protein, typically increased in human vascular restenosis lesions, has been shown to have mitogenic effects. Moreover, it plays a pivotal role in the synthesis of extracellular matrix protein such as collagen, laminin and fibronectin by a variety of cells, which all contribute to the thickening of the neointima. The anti-restenotic effect of targeting TGFβ-1 was also confirmed in another study where a chimeric DNA-RNA HRz significantly decreased VSMC proliferation in vitro and reduced the intima/media ratio down to 30% of control in the rat (Ando et al. 2004).

Despite the significant efficacies of the tested HRzs, complete prevention of neointima formation was not achieved. Among the several variables which can explain this observation, one may be represented by the choice of the target gene.

It is possible that the blockade of a single pathway, i.e., PDGF-A or TGFβ-1 cascade, is not sufficient to prevent VSMC proliferation because of the existence of alternative/ redundant signaling pathways. In this regard, a PDGF-A targeted HRz successfully reduced the proliferation of VSMC stimulated by $TGF\beta_1$, but only modestly, after angiotensin II stimulation (Hu et al. 2001a).

The inhibition of genes shared by multiple pathways and directly involved in the control of cell cycle progression was also explored. A chimeric DNA-RNA HRz against the mRNA of the proliferating cell nuclear antigen (PCNA) achieved a reduction of about 30% in neo-intimal thickening compared to controls, in a pig model of artery restenosis (Frimerman et al. 1999). PCNA was chosen as target as it is a cofactor for DNA polymerase and it is required for DNA synthesis and S-phase progression (Fairman 1990). Moreover, it forms complexes with different cyclins and cyclin-dependent kinases, pivotal regulators of cell cycle progression. Notably, the authors evaluated the anti-restenotic effect of the PCNA-HRz in the presence of a stent. This experimental design is particular relevant as it perfectly matches the condition of the common clinical practice where PTCA is followed by stent implantation. The reported results are also particularly interesting considering that the delivery system used (cationic liposomes) are not ideal for in vivo delivery. It is reasonable to assume that optimized delivery systems (i.e., polymeric mediated delivery) may further improve the efficacy.

Another cell cycle related gene targeted by HRz is the proto-oncogene *c-myb*. High expression level of this protein has been correlated to increased cell proliferation rates. Conversely, reduced *c-myb* expression has been associated with cell differentiation. Moreover, in VSMCs, the expression level of *c-myb* has been directly correlated to cell proliferation rate (Brown et al. 1992). In rat cultured VSMC, a HRz targeted against the mRNA of *c-myb* (Jarvis et al. 1996a,b), decreased cell proliferation down to 14% of controls in vitro and reduced the neo-intimal/media ratio down to 50% of controls in a rat model (Macejak et al. 1999). In this last case, the authors delivered the HRz by an adenoviral vector, a commonly used strategy to transfer genes for gene therapy purposes. The reason for the partial success of this approach can, at least in part, depend on the fact that the adenoviral vector may not represent the optimal delivery system. In vitro it has been show that adenoviral vector can efficiently transduce both VSMC and endothelial cells (Grassi et al. 2006b). This implies that the anti-proliferative HRz can be up-taken also by the endothelium with consequent retardation in endothelium growth, i.e., in the so-called re-endothelization of the injured artery, a fact known to favor restenosis (Serruys and Daemen 2007).

We proposed a novel anti-proliferative approach based on the specific knockdown of two pivotal cell cycle promoting genes which regulate G1/S transition of the cell cycle, i.e., cyclin E and the transcription factor E2F1 (Dyson 1998; Ohtsubo et al. 1995). The inhibition of the expression of these genes is particularly attractive because of the existence of a feed-forward loop between them which amplifies the G1 to S phase promoting signals (Geng et al. 1996). Therefore, the downmodulation of the interaction between these two genes might augment the proliferation inhibition effect. Additionally, the fact that they are synthesized de novo at each new G1 phase and then rapidly degraded at the protein and mRNA levels with the progression of the cell cycle, implies that no active protein and mRNA is left for the next cycle. Thus, their knockdown during G1 phase may efficiently reduce the intracellular level preventing their biological functions. Finally, the observation (O'Sullivan et al. 2003) that, compared to normal VSMCs, cyclin E is over-expressed in in-stent restenotic VSMCs, points towards the central role of cyclin E in sustaining VSMC proliferation, and makes it an ideal target to prevent in-stent stenosis. In cultured human VSMC (Grassi et al. 2001, 2005), we observed a dose- and time-dependent decrease in the amount of S phase cells which, 2 days after transfection and at HRz concentration of 420 nM, corresponded to one-fifth/one-tenth of controls. Moreover, the concomitant targeting of cyclin E and E2F1 resulted in a more pronounced inhibition of VSMC proliferation compared to the independent targeting of each of the two genes, at the same HRz concentration.

Alternative approaches to downregulate VSMC include the promotion of cell death rate, instead of the reduction of cell proliferation. Over-expression of Bcl-2 protein, first identified as an oncogene, inhibits apoptosis induced by a variety of circumstances including growth factor withdrawal, DNA damage and conflicting sub-cellular signal events. Thus, the downregulation of its expression may in principle be useful to favor VSMC death diminishing their accumulation within the neointima. In this regard, a HRz targeted against the mRNA of the Bcl-2 greatly increased apoptotic cell rate compared to controls in cultured rat cells (Perlman et al. 2000). In vivo, the neo-intimal/media ratio was reduced to about of 50% of controls with a concomitant reduction of the cell number within arterial media layer. Despite the effectiveness of this approach, it cannot be excluded that it might induce a weakening of the arterial wall possibly resulting in the formation of secondary artery lesions. A tight regulation of the expression of the pro-apoptotic HRz should be therefore be considered in order to keep an appropriate apoptotic rate.

The concomitant prevention of different patho-biological events which concur to the development of artery re-occlusion, such as VSMC proliferation, migration and deposition of extracellular matrix components, has also been explored. An attractive target is represented by the leukocyte-type 12-lipoxygenase (12-LOX) mRNA whose protein mediates the stimulatory growth effect of Ang II and the chemotactic effect of platelet derived growth factor (PDGF)-BB on VSMCs. A chimeric DNA-RNA HRz successfully downmodulated leukocyte 12-LOX expression in cultured porcine VSMC (Gu et al. 1995). Moreover, an analogous HRz, significantly reduced, in cultured rat VSMC, the PDGF-induced cell migration and deposition of fibronectin, a key matrix protein (Gu et al. 2001). The same HRz decreased the neo-intimal/media ratio to 40% of controls in a rat carotid artery model of restenosis. It should be noted that 12-LOX is also expressed, in addition to VSMC, in monocytes/macrophages, and that 12-LOX products mediate monocytes adhesion to endothelial cells. This implies that the anti-12-LOX HRz may have not only downregulated VSMCs proliferation/migration but also the adhesion of inflammatory cells (monocytes) to endothelial cells. In turn, this may have reduced, in the injured area, the inflammation, a relevant factor which promotes and sustain artery restenosis.

Whereas all the different HRzs described here efficiently reduced VSMC proliferation ex vivo, in vivo (animal models) their effectiveness was reduced. A relevant contribution to this discrepancy may depend on the sub-optimal delivery systems available so far. The major challenge for the future will be the identification of efficient and controllable delivery systems for local HRz release. Moreover, an optimized and standardized HRz release system will allow the comparison of the efficacies of different HRzs and will contribute to select the best candidate genes to knock out.

4 HRzs and Small Interfering RNAs in the Cardiovascular Field

The potential applications of HRz in the cardiovascular field are not limited to artery-restenosis since other vascular and non-vascular pathologies might benefit from a HRz-based approach. A HRz has been successfully directed against the mRNA of the endothelial 12/15 lipoxygenase (12/15LO), a protein which favors monocytes binding to endothelial cells (Hatley et al. 2003). This is considered an early inflammatory event which may contribute to vessel damage, especially in diabetics. It follows that the downregulation of 12/15LO expression by HRzs may be of therapeutic benefit, as this strategy would reduce monocyte adhesion to vascular endothelial cells thus attenuating inflammation-dependent damage of the vessel. These findings confirm a previous report (see above; Gu et al. 2001) of anti-restenotic and anti-inflammatory effects achieved by downregulating the expression of 12LO. Another HRz has been directed against the tissue factor (TF), a glycoprotein bound to the membrane of vascular cells that initiates the clotting cascade (Cavusoglu et al. 2002) causing vessel occlusion with dangerous ischemic consequences. The reduction of the TF mRNA levels in cultured vascular smooth muscle cells by HRzs indicated that it is potentially feasible to think of an anti-thrombotic strategy based on the reduction of the expression levels of the pro-thrombotic protein TF. Another potentially interesting application of HRzs deals with their ability to downmodulate the expression of the mannose 6-phosphate/insulin-like growth factor-II receptor (M6P/IGF2R) in cardiomyocytes (Chen et al. 2004). M6P/IGF2R is a multi-functional protein implicated in the regulation of cardiomyocytes growth and apoptosis. The regulation of its expression levels may thus represent a novel approach to favor cardiomyocytes survival, essential for the treatment of myocardial pathological conditions such as myocardial infarction and heart failure. Finally, HRzs have also been used to knockdown gene expression as a strategy to understand the contribution of a given gene product in cardiac-related biochemical pathways (Montagnani et al. 2002; Wang et al. 2002; Katsuyama et al. 2002).

Up to the end of the last century, HRzs represented one of the most commonly used nucleic acid molecules potentially able to down-regulate gene expression in the cardiovascular system as well as in other human organs. From the year 2000 on, a novel class of nucleic acid molecules named small interfering RNAs (siRNAs) become more and more popular (see Agostini et al. 2006 for a review) to knock-down gene expression. siRNA are short double-stranded RNA molecules able to drive, in a sequence-specific manner, the degradation of virtually any RNA when bound to a cellular protein complex named RNA-induced silencing complex (RISC). The specificity of action is given by the antisense strand, present in the RNA duplex, which directs RISC to the complementary sequence on the target RNA thus triggering its catalysis. Despite the relatively recent identification of siRNAs, several potential applications as therapeutics for different human pathological conditions have been shown (Scherer and Rossi 2004). With regard to the cardiovascular field (Agostini et al. 2006), three main research topics can be identified: (1) studies devoted to ameliorate and to understand the mechanisms regulating cardiomyocytes functionality; (2) studies aimed at the prevention of viral myocarditis; and (3) studies dealing with vascular pathologies. In the first case, studies were focused on the development of novel approaches to treat heart failure and to ameliorate cardiac damages following hypoxia-ischemia. In the second case, attention was put on the attenuation of cardiac infection sustained by coxsackieviruses of group B, the most relevant etiological cause of viral myocarditis. Finally, investigations related to the vascular system focused on the identification of novel approaches to treat atherosclerosis, both by interfering with the atherosclerotic process itself and by improving the outcome of the technique used to re-vascularize vessels occluded by atherosclerotic plaques.

Despite the different mechanism of action, both HRzs and siRNA action results in the destruction of the target RNA in a sequence-specific fashion. This implies that, in principle, the use of one or the other nucleic acid molecule for gene expression knockdown purposes is equivalent. So far, however, it is not completely clear which, among the two kind of molecule, is the most convenient. Few studies have been conducted to compare the knockdown efficiency of HRzs and siRNAs (Miyagishi et al. 2003) targeted against the same substrate RNA. We have compared, in VSMCs, the anti-proliferative effects of two HRzs (Grassi et al. 2005) and two siRNAs (Dapas et al. 2007) all directed against two pivotal cell cycle promoting genes, i.e., cyclin E1 and the transcription factor E2F1. Under optimized conditions, we did not detect significant differences between the anti-proliferative effects of HRz (evaluated 2 days after treatment at a concentration of 220 nM) and siRNA (evaluated 3 days after treatment at a concentration of 220 nM). In progress is the evaluation of the reduction of the intracellular levels of the target mRNAs induced by HRzs and siRNAs. The reported examples are obviously not enough to draw final conclusions about the efficacy of HRs and siRNAs. Despite this fact, it is becoming evident that the selection of efficient siRNAs represents an easier task compared to the selection of efficient HRzs. Computer-based approaches to select active siRNA (Mercatanti et al. 2002) are now highly efficient (more than 75%

success). In contrast, computer-based approaches to select active HRzs are far less reliable and prolonged experimental work is required to select active molecules.

5 Conclusion

HRzs base the biological effect on the sequence-specific recognition of the target RNA leading to its destruction. Thus, they can be used to reduce the intracellular level of a specific RNA coding for a protein which affects cellular metabolism or environment causing disease. The data presented indicate that they have the potential to become of therapeutic value for human pathologies such as artery restenosis. However, despite the encouraging results achieved, HRz technology can and should be optimized to bring it closer to clinical practice.

Acknowledgments We wish to thank the "Fondazione Cassa di Risparmio of Trieste," "Fondazione Sostegno delle Strutture Cardiovascolari, Mirano, Venezia," "Fondazione Benefica Kathleen-Foreman Casali of Trieste and Fondo Trieste 2006.

References

Agostini F, Dapas B, Farra R et al (2006) Potential application of small interfering RNAs in the cardiovascular field. Drug Future 31:513–525

American Heart Association (2001) Heart and stroke statistical update

Ando H, Fukuda N, Kotani M et al (2004) Chimeric DNA–RNA hammerhead ribozyme targeting transforming growth factor-beta 1 mRNA inhibits neointima formation in rat carotid artery after balloon injury. Eur J Pharmacol 483:207–214

Balicki D, Beutler E (2002) Gene therapy of human disease. Medicine (Baltimore) 81:69–86

Bertrand E, Castanotto D, Zhou C et al (1997) The expression cassette determines the functional activity of ribozymes in mammalian cells by controlling their intracellular localization. RNA 3:75–88

Birikh KR, Berlin YA, Soreq H et al (1997a) Probing accessible sites for ribozymes on human acetylcholinesterase. RNA 3:429–437

Birikh KR, Heaton PA, Eckstein F (1997b) The structure, function and application of the hammerhead ribozyme. Eur J Biochem 245:1–16

Brown KE, Kindy MS, Sonenshein GE (1992) Expression of the c-myb proto-oncogene in bovine vascular smooth muscle cells. J Biol Chem 267:4625–4630

Cairns MJ, Hopkins TM, Witherington C et al (1999) Target site selection for an RNA-cleaving catalytic DNA. Nat Biotechnol 17:480–486

Califf RM (1995) Restenosis: the cost to society. Am Heart J 130:680–684

Cavusoglu E, Chen I, Rappaport J et al (2002) Inhibition of tissue factor gene induction and activity using a hairpin ribozyme. Circulation 105:2282–2287

Chen Z, Ge Y, Kang J (2004) Down-regulation of the M6P/IGF-II receptor increases cell proliferation and reduces apoptosis in neonatal rat cardiac myocytes. BMC Cell Biol 5

Colombo A, Moses JW, Morice MC, Ludwig J, Holmes DR Jr, Spanos V, Louvard Y, Desmedt B, Di Mario C, Leon MB (2004) Randomized study to evaluate sirolimus-eluting stents implanted at coronary bifurcation lesions. Circulation 109:1244–1249

Dahm SC, Uhlenbeck OC (1991) Role of divalent metal ions in the hammerhead RNA cleavage reaction. Biochemistry 30:9464–9469

Dapas B, Farra R, Giansante C et al (2007) Cyclin E1, E2 and E2F1 targeted siRNAs down regulate the growth of coronary smooth muscle cell, thus showing their potential use in the prevention of in-stent restenosis. Proceedings of the Ireland Society of Gene and Cellular Therapy Conference, Galway, 18 May 2007

Dyson N (1998) The regulation of E2F by pRB-family proteins. Genes Dev 12:2245–2262

Edelman ER, Rogers C (1998) Pathobiologic responses to stenting. Am J Cardiol 81: 4E–6E

Fairman MP (1990) DNA polymerase delta/PCNA: actions and interactions. J Cell Sci 95(Pt 1):1–4

Ferns GA, Avades TY (2000) The mechanisms of coronary restenosis: insights from experimental models. Int J Exp Pathol 81:63–88

Fischman DL, Leon MB, Baim DS et al (1994) A randomized comparison of coronary-stent placement and balloon angioplasty in the treatment of coronary artery disease. Stent Restenosis Study Investigators. N Engl J Med 331:496–501

Frimerman A, Welch PJ, Jin X et al (1999) Chimeric DNA–RNA hammerhead ribozyme to proliferating cell nuclear antigen reduces stent-induced stenosis in a porcine coronary model. Circulation 99:697–703

Geng Y, Eaton EN, Picon M et al (1996) Regulation of cyclin E transcription by E2Fs and retinoblastoma protein. Oncogene 12:1173–1180

Ghosh S, Parvez MK, Banerjee K et al (2002) Baculovirus as mammalian cell expression vector for gene therapy: an emerging strategy. Mol Ther 6:5–11

Grassi G, Marini JC (1996) Ribozymes: structure, function, and potential therapy for dominant genetic disorders. Ann Med 28:499–510

Grassi G, Grassi M, Platz J et al (2001) Selection and characterization of active hammerhead ribozymes targeted against cyclin E and E2F1 full-length mRNA. Antisense Nucleic Acid Drug Dev 11:271–287

Grassi G, Grassi M, Kuhn A et al (2002) Determination of hammerhead ribozyme kinetic constants at high molar ratio ribozyme-substrate. J Math Biol 45:261–277

Grassi G, Dawson P, Guarnieri G et al (2004) Therapeutic potential of hammerhead ribozymes in the treatment of hyper-proliferative diseases. Curr Pharm Biotechnol 5:369–386

Grassi G, Schneider A, Engel S et al (2005) Hammerhead ribozymes targeted against cyclin E and E2F1 co-operate to down regulate coronary smooth muscle cells proliferation. J Gene Med 7:1223–1234

Grassi G, Crevatin A, Farra R et al (2006a) Rheological properties of aqueous Pluronic-alginate systems containing liposomes. J Colloid Interface Sci 301:282–290

Grassi G, Kohn H, Dapas B et al (2006b) Comparison between recombinant baculo- and adenoviral-vectors as transfer system in cardiovascular cells. Arch Virol 151:255–271

Gruntzig AR, Senning A, Siegenthaler WE (1979) Nonoperative dilatation of coronary-artery stenosis: percutaneous transluminal coronary angioplasty. N Engl J Med 301:61–68

Gu JL, Veerapanane D, Rossi J et al (1995) Ribozyme-mediated inhibition of expression of leukocyte-type 12-lipoxygenase in porcine aortic vascular smooth muscle cells. Circ Res 77:14–20

Gu JL, Pei H, Thomas L et al (2001) Ribozyme-mediated inhibition of rat leukocyte-type 12-lipoxygenase prevents intimal hyperplasia in balloon-injured rat carotid arteries. Circulation 103:1446–1452

Hart CE, Forstrom JW, Kelly JD et al (1988) Two classes of PDGF receptor recognize different isoforms of PDGF. Science 240:1529–1531

Haseloff J, Gerlach WL (1988) Simple RNA enzymes with new and highly specific endoribonuclease activities. Nature 334:585–591

Hatley ME, Srinivasan S, Reilly KB et al (2003) Increased production of 12/15 lipoxygenase eicosanoids accelerates monocyte/endothelial interactions in diabetic db/db mice. J Biol Chem 278:25369–25375

Heidenreich O, Eckstein F (1992) Hammerhead ribozyme-mediated cleavage of the long terminal repeat RNA of human immunodeficiency virus type 1. J Biol Chem 267:1904–1909

Heidenreich O, Benseler F, Fahrenholz A et al (1994) High activity and stability of hammerhead ribozymes containing 2′-modified pyrimidine nucleosides and phosphorothioates. J Biol Chem 269:2131–2138

Hu WY, Fukuda N, Kishioka H et al (2001a) Hammerhead ribozyme targeting human platelet-derived growth factor A-chain mRNA inhibited the proliferation of human vascular smooth muscle cells. Atherosclerosis 158:321–329

Hu WY, Fukuda N, Nakayama M et al (2001b) Inhibition of vascular smooth muscle cell proliferation by DNA-RNA chimeric hammerhead ribozyme targeting to rat platelet-derived growth factor A-chain mRNA. J Hypertension 19:203–212

Jarvis TC, Alby LJ, Beaudry AA et al (1996a) Inhibition of vascular smooth muscle cell proliferation by ribozymes that cleave c-myb mRNA. RNA 2:419–428

Jarvis TC, Wincott FE, Alby LJ et al (1996b) Optimizing the cell efficacy of synthetic ribozymes. Site selection and chemical modifications of ribozymes targeting the proto-oncogene c-myb. J Biol Chem 271:29107–29112

Katsuyama M, Fan C, Yabe-Nishimura C (2002) NADPH oxidase is involved in prostaglandin F2alpha-induced hypertrophy of vascular smooth muscle cells: induction of NOX1 by PGF2alpha. J Biol Chem 277:13438–13442

Kost TA, Condreay JP (2002) Recombinant baculoviruses as mammalian cell gene-delivery vectors. Trends Biotechnol 20:173–180

Kotani M, Fukuda N, Ando H et al (2003) Chimeric DNA–RNA hammerhead ribozyme targeting PDGF A-chain mRNA specifically inhibits neointima formation in rat carotid artery after balloon injury. Cardiovasc Res 57:265–276

Kruger K, Grabowski PJ, Zaug AJ et al (1982) Self-splicing RNA: autoexcision and autocyclization of the ribosomal RNA intervening sequence of Tetrahymena. Cell 31:147–157

Lin ZH, Fukuda N, Suzuki R et al (2004) Adenovirus-encoded hammerhead ribozyme to PDGF A-chain mRNA inhibits neointima formation after arterial injury. J Vasc Res 41:305–313

Macejak DG, Lin H, Webb S, Chase J, Jensen K, Jarvis TC, Leiden JM, Couture L (1999) Adenovirus-mediated expression of a ribozyme to c-myb mRNA inhibits smooth muscle cell proliferation and neointima formation in vivo. J Virol 73:7745–7751

Mercatanti A, Rainaldi G, Mariani L et al (2002) A method for prediction of accessible sites on an mRNA sequence for target selection of hammerhead ribozymes. J Computation Biol 9:641–653

Michaelis L, Menten M (1913) Die Kinetic der invertinwirkung. Biochem Z 49:333–369

Miyagishi M, Hayashi M, Taira K (2003) Comparison of the suppressive effects of antisense oligonucleotides and siRNAs directed against the same targets in mammalian cells. Antisense Nucleic Acid Drug Dev 13:1–7

Montagnani M, Ravichandran LV, Chen H et al (2002) Insulin receptor substrate-1 and phosphoinositide-dependent kinase-1 are required for insulin-stimulated production of nitric oxide in endothelial cells. Mol Endocrinol 16:1931–1942

Moreno R, Fernandez C, Alfonso F et al (2004) Coronary stenting versus balloon angioplasty in small vessels: a meta-analysis from 11 randomized studies. J Am Coll Cardiol 43:1964–1972

Moses JW, Leon MB, Popma JJ et al (2003) Sirolimus-eluting stents versus standard stents in patients with stenosis in a native coronary artery. N Engl J Med 349:1315–1323

Nikol S, Isner JM, Pickering JG, Kearney M, Leclerc G, Weir L (1992) Expression of transforming growth factor-beta 1 is increased in human vascular restenosis lesions. J Clin Invest 90:1582–1592

Nilsson J, Sjolund M, Palmberg L et al (1985) Arterial smooth muscle cells in primary culture produce a platelet-derived growth factor-like protein. Proc Natl Acad Sci USA 82:4418–4422

O'Sullivan M, Scott SD, McCarthy N et al (2003) Differential cyclin E expression in human in-stent stenosis smooth muscle cells identifies targets for selective anti-restenosis therapy. Cardiovasc Res 60:673–683

Ohtsubo M, Theodoras AM, Schumacher J et al (1995) Human cyclin E, a nuclear protein essential for the G1-to-S phase transition. Mol Cell Biol 15:2612–2624
Perlman H, Sata M, Krasinski K et al (2000) Adenovirus-encoded hammerhead ribozyme to Bcl-2 inhibits neointimal hyperplasia and induces vascular smooth muscle cell apoptosis. Cardiovasc Res 45:570–578
Platz J, Grassi M, Kuhn A et al (2007) Effects of various promoter derived sequences on the cleavage kinetic of an hammerhead ribozyme directed against cyclin E1 mRNA. Drug Metabol Lett 1:218–225
Pocock SJ, Henderson RA, Rickards AF et al (1995) Meta-analysis of randomised trials comparing coronary angioplasty with bypass surgery. Lancet 346:1184–1189
Press W, Teukolsky S, Vetterling W et al (1992) Numerical recipes in FORTRAN, 2nd edn. Cambridge University Press, Cambridge, USA
Puerta-Fernandez E, Romero-Lopez C, Barroso-DelJesus A et al (2003) Ribozymes: recent advances in the development of RNA tools. FEMS Microbiol Rev 27:75–97
Rossi JJ (1995) Controlled, targeted, intracellular expression of ribozymes: progress and problems. Trends Biotechnol 13:301–306
Ruygrok PN, Webster MW, Ardill JJ et al (2003) Vessel caliber and restenosis: a prospective clinical and angiographic study of NIR stent deployment in small and large coronary arteries in the same patient. Catheter Cardiovasc Interv 59:165–171
Sarkis C, Serguera C, Petres S et al (2000) Efficient transduction of neural cells in vitro and in vivo by a baculovirus-derived vector. Proc Natl Acad Sci USA 97:14638–14643
Scherer L, Rossi JJ (2004) Recent applications of RNAi in mammalian systems. Curr Pharm Biotechnol 5:355–360
Scherr M, Morgan MA, Eder M (2003) Gene silencing mediated by small interfering RNAs in mammalian cells. Curr Med Chem 10:245–256
Serruys PW, Daemen J (2007) Are drug-eluting stents associated with a higher rate of late thrombosis than bare metal stents? Late stent thrombosis: a nuisance in both bare metal and drug-eluting stents. Circulation 115:1433–1439
Serruys PW, de Jaegere P, Kiemeneij F et al (1994) A comparison of balloon-expandable-stent implantation with balloon angioplasty in patients with coronary artery disease. Benestent Study Group. N Engl J Med 331:489–495
Sigwart U, Puel J, Mirkovitch V et al (1987) Intravascular stents to prevent occlusion and restenosis after transluminal angioplasty. N Engl J Med 316:701–706
Stage-Zimmermann TK, Uhlenbeck OC (1998) Hammerhead ribozyme kinetics. RNA 4:875–889
Stone GW, Ellis SG, Cox DA et al (2004) A polymer-based, paclitaxel-eluting stent in patients with coronary artery disease. N Engl J Med 350:221–231
Symons RH (1992) Small catalytic RNAs. Annu Rev Biochem 61:641–671
Tung R, Kaul S, Diamond GA et al (2006) Narrative review: drug-eluting stents for the management of restenosis: a critical appraisal of the evidence. Ann Int Med 144:913–919
Uhlenbeck OC (1987) A small catalytic oligoribonucleotide. Nature 328:596–600
Vaish NK, Heaton PA, Fedorova O et al (1998) In vitro selection of a purine nucleotide-specific hammerheadlike ribozyme. Proc Natl Acad Sci USA 95:2158–2162
Wang Q, Liu M, Mullah B et al (2002) Receptor-selective effects of endogenous RGS3 and RGS5 to regulate mitogen-activated protein kinase activation in rat vascular smooth muscle cells. J Biol Chem 277:24949–24958
Warashina M, Kuwabara T, Kato Y et al (2001) RNA-protein hybrid ribozymes that efficiently cleave any mRNA independently of the structure of the target RNA. Proc Natl Acad Sci USA 98:5572–5577
Yamamoto K, Morishita R, Tomita N et al (2000) Ribozyme oligonucleotides against transforming growth factor-beta inhibited neointimal formation after vascular injury in rat model: potential application of ribozyme strategy to treat cardiovascular disease. Circulation 102:1308–1314
Zuker M, Jacobson AB (1998) Using reliability information to annotate RNA secondary structures. RNA 4:669–679

Applications of Ribozymes and Pyrrole–Imidazole Polyamides for Cardiovascular and Renal Diseases

E.-H. Yao and N. Fukuda(✉)

Contents

Abstract Ribozymes are RNA molecules that cleave a phosphodiester bond in the appropriate target RNAs in a sequence-specific manner, thereby inhibiting the expression of specific gene products. Ribozymes have progressed from objects of scientific study to potential therapeutic agents for treatment of both acquired and inherited diseases. We have developed ribozymes for therapy of arterial proliferative diseases, such as coronary artery restenosis after angioplasty or stent implantation, hypertensive vascular diseases, atherosclerosis, and progressive renal diseases. We designed nuclease resistant chimeric DNA–RNA hammerhead ribozymes targeting platelet-derived growth factor (PDGF) A-chain and transforming growth factor (TGF)-β1, which effectively and specifically inhibited the exaggerated growth of vascular smooth muscle cells from hypertensive rats in vitro, the arterial stenosis after injury and the renal sclerosis in hypertensive rats in vivo. Pyrrole–imidazole (PI) polyamides are novel gene-silencing compounds, which bind to minor groove of double-strand DNA in the sequence-specific manner to suppress target gene expression. We have developed PI polyamide targeting TGF-β1 promoter and lectin-like oxidized low density lipoprotein receptor-1

N. Fukuda
Department of Medicine, Division of Nephrology and Endocrinology, Nihon University School of Medicine, Tokyo, Japan and Advanced Research Institute of Science and Humanities, Nihon University School of Medicine, Tokyo, Japan
e-mail: fukudan@med.nihon-u.ac.jp

V.A. Erdmann et al. (eds.) *RNA Technologies in Cardiovascular Medicine and Research*,
doi: 10.1007/978-3-540-78709-9,

(LOX-1). The polyamide to TGF-β1 inhibited expression of TGF-β1 gene and protein in vitro and in vivo. Prolonged treatments with the polyamide to TGF-β1 markedly improved the progressive renal dysfunction without any side effects in hypertensive rats. PI polyamide to LOX-1 significantly suppressed the apoptosis of endothelial cells in vitro and the arterial stenosis after angioplasty in vivo. These findings suggest that ribozymes and PI polyamides will be effective gene suppressive agents for cardiovascular and renal diseases.

Abbreviations Ang II: Angiotensin II; Im: *N*-methylimidazole; LOX-1: Lectin-like oxidized low-density lipoprotein receptor-1; PDGF: Platelet-derived growth factor; PI: Pyrrole–Imidazole; Py: *N*-methylpyrrole; SHR: Spontaneously hypertensive rats; siRNA: Small interference RNA; TGF: Transforming growth factor; VSMC: Vascular smooth muscle cell

1 Introduction

It has been estimated that cardiovascular diseases will be the leading cause of mortality in the world within the next decade. These pathologies, such as hypertensive vascular diseases, artery restenosis after angioplasty, vascular bypass graft occlusion, are resistant to current therapeutic approaches, and remain as major clinical problems. The coronary artery restenosis occurs in approximately 30% of patients (Fischman et al. 1994; Serruys et al. 1994), which markedly limit the stent implantation for coronary arterial diseases. Various drugs and devices have been tested to prevent the restenosis. On the other hand, there are still no effective treatments for chronic kidney diseases which progress to end-stage renal diseases.

Application of gene therapy to regulate expression of disease-related genes has important therapeutic potential. Gene therapy technologies can be classified as the supplement of genes for diseases caused by gene deficient or gene abnormality, and the suppression of gene expression for diseases caused by increases in specific molecules. The latter is by the nucleic acid agents such as antisense oligonucleotides, decoys, ribozymes, and small interference RNAs (siRNA). These medicines seem to represent potential therapeutic agents for human diseases, although gene therapy by these strategies is still in its infancy (Scherer and Rossi 2003), but is considered to be a promising therapeutic tool for sever cardiovascular diseases (Isner 2002). The nucleic acid agents are expected to be novel therapeutics since they can be designed freely and applied to any diseases. RNA was traditionally viewed as a molecule that only transfers genetic information from the nucleus to the cytoplasm. Twenty-five years ago, it was discovered that RNA has enzymatic properties, and catalytic RNAs are called ribozymes. A ribozyme is an RNA zyme which contains catalytic domain to catalyze RNA splicing and cleavage reactions. RNA catalysis was first described by Guerrier-Takada and Altman (1984) with the discovery of RNase P, and by Cech et al. (1981) with the discovery of the group I intron. RNA property of cleaving corresponding mRNA can be used to develop the ribozymes as therapeutic agents to block expression of deleterious proteins (Usman and Stinchcomb 1996). Ribozymes gain their target specificity from Watson-Crick base-pairing between nucleotides of the ribozyme binding-arm and those that flank the cleavage site of the target RNA. The mechanism of cleavage involves attack of the 2′-OH of nucleotides that is 5′ to the scissile bond in the target, thus destabilizing the phosphate backbone of the target RNA. Upon cleavage, resultant products dissociate from the ribozyme complex, allowing the ribozyme to attack a new target. The cleavage event renders the mRNA untranslatable and leads to further degradation by cellular ribonucleases. Since these nucleic acid agents are easily degraded by nucleases, suitable chemical modifications or drug-delivery systems including vectors are required for their therapeutic applications. Therefore, we have developed pyrrole–imidazole (PI) polyamides as a novel gene-suppressing agent. PI polyamides are small synthetic molecules composed of aromatic rings *N*-methylpyrrole (Py) and *N*-methylimidazole (Im) amino acid (White et al. 1997; Gottesfeld et al. 1997). Such synthetic polyamides can bind to specific base pairs in the minor groove of the DNA double helix with high affinity and specificity, and block binding of transcription factors to the element of gene promoter. Gottesfeld et al. (1997) first reported an eight-ring PI polyamide designed to target to a specific region of the transcription factor TFIIIA binding site inhibited 5S RNA gene expression in *Xenopus* kidney cells, suggesting that the PI polyamide can inhibit the transcription of specific genes. So far, various types of sequence-specific DNA-binding PI polyamide have been developed to regulate gene expression by targeting to the specific region in the gene promoter (Murty and Sugiyama 2004). Thus, PI polyamides can be feasible gene suppressing agents for severe diseases which cannot be rescued with current medicines.

We have developed several ribozymes and PI polyamides targeting transforming growth factor-β1 (TGF-β1), platelet-derived growth factor (PDGF) A-chain or lectin-like oxidized low-density lipoprotein receptor-1 (LOX-1) as gene therapy agents for arterial proliferative diseases and progressive renal diseases.

2 Ribozymes

Gene coding sequences (exons) are interrupted by non-coding parts called introns. Transcripts of such genes undergo cleavage-ligation reactions to produce mature functional mRNAs to direct peptide synthesis. However, some RNAs can also act as an enzyme. They are ribozymes, RNA molecules that selectively bind to and specifically cleave target mRNA molecules leading to inhibition of gene expressions (Guerrier-Takada and Altman 1984). Ribozymes can suppress expression of several disease-related genes. Ribozymes are composed of RNA backbone which is easily degraded by RNases in vivo (Merdan et al. 2002). To avoid that, we have developed

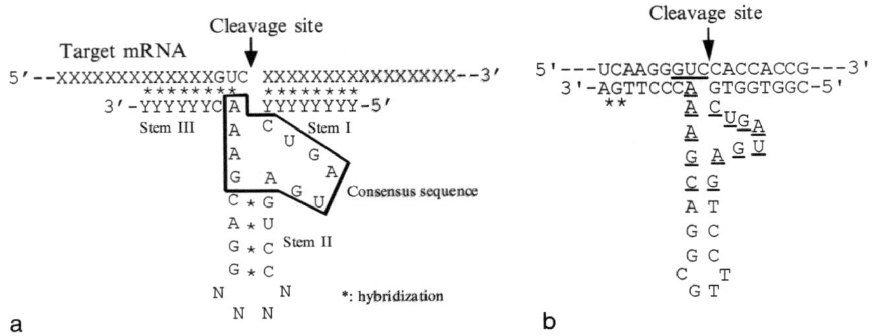

Fig. 1 (**a**) General structures of hammerhead ribozymes. Hammerhead ribozyme consists of antisense arms (stems I and III) and a catalytic consensus sequence with a flanking stem II and loop section. (**b**) Designed structure of chimeric DNA–RNA hammerhead ribozyme to PDGF A-chain mRNA. Ribonucleotides within the DNA are underlined. Phosphoro-thioate linkages are marked by stars

RNase-resistant chimeric DNA–RNA ribozymes for arterial proliferative diseases and renal diseases.

Hammerhead ribozymes have received considerable attention as potential therapeutic tools. They are the smallest ribozyme, approximately 30 nucleotides long, and catalyze site-specific cleavage of phosphodiester bonds (Pieken et al. 1991). Hammerhead ribozyme motif comprises three base-paired helices connected by two single-stranded regions, which contain the largely invariant catalytic domain. Hammerhead ribozyme consists of antisense arms (stems I and III) and a catalytic consensus sequence with a flanking stem II and loop section (Fig. 1a). The catalytic consensus sequence cleaves the target RNA. The target sequence for cleavage usually consists of nucleotide sequence NUX, where N is any base and X can be either C, U, or A, but not G. Usually GUC is recommended for the NUX sequence (Shimayama et al. 1995). The cleavage reaction occurs at the 3′-end of the NUX sequence. Once the target is cleaved, the ribozyme dissociates from the cleaved transcript and repeats the process with another RNA molecule. Generally, cleavage susceptible sequences are determined by predictions of the secondary structure of the target.

2.1 *Design and Delivery Techniques*

For therapeutical application of a ribozyme, one has to determine several important factors, such as target gene and cleavage site selection. The position of the NUX target site within the mRNA has to be selected first. Efficiency of the ribozyme is dependent on the target sequence. The rate of cleavage of different NUX sequences can vary by more than 100-fold, and nucleotide sequence flanking the cleavage site is also important (Zoumadakis et al. 1994). Function of mRNA is dependent on secondary structures. Ribozymes can significantly change mRNA secondary structure by cleavage at NUX sequences located at the 3′end of loop structures. If NUX sequence is located in the double-stranded stem structure of target mRNA, the ribozyme cleavage does not suppress mRNA function. Thus, the ribozyme cleavage site should be located in the loop structures of target mRNAs. RNA folding programs can be used to help determine potential cleavage sites. For example, we survey mRNA targets for GUC cleavage sites in single stranded chain using *GENETYX-MAC: Secondary Structure and Minimum Free Energy*. Biochemical analysis, such as nuclease mapping or chemical probing, can be performed to monitor the accessibility of ribozyme designing (Kronenwett et al. 1996). In addition, to determine whether the designed ribozyme actually cleaves the target RNA, in vitro cleavage experiments should be performed prior to in vivo experiments. All-RNA ribozyme and target RNA were synthesized using T7 RNA polymerase and synthetic DNA templates as described previously (Milligan et al. 1987), and then cleavage reactions (Fig. 2) were performed (Saxena and Ackerman 1990).

Delivery systems will play a critical role in the realization of the therapeutic potential of ribozymes. Currently available delivery methods of ribozymes can be classified into exogenous delivery techniques, and endogenous delivery systems.

In exogenous delivery, in vitro synthesized ribozyme is directly applied to cells in culture medium and rapidly degraded. Some delivery reagents, such as lipofectin or polyethylenimine, should be used to prevent synthetic ribozymes from degradation. Lipofectin-complexed molecules are preferentially transported to the

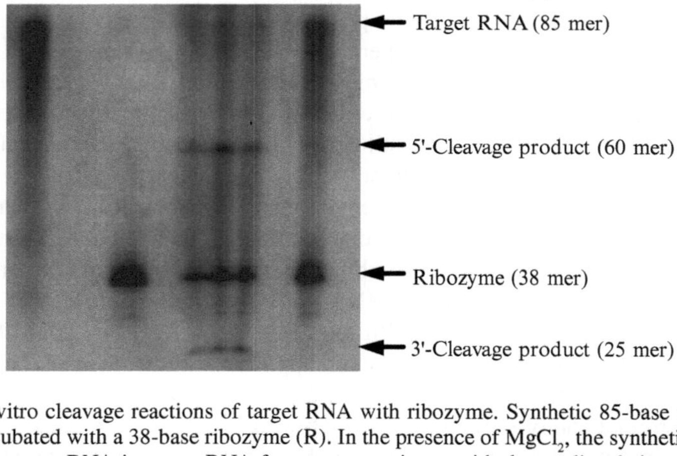

Fig. 2 In vitro cleavage reactions of target RNA with ribozyme. Synthetic 85-base target RNA (T) was incubated with a 38-base ribozyme (R). In the presence of $MgCl_2$, the synthetic ribozyme cleaved the target RNA into two RNA fragments consistent with the predicted sizes (60 and 25 bases)

cytoplasm. However, delivery of ribozymes to the nucleus may be more efficient. In the previous studies, we demonstrated that polyethylenimine can protect the ribozyme from degradation by nucleases in its delivery into tissues in vivo (Kotani et al. 2003).

In endogenous delivery approach, the constructed ribozyme-expressing vector is delivered by transfection or by infection into cells or organs. Ribozyme template DNA sequences are inserted into the untranslated regions transcribed by RNA polymerase II, such as the SV40 early promoter (Cameron and Jennings 1994) or actin gene (Sarver et al. 1990). The RNA polymerase III (pol III) promoters of the U6 small nuclear RNA140 or from certain tRNAs have also been used to express ribozymes (Thompson et al. 1995). Ribozymes are expressed from viral long terminal repeat promoters or from introduced pol II or pol III promoters. Retroviral vectors are relatively efficient and safe and are integrated stably in the host genome of replicating cells. Replication-incompetent adenoviral or adeno-associated viral ribozyme expression vectors enable transient expression of ribozymes because these vectors are not integrated into the host genome. One significant advantage of viral vectors is their extremely high transfection efficiency in a variety of tissues.

2.2 *Modification of Ribozymes*

For application of the ribozyme therapy, high catalytic efficiency, high stability, and adequate levels of ribozymes in the organs are needed. Several approaches can be used for protection of ribozymes from degradation. One widely employed strategy is chemical modifications of synthetic ribozymes. A number of modifications can improve stability, specificity, and efficacy of ribozymes. To protect against degradation by host nucleases, ribozymes have developed to contain stabilizing chemical modifications, primarily at the 2′-OH position of the sugar residue in each nucleotide (Chamberlin et al. 2002; Saksmerprome and Burke 2004).

An another approach of the chemical modification to resist degradation by nucleases is the construction of chimeric DNA–RNA hammerhead ribozymes in which deoxyribonucleotides are substituted for ribonucleotides at noncatalytic residues as shown in Fig. 1b. Intracellular stability studies showed that the chimeric DNA–RNA hammerhead ribozymes remain stable for several hours, whereas all-RNA ribozymes are completely degraded within 1 h of incubation. Moreover, the catalytic cleavage rate for a chimeric ribozyme with DNA-only hybridizing arms is higher than for the corresponding all-RNA ribozyme (Taylor et al. 1992). Thus, the chimeric DNA–RNA ribozymes with modified DNA for RNA could be one of the most suitable structures for therapeutic application of the synthetic ribozymes. We have modified the chimeric DNA–RNA ribozymes with two deoxyribonucleotides at the 3′-terminal end of the ribozyme which were modified with phosphorothioate linkages to improve resistance to nucleases (Fig. 1b). We found that these ribozymes were more stable in living cells and organs.

2.3 *Applications of Ribozymes*

2.3.1 Arterial Proliferative Diseases

For application therapies by nucleic acid agents in cardiovascular diseases, the investigation of critical issues for these diseases is important. To define these factors in the arterial proliferative diseases, we investigated mechanisms underlying the exaggerated growth of vascular smooth muscle cells (VSMCs) from spontaneously hypertensive rats (SHR) as neointimal VSMCs in the restenosis artery (Fukuda 1997). SHR-derived VSMCs show synthetic phenotype by which angiotensin II (Ang II) and Ang II-related growth factors such as TGF-β1, PDGF A-chain, and basic fibroblast growth factor are endogenously produced from the cells and induce the exaggerated growth (Kubo et al. 1996; Fukuda et al. 1999a,b; Hu et al. 2000; Satoh et al. 2001). We have found abnormality of TGF-β1 (Fukuda et al. 1995, 1998) and the shortened cell cycle in response to Ang II in SHR-derived VSMCs (Kubo et al. 2000). These findings indicate that the exaggerated growth of SHR-derived VSMCs are associated with Ang II-related growth factors such as TGF-β1 and PDGF A-chain, which can be target molecules to nucleic acid agents for the arterial restenosis after angioplasty.

Restenosis of coronary artery after coronary angioplasty is caused by recoil of the coronary artery after mechanical dilatation and neointima formation with hyperplasia of VSMCs. Restenosis of the coronary artery after stent implantation is caused only by neointima formation. The source of intimal VSMCs was thought to be medial VSMCs of the artery. However, it was reported that the intimal VSMCs originate from bone marrow-derived progenitor cells present in peripheral blood (Sata et al. 2002). Intimal VSMCs show the synthetic phenotype with increased numbers of intracellular organelles and production of Ang II and growth factors such as PDGF, TGF-β1, bFGF, and endothelin that stimulate growth of VSMCs in a paracrine or autocrine manner (Casscells 1992). PDGF is a dimer composed of disulfide-linked A-chain and B-chain (Johnsson et al. 1982). Nilsson et al. (1985) showed that normal growth-arrested VSMCs do not express PDGF mRNA, whereas cultured VSMCs or VSMCs in atherosclerotic plaques express PDGF A-chain mRNA and secrete PDGF-AA protein, indicating that PDGF A-chain contributes to VSMC proliferation in arterial proliferative disease. TGF-β1 plays a pivotal role in the pathogenesis of restenosis after angioplasty (Majesky et al. 1991). It stimulates synthesis of extracellular matrix proteins, including collagen, laminin, and fibronectin, suggesting that TGF-β1 may be a gene therapy target for preventing restenosis. In the previous studies, we have chosen TGF-β1 and PDGF A-chain as the targets for the ribozyme-based gene therapy to treat arterial proliferative diseases including artery neointima after balloon injury.

Nucleic acid agents have been proposed to apply for the arterial stenosis after angioplasty. Recently, ribozymes targeting *c-myb* (Macejak et al. 1999), TGF-β1 (Yamamoto et al. 2000), and leukocyte-type 12-lipoxygenase (Gu et al. 2001) were reported to efficiently inhibit the neointima formation of artery in vivo.

We designed and synthesized a 38-base chimeric DNA–RNA hammerhead ribozyme with two phosphorothioate linkages at the 3′ terminal that cleaves rat PDGF A-chain mRNA at the GUC sequence (Fig. 1b). We examined its effects on growth of VSMCs from SHR, and found that the ribozyme effectively and specifically inhibited the exaggerated growth. This effect was mediated by degradation of the rat PDGF A-chain mRNA resulting in reduced production of rat PDGF-AA protein synthesis (Hu et al. 2001a,b). To evaluate action of chimeric DNA–RNA hammerhead ribozyme specific to PDGF A-chain on the stenosis after angioplasty, we examined the effects of that ribozyme on neointima formation in rat carotid artery after balloon injury in vivo. FITC-labeled ribozyme delivered with polyethylenimine (Aigner et al. 2002) to the injured artery was taken up into the midlayer smooth muscle of the carotid artery until 24 h after balloon injury, suggesting efficient local delivery of ribozyme (Fig. 3). The ribozyme reduced neointima formation by 55% (Fig. 4), and inhibited PDGF A-chain mRNA expression and synthesis of PDGF-AA protein in injured vessels, suggesting that ribozyme blocks neointima formation through degradation of PDGF A-chain mRNA (Kotani et al. 2003).

We have constructed a replication-deficient recombinant adenovirus that expresses a ribozyme specific for PDGF A-chain mRNA for endogenous delivery of ribozyme to a tissue. Transfection of this vector significantly reduced proliferation of VSMCs from SHR in a dose-dependent manner (Hu et al. 2002). The adenovirus-encoded ribozyme to PDGF A-chain reduced expression of PDGF A-chain mRNA and showed prolonged inhibition of proliferation of VSMCs in comparison to chimeric DNA–RNA hammerhead ribozyme. Adenovirus-encoded ribozyme to PDGF A-chain reduced the neointima formation by 68%, which is greater than the efficacy of the chimeric DNA–RNA ribozyme to PDGF A-chain. Decreased expression of PDGF A-chain mRNA and protein in rat carotid artery was observed at 14 days after balloon injury, suggesting sustained suppression of expression of PDGF A-chain due to expression of the ribozyme encoded by the adenovirus vector system (Lin et al. 2004).

We have designed and examined effects of the chimeric DNA–RNA hammerhead ribozyme targeting TGF-β1 mRNA on neointima formation (Teng et al. 2000; Su et al. 2000). This chimeric ribozyme reduced neointima formation in rat carotid artery by 65% (Ando et al. 2004). We also observed a reduction in expression of fibronectin mRNA, suggesting that the ribozyme somehow inhibits extracellular matrix formation. Because neointima contains both vascular smooth muscle and extracellular matrix (Majesky 1994), TGF-β1 may influence restenosis through accumulation of extracellular matrix and not increased proliferation of VSMCs. Taken together, these findings suggest that ribozymes targeting PDGF A-chain and TGF-β1 are viable strategies to prevent coronary artery restenosis after angioplasty.

Because of an instability of ribozyme in the body, its local delivery and application for short-term diseases such as arterial stenosis after angioplasty or stent implantation are better than systematic administrations and for chronic diseases. The in-stent restenosis occurs up to 50% in high-risk patients undergoing coronary stenting. Since many drugs and devices have been tried and failed to prevent the

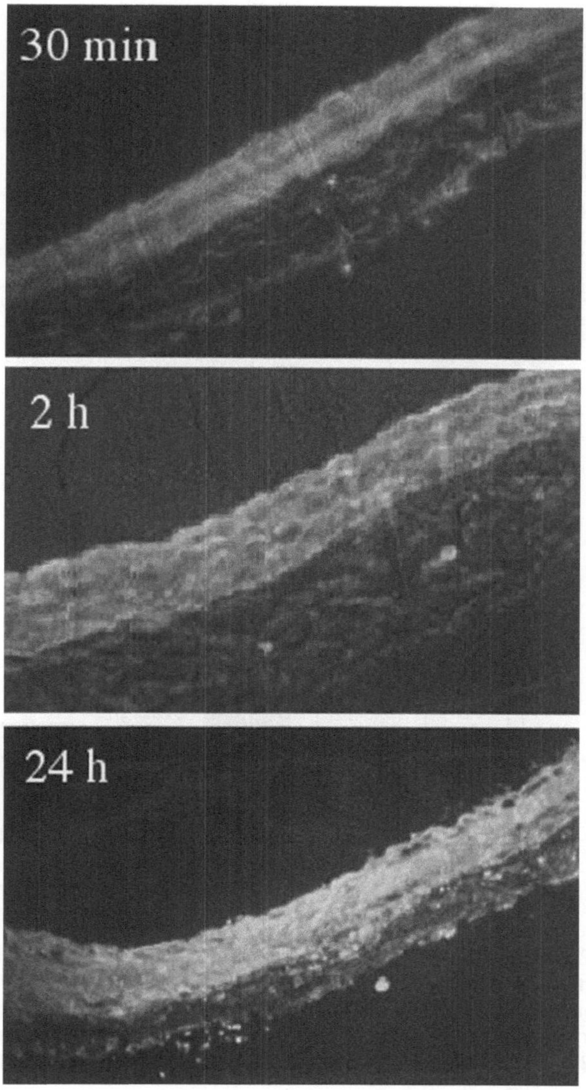

Fig. 3 Uptake of FITC-labeled chimeric DNA–RNA hammerhead ribozyme targeting PDGF A-chain into carotid artery after balloon injury. Five micrograms of FITC-labeled ribozyme was incubated within the artery lumen for 10 min. FITC-labeled ribozyme was localized in the midlayer of smooth muscle in the carotid artery by 30 min after injury, and levels increased until 24 h after balloon injury

in-stent restenosis, DES were introduced and showed potent reduction of the in-stent restenosis. DES with sirolimus or pacritaxel has been widely used to prevent the in-stent restenosis in the clinical fields (Regar et al. 2002; Colombo et al. 2003; Park et al. 2003; Cohen et al. 2004).

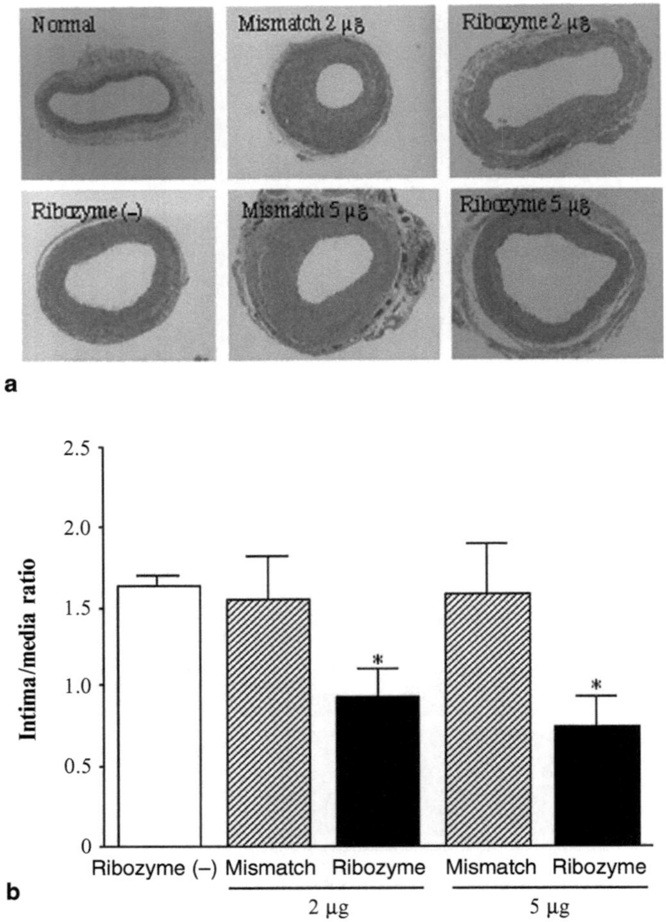

Fig. 4 Effect of chimeric DNA–RNA ribozyme targeting PDGF A-chain on neointima formation in rat carotid artery 2 w after balloon injury. (**a**) Specimens were cross-sectioned at 3 mm and stained with hematoxylin and eosin. (**b**) Intimal and medial cross-sectional areas of four cross sections of artery obtained from each rat were measured. Neointima of rat carotid artery after treatment with 2 or 5 μg of ribozyme was 44 and 55% less, respectively, than that receiving no treatment

Since sirolimus stops the cell cycle progression (Murakami et al. 2004), the sirolimus-coated DES can prevent the in-stent restenosis with complete suppression of VSMCs hyperplasia. However, since sirolimus also prevents the re-endothelialization inner side of the metal stent, complications of the subacute thrombosis or late thrombosis have recently been reported in patients implanted the sirolimus-coated DES (Iakovou et al. 2004; Werner et al. 2004). These clinical recognitions led to development of the second-generation DESs that do not induce the late thrombosis.

Ribozymes can be administered via a DES. However, binding of nucleotides to metal surfaces is poor and, therefore, improvement of ribozyme binding capacity would be helpful. Negatively charged nucleotides that can interact with a positively charged stent. Hydrogel on the balloon and stent are better for binding nucleic acids to a stent. Frimerman et al. (1999) demonstrated that chimeric DNA–RNA hammerhead ribozyme to proliferating cell nuclear antigen reduced stent-induced stenosis in a pig model of coronary desease. We have confirmed that the chimeric DNA–RNA ribozymes targeting PDGF A-chain and TGF-β1 strongly and specifically inhibited the arterial stenosis after angioplasty (Teng et al. 2000; Kotani et al. 2003). The adenovirus-encoded hammerhead ribozyme to PDGF A-chain inhibited arterial thrombosis after the injury and induced the re-endothelialization (Lin et al. 2004). Since TGF-β1 inhibits proliferation of endothelial cells, the suppression of TGF-β1 by ribozyme certainly induces the endotheliarization. Thus, local delivery of the ribozymes targeting PDGF A-chain and TGF-β1 via DES will be a clinical useful approach not only to prevent the instent restenosis but also to induce the endothelialization in coronary artery.

2.3.2 Renal Diseases

Although a variety of drugs have been used to treat the progressive renal diseases experimentally and clinically, none of them has been shown to effectively protect the kidney from progressing to the end-stage renal disease. Thus, gene therapy is now being proposed for the progressive renal diseases. TGF-β1 is a multifunctional regulatory protein that can stimulate cellular proliferation and strongly induces the extracellular matrix formation (Roberts et al. 1986; Sporn et al. 1986; Massague 1987). Thus, TGF-β1 involves in pathogenesis glomerulosclerosis and glomerulonephritis with mesangial proliferation and increases in synthesis of extracellular matrix such as fibronectin and collagen (Morales and Roberts 1988; Okuda et al. 1990; Yamamoto et al. 1996). Dahly et al. (2002) injected anti-TGF-β1 antibody into salt-loaded Dahl-S rats and found a significant decrease in proteinuria and degree of glomerulosclerosis and renal interstitial fibrosis. In addition, PDGF is a potent stimulator for renal mesangial cells. We demonstrated contributions of TGF-β1 (Tahira et al. 2002) and PDGF A-chain (Kishioka et al. 2001) in kidney of stroke-prone spontaneously hypertensive rats (SHR-SP) which show severe hypertension and renal damages.

We designed chimeric DNA–RNA hammerhead ribozyme targeting TGF-β1, and examined its effects on hypertensive renal sclerosis in SHR-SP and Dahl salt-sensitive rats, which show increased expression of TGF-β1 in renal cortex. Intrapertonially administered the chimeric ribozyme distributed into glomerulus in vivo, and significantly inhibited expression of TGF-β1 as well as extracellular matrix molecules such as type I collagen and fibronectin. One intraperitorial injection of the chimeric ribozyme to TGF-β1 markedly ameliorated wall thickness

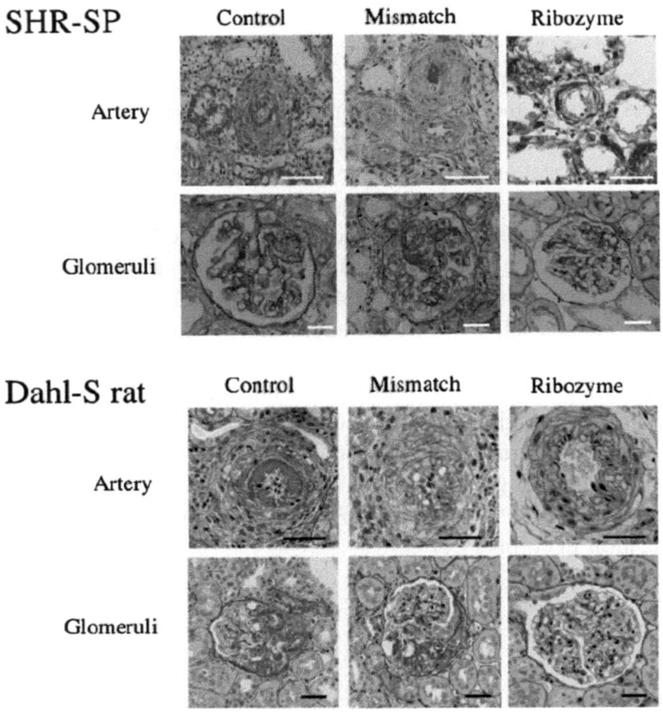

Fig. 5 Effects of chimeric DNA–RNA ribozyme to TGF-β1 on arterial thickness and gromelurosclerosis in renal cortex of salt-loaded SHR-SP and Dahl-S rats treated with. After 4 weeks of salt loading, rats received an intraperitoneal injection of the chimeric ribozyme (200 μg/body weight). Intraperitoneal administration of chimeric DNA–RNA ribozyme to TGF-β1 markedly ameliorated thickening of the capillary artery wall and glomerulosclerosis in kidneys of salt-loaded SHR-SP and Dahl-S rats

of renal capillary artery and glomerulosclerosis in hypertensive rats (Fig. 5), (Tahira et al. 2007). These findings suggest that chimeric DNA–RNA hammerhead ribozyme targeting TGF-β1will be feasible for the treatment of progressive renal diseases.

2.3.3 Clinical Trials of Ribozymes

Besides basic studies to develop ribozyme therapies, several clinical trials of ribozyme have been reported, although they are not yet in the field of cardiovascular diseases. A retrovirally expressed ribozyme that targets HIV Tat is currently in phase II testing for patients with AIDS-related lymphoma. ANGIOZYME, directed against mRNA for Flt-1 (VEGF-R1), the high-affinity receptor for VEGF, is currently in phase II trials for several tumor types. HEPTAZYME, a ribozyme targeting 5′-untranslated region (5′-UTR) of the hepatitis C virus (HCV) RNA genome, has recently passed through a phase I/II clinical trial in patients with chronic hepatitis C (Macejak et al. 2000). These studies strongly suggest that ribozyme-based strategy would be a useful tool of gene therapy for human diseases.

2.4 Specificity of Ribozymes

To evaluate the specificity of our engineered ribozyme targeting PDGF A-chain mRNA, we used microarray technology to examine levels of specific transcripts in injured vessels (Kotani et al. 2003). The transcripts were used as gene-specific hybridization targets to quantitative expression of the corresponding genes. In the microarray analysis, the injured vessels showed increases in 525 transcripts for metabolic enzymes, growth factors, cytokines, cell cycle regulators, and transcription factors. Among the transcripts, several factors of intracellular signaling systems, kinases, and cell cycle-related peptides were completely inhibited by treatment of the chimeric DNA–RNA hammerhead ribozyme targeting PDGF A-chain mRNA in the injured vessel. Some over-expressed transcripts of metabolic enzymes were completely inhibited by the ribozyme. Thus, the inhibition of neointima formation by the ribozyme coincides with suppression of cell cycle, metabolism, and phenotype of VSMCs. Therefore, suppression of expression of intracellular signaling, kinase, and cell cycle-related peptide mRNAs may be secondary to inhibition of neointima formation by the ribozyme. The ribozyme targeting PDGF A-chain mRNA completely inhibited expression of only that mRNA, indicating that inhibition of PDGF A-chain expression was specific (Kotani et al. 2003). Thus, the major advantage of ribozymes is that they can sequence-specifically cleave multiple target mRNA molecules, whereas antisense nucleotides do not cleave the target molecules and act only at an equimolar ratio.

Although ribozymes have been applied to inhibit disease-related genes in numerous studies, several problems may limit the application of ribozyme therapies. Since a major limitation in the development of ribozymes is degradation by RNases, the chemical modifications are necessary to increase their stability in vivo. However, such modifications should not affect the ribozymes cleavage activity. Mechanistically, various studies reveal that even a slight structural change of ribozymes can dramatically affect their chain folding and alter the binding properties of their catalytic core (Lonnberg and Lonnberg 2005). Synthetic ribozymes should also overcome several other problems, including accessibility to target mRNAs, efficient delivery to target cells and preserving their activity, and non-specific effects. Thus, the endogenous delivery of ribozymes seems to be more feasible. Since ribozymes cannot be administered orally or delivered locally to the appropriate targets, such as kidney in human, more efficient administration methods, for example, tissue-specific ribozyme expression vectors, should be developed.

3 Pyrrole–Imidazole (PI) Polyamides

Since nucleic acid agents including antisense oligonucleotides, ribozymes and siRNA have disadvantage of their degradation by nucleases in vivo, we moved to develop PI polyamide as a novel gene expression suppressive agent. PI polyamides were firstly identified from antibiotics duocarmycin A and distamycin A (Trauger et al. 1996; White et al. 1997, 1998). PI polyamides are chemical small synthetic molecules composed of the aromatic rings of *N*-methylpyrrole (Py) and *N*-methylimidazole (Im) amino acids which recognize and bind DNA in a sequence specific manner (Trauger et al. 1996; White et al. 1997). Gene promoter sequences play a vital role in initiating transcription. They offer recognition sites for the transcription factors and RNA polymerase. Binding of transcription factors to the DNA response elements in the region of gene promoter is required for initiation of transcription. PI polyamides can bind within the minor groove of DNA and block binding of transcription factors to inhibit gene transcription. Aromatic Py and Im can be coupled and adopt a U-shape conformation in the presence of γ-aminobutyric acid. Such synthetic polyamides can bind to base pairs in the minor groove of the DNA double helix with high affinity and specificity. Base pair specificity is dependent on the pairing of Py and Im. The Py/Im pair targets C-G base pairs. Im/Py recognizes G-C base pairs, and a Py/Py binds both A-T and T-A base pairs (Trauger et al. 1996; White et al. 1997). It was recently reported that the A-T degeneracy can be overcome by replacing one pyrrole ring of the Py/Py pair with 3-hydroxypyrrole (Hp); Hp/Py preferentially binds T-A pairs (Gottesfeld et al. 1997), (Fig. 6). Designed PI polyamides block binding of transcription factors to inhibit the gene expression by docking to minor groove in the promoter lesion. Thus PI polyamide could be a gene silencer that inhibits target gene expression in stimulatory condition as diseases, which does not knock down the gene expressions.

We evaluated the binding affinity and specificity of polyamides for target DNA by gel mobility shift and BiaCore analysis. In our previous studies, we have shown that PI polyamide bound to the appropriate double-stranded DNA but not to mutated DNA. Fast binding of PI polyamide to target sequence occurred relative to that of mismatch polyamides to make match binding being able to reach equilibrium at high concentrations. Dissociation equilibrium constant of the designed PI polyamide was 700 times higher than it in mismatch polyamides. These findings indicate that PI polyamide can strongly and specifically bind to target DNA of the promoter activity (Fig. 7) (Matsuda et al. 2006).

Since polyamides are completely resistant to nucleases, they can be delivered into organs without delivery systems. Fluorescent-labeled polyamides were prepared to analyze the intracellular distribution of these molecules in several cell lines. Polyamide-dye conjugates were observed to enter the nuclei of living cells (Dickinson et al. 1998; Belitsky et al. 2002). These studies indicated that polyamides are cell permeable and localize in the nucleus in the cultured cells, and can access their target sites in the genes. We have intravenously injected fluorescent-labeled PI polyamide targeting TGF-β1 in rats and observed that polyamides

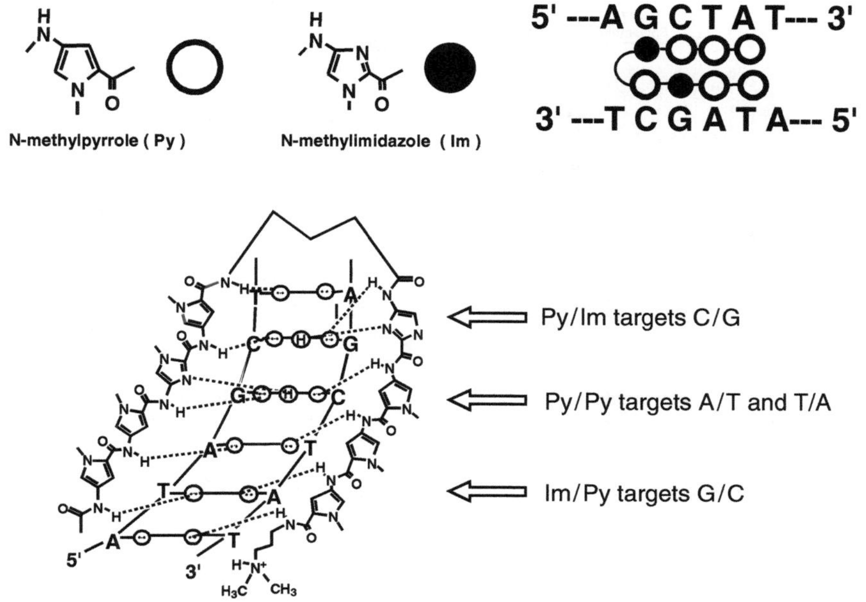

Fig. 6 Rule for recognition of double strand DNA by pyrrole (Py)-imidazole (Im) polyamide. DNA recognition by Py-Im polyamides depends on a code of side-by-side pairing of Py and Im in the minor groove; pairing of Im opposite Py (Im/Py) targets a G-C base pair, and Py/Im targets a C-G base pair. Py/Py either T-A or A-T base pair. The binding constants and specificity of Py-Im polyamides are comparable to those of transcription factors

were up taken into several organs, including aorta without any vectors, and excreted into urine detected by high performance liquid chromatography (Fig. 8) (Matsuda et al. 2006).

For development of synthetic PI polyamides as a gene silencer, various types of sequence-specific DNA-binding PI polyamide have been designed (Murty and Sugiyama 2004). Gottesfeld et al. (1997) firstly reported an eight-ring PI polyamide designed to target to a specific region of the transcription factor TFIIIA binding site inhibited 5S RNA gene expression in *Xenopus* kidney cells. To block activity of the human immunodeficiency virus type 1, two polyamides were designed to bind two transcription factor binding sites, and inhibits virus replication by >99% (Dickinson et al. 1998).

Because TGF-β1 plays an important role in the cardiovascular and renal diseases, we have developed PI polyamides targeting TGF-β1. We synthesized a PI polyamide targeting the promoter of human TGF-β1 adjacent to the fat-specific element 2. Gel mobility shift assays showed that the synthetic PI polyamides bound to the corresponding double-stranded oligonucleotides, whereas the mismatched polyamides did not. FITC-conjugated PI polyamide has been detected in the nuclei

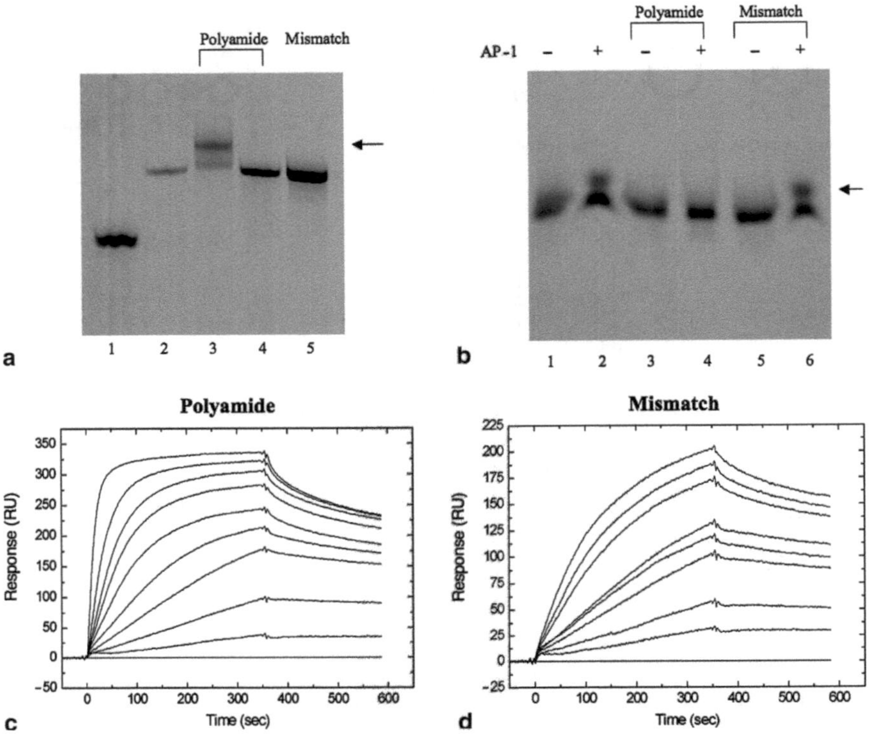

Fig. 7 Binding and specificity of polyamide to double-stranded DNA and inhibition of AP-1 complex binding. (**a**) Polyamide bound the appropriate 21-bp double-stranded DNA but did not bind the 2-bp mutated DNA, whereas Mismatch did not show binding to appropriate DNA. (**b**) A single mobility band was observed when the DNA were incubated with AP-1 or treatment with Mismatch. Polyamide inhibited AP-1 binding to target DNA. (**c,d**) Kinetics of Polyamide and Mismatch bindings with target double-strand DNA obtained from fitting resulting sensorgrams (Biacore assay). Fast binding of Polyamide to target sequence occurred relative to that of Mismatch to make match binding being able to reach equilibrium at high concentrations

of VSMCs. PI polyamide significantly decreased the activity of the TGF-β1 promoter (Lai et al. 2005). In cultured human VSMCs, this polyamide inhibited expression of the TGF-β1 mRNA and protein. We also synthesized a PI polyamide targeting AP-1 binding site of rat TGF-β1 promoter. This polyamide completely inhibited the increases in proteinuria and albuminuria in salt-loaded Dahl-S rats along with suppression of TGF-β1 expression in glomeruli and nephrotubuli (Fig. 9) (Matsuda et al. 2006).

In addition, we recently designed and synthesized PI polyamides targeting rat oxidatively modified low-density lipoprotein (OxLDL) gene promoter. OxLDL leads to endothelial dysfunction, and it is implicated in the atherosclerosis. LOX-1 is the major OxLDL receptor, which is a membrane protein and is expressed in

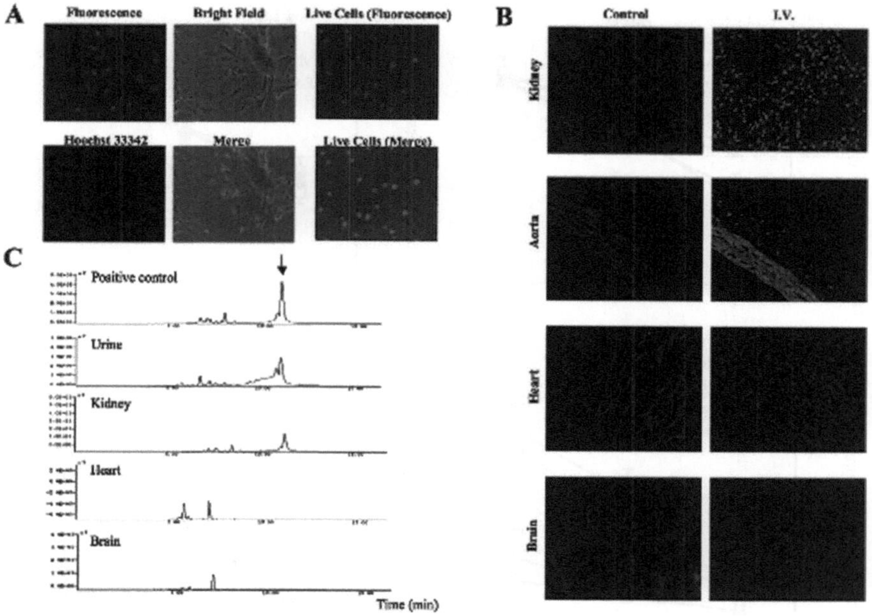

Fig. 8 Distribution of FITC-labeled PI polyamide in vitro and in vivo. (**a**) Mesangial cells from Wistar rats were incubated with FITC-labeled polyamide. (**b**) FITC-labeled polyamide was injected into Wistar rats intravenously. Twenty-four hours after injection, the kidneys, aorta, heart, and brain were removed, and frozen specimens were made. (**c**) HPLC analysis for the fluorescein-labeled polyamide in the urine, renal cortex, aorta, heart, and brain. Arrow indicates of polyamide positive control

vascular endothelial cells and vascular-rich organs (Sawamura et al. 1997). The designed PI polyamide targets AP-1 binding site of human or rat LOX-1 gene promoter. PI polyamide targeting human LOX-1 promoter suppressed the apoptosis of human endothelial cells. In a rat model, PI polyamide targeting rat LOX-1 promoter significantly inhibited neointimal hyperplasia of carotid artery after balloon injury. These findings suggest that this polyamide would be useful for the treatment of LOX-1-associated diseases such as atherosclerosis, arterial stenosis after angioplasty or instent restenosis.

While many researchers have reported that ribozymes and PI polyamides act as effective gene suppressing agents to control disease-related genes, no study has compared the efficacy of ribozymes and PI polyamides. For the potential applications of gene therapy, PI polyamides may have several advantages over ribozymes: (1) PI polyamides are resistant to the biological degradation induced by nucleases, thus they are more stable than ribozymes in vivo, (2) no cleavage reaction for the targeted gene occurs, and (3) vector-assisted delivery systems are not needed, thus the therapeutic application of PI polyamides to human diseases is not limited by delivery systems.

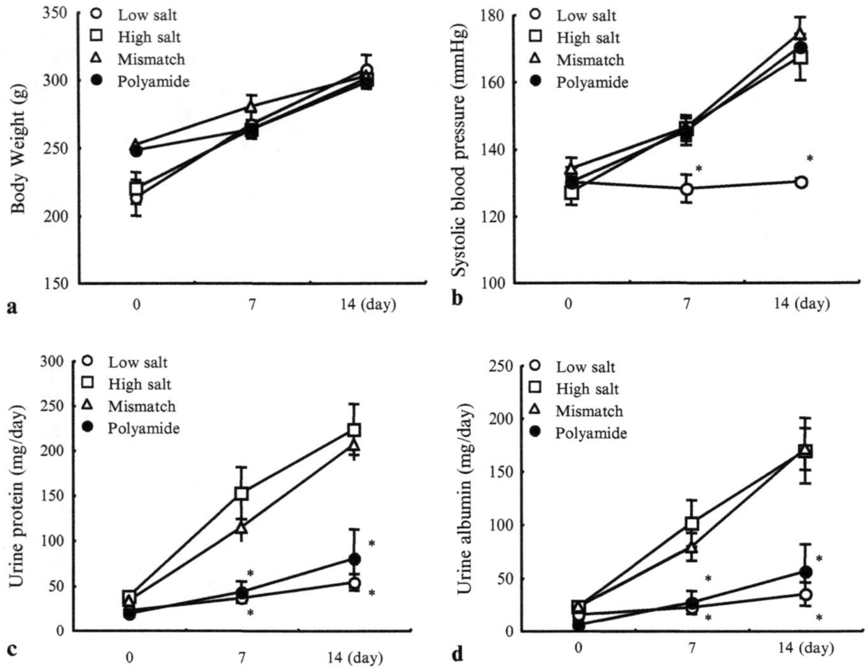

Fig. 9 Changes in body weight (**a**), systolic blood pressure (**b**), urinary protein (**c**), and urinary albumin (**d**) in Dahl-salt sensitive (Dahl-S) rats before and after treatments with PI polyamide targeting TGF-β1 promoter (Polyamide). Dahl-S rats fed a high-salt received intravenous injection of 1 mg Polyamide (filled circles) or mismatch polyamide (Mismatch, open triangles) every 2 days seven times diet for 2 wk

4 Conclusions

Our long-term goal is the control of disease-related gene expression in human. Now this goal can be pursued, because ribozymes or PI polyamides are considered to be a promising therapeutic tool for several human diseases that are resistant to current approaches. Recently, completed sequences of the human genome and genomes of several pathogenic organisms provide a large number of potential gene targets for developing ribozymes or PI polyamides. The in-stent restenosis of coronary artery and the end-stage renal diseases still remain major clinical issues, which waste large amounts of medical resources. These medical expenses for both hospitals and patients would be reduced if ribozymes or PI polyamides could be used to prevent restenosis and renal dysfunction in the future. Thus, we anticipate that ribozymes or PI polyamides may be emerging as new and broadly useful therapeutic agents for many diseases.

Chimeric DNA–RNA hammerhead ribozymes and PI polyamides have been employed to treat cardiovascular and renal diseases, including prevention of restenosis or progressive renal diseases, and encouraging results have been reported, although several problems still remain to be resolved. Ribozymes and PI polyamides will be potential therapeutic tools to control disease-related genes in the field of cardiovascular and renal diseases. It is anticipated that the ribozymes or the PI polyamides will play an important role in the treatment of cardiovascular and renal diseases in the future.

Acknowledgments We thank Drs. T. Ueno, H. Kishioka, C. Satoh, M. Nakayama, S. Kunimoto, R. Suzuki, H. Ando, and M. Kotani in Department of Medicine, Nihon University School of Medicine; Prof. K. Matsumoto, K. Kanmatsuse, H. Mugishima in Nihon University School of Medicine provided direction for the gene therapy projects; Drs. W.Y. Hu, J. Su, Y. Lai, Z. Lin as Research Fellow of Fujian Medical School, P.R. China for experimental efforts on the gene therapy. Prof. H. Sugiyama in Kyoto University for help with the PI polyamide studies.

References

Aigner A, Fischer D, Merdan T et al (2002) Delivery of unmodified bioactive ribozymes by an RNA-stabilizing polyethylenimine (LMW-PEI) efficiently down-regulates gene expression. Gene Ther 9:1700–1707

Ando H, Fukuda N, Kotani M et al (2004) Chimeric DNA–RNA hammerhead ribozyme targeting TGF-β1 mRNA inhibits neointima formation in rat carotid artery after balloon injury. Eur J Pharmacol 483:207–214

Belitsky JM, Leslie SJ, Arora PS et al (2002) Cellular uptake of *N*-methylpyrrole/*N*-methylimidazole polyamide-dye conjugates. Bioorg Med Chem 10:3313–3318

Cameron FH, Jennings PA (1994) Multiple domains in a ribozyme construct confer increased suppressive activity in monkey cells. Antisense Res Dev 4:87–94

Casscells W (1992) Migration of smooth muscle and endothelial cells. Critical events in restenosis. Circulation 86:723–729

Cech TR, Zaug AJ, Grabowski PJ (1981) In vitro splicing of the ribosomal RNA precursor of tetrahymena: involvement of a guanosine nucleotide in the excision of the intervening sequence. Cell 27:487–496

Chamberlin SI, Merino EJ, Weeks KM (2002) Catalysis of amide synthesis by RNA phosphodiester and hydroxyl groups. Proc Natl Acad Sci USA 99:14688–14693

Cohen DJ, Bakhai A, Shi C et al (2004) Cost-effectiveness of sirolimus-eluting stents for treatment of complex coronary stenoses: results from the sirolimus-eluting balloon expandable stent in the treatment of patients with de novo native coronary artery lesions (SIRIUS) trial. Circulation 110:508–514

Colombo A, Drzewiecki J, Banning A et al (2003) Randomized study to assess the effectiveness of slow- and moderate-release polymer-based paclitaxel-eluting stents for coronary artery lesions. Circulation 108:788–794

Dahly AJ, Hoagland KM, Flasch AK et al (2002) Antihypertensive effects of chronic anti-TGF beta antibody therapy in Dahl S rats. Am J Physiol Regul Integr Comp Physiol 283: R757–R767

Dickinson LA, Gulizia RJ, Trauger JW et al (1998) Inhibition of RNA polymerase II transcription in human cells by synthetic DNA-binding ligands. Proc Natl Acad Sci USA 95:12890–12895

Fischman DL, Leon MB, Baim DS et al (1994) A randomized comparison of coronary-stent placement and balloon angioplasty in the treatment of coronary artery disease. Stent Restenosis Study Investigator. N Engl J Med 331:496–501

Frimerman A, Welch PJ, Jin X et al (1999) Chimeric DNA–RNA hammerhead ribozyme to proliferating cell nuclear antigen reduces stent-induced stenosis in a porcine coronary model. Circulation 99:697–703

Fukuda N (1997) Molecular mechanisms of the exaggerated growth of vascular smooth muscle cells in hypertension. J Atheroscler Thromb 4:65–72

Fukuda N, Kubo A, Izumi Y et al (1995) Characteristics and expression of transforming growth factor-beta receptor subtypes on vascular smooth muscle cells from spontaneously hypertensive rats. J Hypertens 13:831–837

Fukuda N, Hu W-Y, Kubo A et al (1998) Abnormal regulation of transforming growth factor-beta receptors on vascular smooth muscle cells from spontaneously hypertensive rats by angiotensin II. Hypertension 31:672–677

Fukuda N, Satoh C, Hu W-Y et al (1999a) Production of angiotensin II by homogeneous cultures of vascular smooth muscle cells from spontaneously hypertensive rats. Arterioscler Thromb Vasc Biol 19:1210–1217

Fukuda N, Hu W-Y, Satoh C et al (1999b) Contribution of synthetic phenotype on the enhanced angiotensin II-generating system in vascular smooth muscle cells from spontaneously hypertensive rats. J Hypertens 17:1099–1107

Gottesfeld JM, Neely L, Trauger JW et al (1997) Regulation of gene expression by small molecules. Nature 387:202–205

Gu JL, Pei H, Thomas L et al (2001) Ribozyme-mediated inhibition of rat leukocyte-type 12-lipoxygenase prevents intimal hyperplasia in balloon-injured rat carotid arteries. Circulation 103:1446–1452

Guerrier-Takada CG, Altman S (1984) Catalytic activity of an RNA molecule prepared by transcription in vitro. Science 223:285–286

Hu WY, Fukuda N, Satoh C et al (2000) Phenotypic modulation by fibronectin enhances the angiotensin II-generating system in cultured vascular smooth muscle cells. Arterioscler Thromb Vasc Biol 20:1500–1505

Hu WY, Fukuda N, Nakayama M et al (2001a) Inhibition of vascular smooth muscle cell proliferation by DNA-RNA chimeric hammerhead ribozyme targeting to rat platelet-derived growth factor A-chain mRNA. J Hypertens 19:203–212

Hu WY, Fukuda N, Kishioka H et al (2001b) Hammerhead ribozyme targeting human platelet-derived growth factor A-chain mRNA inhibited the proliferation of human vascular smooth muscle cells. Atherosclerosis 158:321–329

Hu WY, Fukuda N, Kotani M et al (2002) Adenovirus-mediated transfer of ribozyme targeting platelet-derived growth factor A-chain mRNA inhibits growth of vascular smooth muscle cells from spontaneously hypertensive rats. J Cardiovasc Pharmacol 39:858–865

Iakovou I, Sangiorgi GM, Stankovic G et al (2004) Effectiveness of sirolimus-eluting stent implantation for treatment of in-stent restenosis after brachytherapy failure. Am J Cardiol 94:351–354

Isner JM (2002) Myocardial gene therapy. Nature 415:234–239

Johnsson A, Heldin CH, Westermark B et al (1982) Platelet-derived growth factor: identification of constituent polypeptide chains. Biochem Biophys Res Commun 104:66–74

Kishioka H, Fukuda N, Watanabe Y et al (2001) Effects of PDGF A-chain antisense oligodeoxynucleotides on growth of cardiovascular organs in stroke-prone spontaneously hypertensive rats. Am J Hypertens 14:439–445

Kotani M, Fukuda N, Ando H et al (2003) Chimeric DNA-RNA hammerhead ribozyme targeting to PDGF A-chain mRNA specifically inhibited neointima formation of rat carotid artery after balloon injury. Cardiovasc Res 57:265–276

Kronenwett R, Haas R, Sczakiel G (1996) Kinetic selectivity of complementary nucleic acids: *bcr-abl*-directed antisense RNA and ribozymes. J Mol Biol 259:632–644

Kubo A, Fukuda N, Soma M et al (1996) Inhibitory effect of an angiotensin II type 1 receptor antagonist on growth of vascular smooth muscle cells from spontaneously hypertensive rats. J Cardiovasc Pharmacol 27:58–63

Kubo A, Fukuda N, Teng J et al (2000) Angiotensin II regulates the cell cycle of vascular smooth muscle cells from SHR. Am J Hypertens 13:1117–1124

Lai Y, Fukuda N, Ueno T et al (2005) Synthetic pyrrole-imidazole polyamide inhibits expression of the human transforming growth factor-beta1 gene. J Pharmacol Exp Therap 315:571–575

Lin ZH, Fukuda N, Suzuki R et al (2004) Adenovirus-encoded hammerhead ribozyme to PDGF A-chain mRNA inhibits neointima formation after arterial injury. J Vasc Res 41:305–313

Lonnberg T, Lonnberg H (2005) Chemical models for ribozyme action. Curr Opin Chem Biol 9:665–673

Macejak DG, Lin H, Webb S et al (1999) Adenovirus-mediated expression of a ribozyme to c-myb mRNA inhibits smooth muscle cell proliferation and neointima formation in vivo. J Virol 73:7745–7751

Macejak DG, Jensen KL, Jamison S et al (2000) Inhibition of hepatitis C virus (HCV)-RNA-dependent translation and replication of a chimeric HCV Poliovirus using synthetic stabilized ribozymes. Hepatology 31:769–776

Majesky MW (1994) Neointima formation after acute vascular injury. Role of counteradhesive extracellular matrix proteins. Tex Heart Inst J 21:78–85

Majesky MW, Lindner V, Twardzik DR (1991) Production of transforming growth factor beta 1 during repair of arterial injury. J Clin Invest 88:904–910

Massague J (1987) The TGF-β1 family of growth and differentiation factors. Cell 49:437–438

Matsuda H, Fukuda N, Ueno T et al (2006) Development of gene silencing pyrrole-imidazole polyamide targeted to the TGF-β1 promoter for treatment of progressive renal diseases. J Am Soc Nephrol 17:422–432

Merdan T, Kopecek J, Kissel T (2002) Prospects for cationic polymers in gene and oligonucleotide therapy against cancer. Adv Drug Deliv Rev 54:715–758

Milligan JF, Groebe DR, Witherell GW et al (1987) Oligoribonucleotide synthesis using T7 RNA polymerase and synthetic DNA templates. Nucleic Acids Res 15:8783–8798

Morales TI, Roberts AB (1988) Transforming growth factor fi regulates the metabolism of proteoglycans in bovine cartilage organ cultures. J Biol Chem 263:12828–12831

Murakami M, Ichisaka T, Maeda M et al (2004) mTOR is essential for growth and proliferation in early mouse embryos and embryonic stem cells. Mol Cell Biol 24:6710–6718

Murty MS, Sugiyama H (2004) Biology of *N*-methylpyrrole-*N*-methylimidazole hairpin polyamide. Biol Pharm Bull 27:468–474

Nilsson J, Sjolund M, Palmberg L et al (1985) Arterial smooth muscle cells in primary culture produce a platelet-derived growth factor-like protein. Proc Natl Acad Sci 82:4418–4422

Okuda S, Languino LR, Ruoalahti E et al (1990) Elevated expression of transforming growth factor-beta and proteoglycan production in experimental glomerulonephritis. Possible role in expansion of the mesangial extracellular matrix. J Clin Invest 86:453–462

Park SJ, Shim WH, Ho DS et al (2003) A paclitaxel-eluting stent for the prevention of coronary restenosis. N Engl J Med 348:1537–1545

Pieken WA, Olsen DB, Benseler F et al (1991) Kinetic characterization of ribonuclease-resistant 2′-modified hammerhead ribozymes. Science 253:314–317

Regar E, Serruys PW, Bode C et al (2002) Angiographic findings of the multicenter randomized study with the sirolimus-eluting bx velocity balloon-expandable stent (RAVEL): sirolimus-eluting stents inhibit restenosis irrespective of the vessel size. Circulation 106:1949–1956

Roberts AB, Sporn MB, Assoian RK et al (1986) Transforming growth factor type-β rapid induction of fibrosis and angiogenesis in vivo and stimulation of collagen formation in vitro. Proc Nati Acad Sci USA 83:4167–4171

Saksmerprome V, Burke DH (2004) Deprotonation stimulates productive folding in allosteric TRAP hammerhead ribozymes. J Mol Biol 341:685–694

Sarver N, Cantin EM, Chang PS et al (1990) Ribozymes as potential anti-HIV-1 therapeutic agents. Science 247:1222–1225
Sata M, Saiura A, Kunisato A et al (2002) RHematopoietic stem cells differentiate into vascular cells that participate in the pathogenesis of atherosclerosis. Nat Med 8:403–409
Satoh C, Fukuda N, Hu W-Y et al (2001) Role of endogenous angiotensin II in the increased expression of growth factors in vascular smooth muscle cells from spontaneously hypertensive rats. J Cardiovasc Pharmacol 37:108–118
Saxena SK, Ackerman EJ (1990) Ribozymes correctly cleave a model substrate and endogenous RNA in vivo. J Biol Chem 265:17106–17109
Sawamura T, Kume N, Aoyama T et al (1997) An endothelial receptor for oxidized low-density lipoprotein. Nature 386:73–77
Scherer LJ, Rossi JJ (2003) Approaches for the sequence-specific knock-down of mrna. Nat Biotechnol 21:1457–1465
Serruys PW, de Jaegere P, Kiemeneij F et al (1994) A comparison of balloon-expandable-stent implantation with balloon angioplasty in patients with coronary artery disease. Benestent Study Group. N Engl J Med 331:489–495
Shimayama T, Nishikawa S, Taira K (1995) Generality of the NUX rule: kinetic analysis of the results of systematic mutations in the trinucleotide at the cleavage site of hammerhead ribozymes. Biochemistry 34:3649–3654
Sporn MB, Roberts AB, Wakefield LM et al (1986) Transforming growth factor-f: biological function and chemical structure. Science 233:532–534
Su JZ, Fukuda N, Hu W-Y et al (2000) Ribozyme to human TGF-β1 mRNA inhibits the proliferation of human vascular smooth muscle cells. Biochem Biophys Res Commun 19:401–407
Tahira Y, Fukuda N, Endo M et al (2002) Transforming growth factor-beta expression in cardiovascular organs in stroke-prone spontaneously hypertensive rats with the development of hypertension. Hypertens Res 25:911–918
Tahira Y, Fukuda N, Endo M et al (2007) Chimeric DNA-RNA hammerhead ribozyme targeting TGF-β1 mRNA efficiently ameliorated renal injury in hypertensive rats. J Hypertens 25:671–678
Taylor NR, Kaplan BE, Swiderski P et al (1992) Chimeric DNA-RNA hammerhead ribozymes have enhanced in vitro catalytic efficiency and increased stability in vivo. Nucleic Acids Res 20:4559–4565
Teng J, Fukuda N, Hu WY et al (2000) DNA–RNA chimeric hammerhead ribozyme to transforming growth factor-β1 mRNA inhibits the exaggerated growth of vascular smooth muscle cells from spontaneously hypertensive rats. Cardiovasc Res 48:138–147
Thompson JD, Ayers DF, Malmstrom TA et al (1995) Improved accumulation and activity of ribozymes expressed from a tRNA-based RNA polymerase III promoter. Nucleic Acids Res 23:2259–2268
Trauger JW, Baird EE, Dervan PB (1996) Recognition of DNA by designed ligands at subnanomolar concentrations. Nature 382:559–561
Usman N, Stinchcomb DT (1996) Design, synthesis and function of therapeutic hammerhead ribozymes. Nucleic Acids Mol Biol 10:243–264
Werner GS, Emig U, Krack A et al (2004) Sirolimus-eluting stents for the prevention of restenosis in a worst-case scenario of diffuse and recurrent in-stent restenosis. Catheter Cardiovasc Interv 63:259–264
White S, Baird EE, Dervan PB (1997) On the pairing rules for recognition in the minor groove of DNA by pyrrole–imidazole polyamides. Chem Biol 4:569–578
White S, Szewczyk JW, Turner JM et al (1998) Recognition of the four Watson-Crick base pairs in the DNA minor groove by synthetic ligands. Nature 391:468–471
Yamamoto T, Noble NA, Cohen AH et al (1996) Expression of transforming growth factor-beta isoforms in human glomerular diseases. Kidney Int 49:461–469
Yamamoto K, Morishita R, Tomita N et al (2000) Ribozyme oligonucleotides against transforming growth factor-beta inhibited neointimal formation after vascular injury in rat model:

potential application of ribozyme strategy to treat cardiovascular disease. Circulation 102:1308–1314
Zoumadakis M, Neubert WJ, Tabler M (1994) The influence of imperfectly paired helices I and III on the catalytic activity of hammerhead ribozymes. Nucleic Acids Res 22:5271–5278

Part IV
Noncoding, Aptamer and Antisense RNAs

Noncoding RNAs in Human Diseases

M. Szymański(✉) and J. Barciszewski

Contents

Abstract The years following publication of the draft human genome sequence brought about a dramatic change in our understanding of the organization and expression of genetic information. Discoveries of new classes of RNA without protein coding potential and large scale cDNA sequencing projects demonstrated that, contrary to the original view, the cellular functions and the processes associated with growth and development of organisms do not depend solely on the protein-coding genes. It is evident that the size of transcribed noncoding segments of the mammalian genomes significantly exceeds the open reading frames translated into proteins. There is a growing number of genes which yield noncoding RNAs (ncRNA) playing multiple functions in regulation of gene expression. Many of the ncRNAs have been shown to be associated with certain human diseases including neurobehavioral, developmental disorders and cancer.

M. Szymański
Institute of Bioorganic Chemistry of the Polish Academy of Sciences, Noskowskiego 12, 61-704 Poznan, Poland
e-mail: mszyman@ibch.poznan.pl.

V.A. Erdmann et al. (eds.) *RNA Technologies in Cardiovascular Medicine and Research*, 235
doi: 10.1007/978-3-540-78709-9, © Springer-Verlag Berlin Heidelberg 2008

1 Noncoding RNAs

The most unexpected outcome of the human genome sequencing project, followed by other mammalian genomes, was an estimate of a number of genes. Mammalian genomes contain 25,000–30,000 protein-coding genes (Lander et al. 2001; Southan 2004; Venter et al. 2001; Waterston et al. 2002) accounting for approximately 25% of the genome, and their open reading frames constitute less than 2% of nuclear DNA. A much lower than expected number of genes was surprising, primarily because it was assumed that the organization, growth, development and functioning of highly complex organisms requires a large number of different proteins. To some degree the repertoire of proteins encoded by the genome is expanded by alternative splicing which in some cases results in multiple protein variants being produced by one gene. It is estimated that over half of primary transcripts from human genes is subject to alternative splicing (Lareau et al. 2004).

Detailed studies aimed at the determination of the mammalian transcriptome using large scale sequencing of full length cDNA libraries (Okazaki et al. 2002; Carninci et al. 2005) and genome tiling arrays (Cawley et al. 2004; Cheng et al. 2005) suggest that over half of the genome is actually transcribed, and that in addition to protein-coding there is at least an equal number of RNA-coding genes (Fig. 1). These observations are consistent with the hypothesis that the complexity of organisms does not depend on the number of proteins encoded by the genomes but rather on regulatory networks including ncRNAs (Mattick 1994, 2001, 2003). An increase in complexity accompanied by a decrease in the size of protein-coding part of the genomes is evident looking at the sequenced genomes of bacteria, invertebrates and mammals, in which the non-coding portions account for 10, 30–40 and 98% of the genome, respectively.

The noncoding part of the transcriptome consists of untranslated regions of protein-coding genes (introns, 3′- and 5′-UTRs) and independent noncoding RNAs

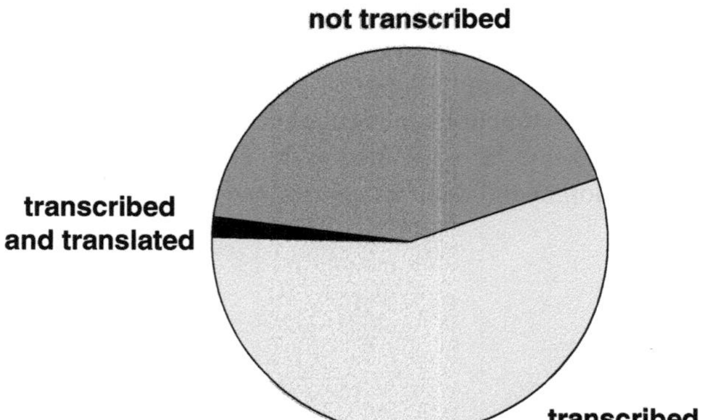

Fig. 1 Coding capacity of human genomic DNA. Over half of genomic DNA is transcribed, but only 1.5–2% encodes proteins. Transcribed and not translated RNAs include untranslated regions of protein-coding genes, introns as well as housekeeping and regulatory ncRNA genes

(ncRNAs). In a broad sense, the term ncRNA applies to any transcript without protein-coding potential, including well-defined classes of housekeeping or infrastructural RNAs (e.g., tRNA, rRNA, snoRNA, snRNA, tmRNA) essential for protein biosynthesis, processing and modifications of RNAs and other basic cellular functions. These RNAs are usually constitutively expressed at constant levels in all cells.

There is, however, another group of highly heterogenous noncoding transcripts which perform regulatory functions and show highly specific expression patterns depending on the environmental conditions, tissue type or developmental stage. RNA-based regulatory mechanisms controlling expression of genes have been shown to be common both in bacteria and in eukaryotes. In mammals, the majority of known regulatory ncRNAs are transcribed by RNA polymerase II (PolII) and possess mRNA-like features: the hypermodified cap structure at the 5′-end and a poly(A) tail at the 3′-end. Another feature shared with protein-coding genes is the usage of the same promoter elements and transcription factors (Cawley et al. 2004). This is also true for the rare examples of ncRNAs transcribed by RNA polymerase III (PolIII) that are primarily used for constitutive transcription of housekeeping genes (tRNA, 5S rRNA, 5.8S rRNA, snRNA). Expression of these genes depends on upstream promoters, specific TFIIIB50 factor and transcription factors used in transcription of PolII-specific genes (Paule and White 2000; Teichmann et al. 2000) that ensures high specificity of expression.

A majority of large PolII-transcribed ncRNAs have been identified through a large-scale sequencing approach using cDNA libraries. However, the analysis of transcriptional activity of human chromosomes using whole-genome tiling arrays demonstrated that 43% of RNAs is not polyadenylated, and that over a half of transcripts show exclusively nuclear localization (Cheng et al. 2005). The cDNA cloning approach does not allow identification of polyA-RNAs whether transcribed by RNA polymerases II or III.

In some cases, the mature, functional ncRNAs are products of processing of introns spliced out of the host genes. Such a mechanism is used in a process of biogenesis of small nucleolar RNAs (snoRNA) which serve as guide molecules in post-transcriptional modification (2′-*O*-methylation and pseudourydylation) of RNAs (Runte et al. 2001). Also, microRNAs that play a key role in translational regulation of many eukaryotic genes are embedded within introns of host genes transcribed by PolII (Rodriguez et al. 2004). Apart from these clearly identifiable groups of ncRNAs other introns may also represent functional by-products constituting elements of regulatory networks and providing information about current gene expression status (Mattick 2003).

2 Functions of ncRNAs

Although in the last few years there was a dramatic growth in the number of identified novel ncRNAs, we are still far from the full understanding of their roles in functioning of the cells. Obviously, the RNA turned out to be much more versatile than was initially believed. It does not only provide templates and decoding machinery for translation of the genetic information from DNA to proteins, but also performs a number of functions that are crucial for many cellular processes. Ribosomal RNAs and snRNA not only provide a scaffold for ribosomal and spliceosomal proteins, but also form catalytic centers of ribosomes and spliceosomes. RNAs play a role of guiding molecules directing RNA modifications and editing. During DNA replication, the synthesis of a new DNA strand is initiated by short RNA primers, and the synthesis of telomeres depends on a presence of an RNA template provided by telomerase RNA. RNA molecules are parts of many other ribonucleoprotein particles which play a role in protein transport (signal recognition particle, SRP). It is now clear that RNAs in contemporary organisms are not only remnants of the ancient "RNA world," but still play a role equally important as that of proteins. One of the recently discovered ncRNAs, HAR1F, may be a key to understanding the evolution of human brain (Pollard et al. 2006). This RNA is expressed during the development of human brain and co-localized with Reelin involved in specifying the structure of the cortex. The involvement of RNAs in processes associated with growth and development of higher organisms suggests that the changes of their expression patterns may be responsible for the evolution of organisms.

Currently, there are relatively few ncRNAs with known cellular functions and even fewer for which mechanisms of action have been unraveled. It seems, however, that a significant fraction of ncRNAs is directly or indirectly involved in controlling expression of genes. RNA-dependent regulatory mechanisms can operate on practically every step of transmission of genetic information. On the transcriptional level, certain RNAs can modulate the activity of transcription factors or RNA polymerase. Post-transcriptional regulation can affect splicing, transport, mRNA stability or translation. RNAs influence transcriptional activity of chromatin by providing signals for setting up epigenetic features (DNA methylation, modification of histone proteins) and changing its structure.

2.1 Transcriptional Regulation

Several mammalian ncRNAs play a role in regulation of the activity of transcription factors and RNA polymerase. One of the RNAs known for over 20 years for which there was no clearly identified function is 7SK RNA. This RNA is transcribed by PolIII and is an inhibitor of positive transcription elongation factor b (P-TEFb), a cofactor of RNA polymerase II required for a transition from abortive to productive elongation (Nguyen et al. 2001; Yang et al. 2001). P-TEFb is also a host cofactor of Tat protein essential for HIV-1 transcription (Kiernan et al. 1999). The action of P-TEFb depends on the activity of its cyclin-dependent kinase 9 (Cdk9) phosphorylating the PolII largest subunit's C-terminal domain. This activity is inhibited by interaction with 7SK RNA and HEXIM1/MAQ1 protein (Yik et al. 2004). The complex dissociates under stress conditions making P-TEFb available for transcription of stress-induced genes (Yik et al. 2003).

Another PolIII transcript implicated in the regulation of PolII activity is mouse B2 RNA transcribed from repetitive short interspersed elements (SINEs) in response to heat shock. The transcriptional activation of B2 SINEs is followed by downregulation of PolII-dependent genes (Allen et al. 2004). B2 RNAs show high affinity for the core PolII and inhibit transcription at the initiation step. Dissociation or removal of B2 RNA from the pre-initiation complex results in reactivation of transcription after the shock (Espinoza et al. 2004). In humans, heat shock and other stress conditions induce transcription from Alu repeats. These abundant PolIII transcripts may therefore play similar role to that of B2 RNAs in mouse (Allen et al. 2004).

One of the first ncRNAs involved in the regulation of transcription identified in mammals was steroid receptor activator RNA (SRA RNA). SRA RNA serves as a coactivator of nuclear receptors of steroid hormones including progestins, estrogens, androgens and glucocorticoids (Lanz et al. 1999). Several tissue-specific isoforms of mature SRA RNA result from alternative splicing of the primary transcript. These variants, ranging in size from ~700 to ~1,500 nucleotides, differ in 5- and 3′-terminal regions flanking a common central core. An interesting feature of the SRA-encoding gene is the fact that, in addition to the noncoding SRA RNA, it also yields a coding mRNA transcribed from the alternative transcription start site (Emberley et al. 2003).

Analysis of the SRA RNA mutants revealed that the core region consists of five essential putative secondary structure elements disruption of which results in reduced transcriptional coactivation (Lanz et al. 2002). The activity of SRA RNA depends on its interaction with SRC-1 (steroid receptor coactivator 1) protein (Lanz et al. 1999) and a hormone induced transcriptional repressors: SHARP, (SMRT/HDAC1 associated repressor protein) (Shi et al. 2001) and SLIRP (SRA stem-loop interacting RNA-binding protein) (Hatchell et al. 2006). Thus, it can function both in activation and repression of hormone-responsive genes and a fine-tuning of their expression may depend on the competition between the two proteins for the RNA component.

The key protein components associated with SRA RNA activity are DEAD-box RNA-binding proteins p68 and p72. They bind SRA RNA and are probably responsible for its association with SRC-1 (Watanabe et al. 2001). The p68/p72 has also been found to form a complex with SRA RNA and a transcription factor MyoD involved in muscle development. Both p68/p72 and SRA RNA are essential for correct gene expression patterns in mouse myoblasts (Caretti et al. 2006).

A noncoding RNA NRON plays a role in the regulation of transcription of the genes activated by the nuclear factor of activated T cells (NFAT). NFAT is a calcium-responsive transcription factor implicated in the development of muscles, vascular and nervous system as well as in T-cell mediated immune response. NRON inactivation using specific shRNAs (short hairpin RNAs) results in a significant increase of NFAT activity. NRON interacts with nuclear import factors (importin-beta family members) essential for NFAT accumulation in the nucleus and subsequent activation of respective genes. Thus NRON acts as a repressor of nuclear trafficking of NFAT, but the precise mechanism is yet unknown (Willingham et al. 2005).

2.2 Post-Transcriptional Regulation

In bacteria, most of the noncoding regulatory RNAs act on the post-transcriptional step of gene expression. Complementary interaction between ncRNA and a target mRNA results in either repression or stimulation of translation (Storz et al. 2005). In mammalian genomes, there is a significant fraction of expressed regions showing evidence of overlapping sense-antisense transcription (Kapranov et al. 2005; Katayama et al. 2005). The discovery of RNA interference (RNAi) mechanism prompted speculations that the sense–antisense pairs of transcripts could serve as substrates for Dicer to generate siRNAs (short interfering RNA) targeting either of the transcripts. However, there is no evidence of such a mechanism in mammals and sense–antisense pairs are not detectable in the cytoplasm (Faghihi and Wahlestedt 2006).

There are relatively few well-studied examples of functional relationship between sense and antisense transcripts. Transcription of the antisense noncoding gene was shown to interfere with expression of the HOXA11 gene (Chau et al. 2002), and a similar mechanism may be responsible for the suppression of transcription of the β isoform of the cardiac myosin heavy chain (Haddad et al. 2003). Antisense transcripts can also regulate splicing of the overlapping genes. Expression of the two isoforms of thyroid hormone receptor α (TRα1 and TRα2) is regulated by the antisense transcript encoding a nuclear receptor, RevErb. RevErb mRNA overlaps sequences coding for TRα2 and suppressing alternative splicing. This results in preferential expression of TRα1 (Hastings et al. 2000). Possible regulatory mechanisms involving overlapping antisense transcripts may also involve cytoplasmic export of the mRNA.

An interesting example of the post-transcriptional regulation of gene expression involves the expressed processed pseudogene of Makorin1 (*Makorin-p1*). Both Makorin1 gene and expressed pseudogene share the *cis*-acting RNA decay element which is responsible for their degradation. Expression of the pseudogene results in decreased rate of turnover of the coding mRNA and its deletion is lethal (Hirotsune et al. 2003). In human and mouse genomes there are 5,206 and 3,428 processed pseudogenes, respectively (Adel et al. 2005), some of which may play a similar role to that of *Makorin-p1* (Yano et al. 2004).

In the neurons of rodents and primates, translation of mRNA is regulated by PolIII transcripts BC1 and BC200 (Martignetti and Brosius 1995) that form ribonucleoprotein particles found in cell bodies and in dendrites (Tiedge et al. 1991, 1993). BC1 RNA plays a role of translational repressor interfering with the formation of a stable 48S preinitiation complex. BC1 specifically inhibits a helicase eIF4A and forms stable complexes with the poly(A)-binding protein (PABP) (Wang et al. 2002). At synapses, BC1 RNA also interacts with FMRP (fragile X mental retardation associated protein) and this association is essential for translational repression of a subset of mRNAs. BC1 probably facilitates FMRP binding to the mRNA (Zalfa et al. 2003).

MicroRNAs constitute a class of ncRNAs that have attracted most attention in recent years. Small 20-nucleotides-long microRNAs are ubiquitously expressed in animals, plants and viruses. Animal microRNAs are primarily involved in the regulation of translation. They bind to partially complementary sequences localized within 3′-UTRs of target mRNAs which results in translational repression, but the details of this mechanism remains obscure (Bartel 2004). According to one model, the binding of microRNA to target mRNA stops protein biosynthesis after the initiation step (Petersen et al. 2006). It has, however, been demonstrated that microRNAs are active only against mRNAs with the 7-methyl guanosine cap. This suggests that the inhibition takes place at the initiation step and may involve recognition of the cap by eIF4E (Pillai et al. 2005). The latter model is supported by recent studies which implicate one of the Argonaute family proteins (Ago2) in the process. Ago2 possess an element similar to the cap-binding domain of the eIF4E and are able to bind cap of microRNA-bound mRNA. This competition with eIF4A results in decreased translation (Kiriakidou et al. 2007). It was also suggested that inhibition of initiation step may be due to the recruitment of eIF6 by the microRNA–mRNA complex which in turn blocks joining of the ribosomal subunits and prevents translation (Chendrimada et al. 2007).

Although, the majority of animal microRNAs play a role of translational repressors, in mammals there are known examples of the induced cleavage of target mRNAs. In mouse, mir-196 is responsible for degradation of the mRNA encoding Hox-B8 protein (Yekta et al. 2004) and the Epstein-Barr virus miR-BART2 targets mRNA for viral DNA polymerase (Pfeffer et al. 2004).

A fundamental difference between the microRNA-induced cleavage of and translational repression is the extent of complementarity between microRNA and target mRNA. The mRNA cleavage by Dicer requires that the microRNA is fully complementary to the target mRNA, and a single binding site for one miRNA is sufficient for the gene expression silencing. In the case of translational repression, the miRNA binding sites within the mRNA's 3′-UTR are not fully complementary, but silencing requires interaction with more than one microRNA.

These single-stranded regions within miRNA–mRNA complex are crucial for miRNA's activity (Vella et al. 2004). Most of the microRNAs seem to be involved in the regulation of a small subset of genes, but there are miRNAs for which binding sites have been found in hundreds of mRNAs, and some of them are probably regulated by more than one microRNA. The miRNA target sites conserved in other mammals have been found in over 2,000 human genes (John et al. 2004).

2.3 Epigenetic Regulation

One of the most interesting aspects of ncRNAs activity is their involvement in epigenetic processes. In mammals, ncRNAs play a key role in silencing of genes within imprinted clusters and on the inactive X chromosomes as well as DNA methylation.

Genomic imprinting is a process whereby the expression of genes is restricted to either the paternal or maternal allele. In mouse, there are about 80 genes that show imprinted expression. Imprinted genes are usually arranged in clusters. Interestingly, one of the genes within a cluster always specifies an ncRNA, and the protein and RNA genes show reciprocal imprinted expression patterns (Morison et al. 2005). The precise role of ncRNAs in most cases remains unknown, but there are two well-studied cases in which the expression of ncRNA-coding gene was shown to be essential for imprinted expression of other genes.

The cluster of imprinted genes associated with Beckwith-Wiedemann syndrome (BWS) on human chromosome 11p15 contains 13 maternally and 4 paternally expressed genes, and the aberrant expression or mutations of its genes is responsible for BWS and several human cancers. One of the imprinting control elements is a differentially methylated CpG island (*KvDMR1*) within intron 10 of the maternally expressed *KCNQ1* (*KvLQT1*) gene. *KvDMR1* is methylated on the maternal allele. The unmethylated paternal allele gives rise to an ncRNA (LIT1/Kcnq1ot1) antisense to the KCNQ1 expressed in most human tissues. Absence of maternal allele's methylation and deletions of *KvDMR1* are the most frequent defects found in BWS patients (DeBaun et al. 2002). The expression of LIT1 RNA is required for silencing of paternal alleles of several maternally expressed genes (Horike et al. 2000). The silencing is bidirectional (Thakur et al. 2004) and its extent depends on the length of LIT1 RNA (Mancini-Dinardo et al. 2006; Kanduri et al. 2006). This process involves recruitment of Polycomb group complex Eed-Ezh2, and histone H3 methylations responsible for transcriptional repression, but it is unclear whether it depends on the opening of the chromatin by transcription or on the structural features of LIT1 RNA (Umlauf et al 2004).

Similar bidirectional silencing of genes was observed for imprinted genes on mouse chromosome 17. The maternally methylated imprinting control region within intron 2 of insulin-like growth factor type-2 receptor (*Igf2r*) gene regulates maternal expression of *Igf2r*, *Slc22a2* and *Slc22a3* genes (Wutz et al. 1997). Unmethylated paternal allele is transcribed as an unspliced, antisense 108-kb-long Air RNA (antisense *Igf2r*). Air RNA overlaps 30 kb of *Igf2r* gene, its promoter and extends over 70 kb upstream. Its expression is required for the silencing of all maternally expressed genes within the cluster, including *Slc22a2* and *Slc22a3* located 110 and 155 kb downstream from *Igf2r*, respectively (Sleutels et al. 2002, 2003).

Thus, both LIT1 and Air RNAs serve as triggers of bidirectional silencing. Interestingly, the silencing does not affect all the genes within the cluster, including ncRNA gene itself. Several models of silencing mechanism have been proposed, but there is no clear-cut evidence in favor of either. It is also possible that ncRNA-dependent silencing is not restricted to imprinted clusters, but may be more widespread (Pauler et al. 2007).

NcRNA is also fundamental for the silencing transcriptional silencing of genes in the process of X-chromosome inactivation. This phenomenon shares certain characteristics with the silencing of imprinted genes induced, but it extends to the whole X chromosome. The key element responsible for X-chromosome inactivation is Xic (X-inactivation center) locus from which a long, spliced and polyadenylated ncRNA Xist (X-inactive specific transcript) is transcribed. The expression of Xist RNA is critical for initiation of inactivation (Newall et al. 2001) but not for the maintenance of the silent state (Csankovszki et al. 2001). Xist RNA becomes associated with the chromatin which results in exclusion of PolII and its transcription factors creating a transcriptionally silent nuclear compartment (Chaumeil et al, 2006). This is followed by chromatin modification by methylation of the CpG islands in DNA and changes in histone composition (Chadwick and Willard 2004; Rougeulle et al. 2004). The protein factors implicated in these modifications include Polycomb group (PcG) repressive complexes 1 (PRC1) and 2 (PRC2) (Plath et al. 2004; de Napoles et al. 2004; Fang et al. 2004).

In contrast to RNAs involved in the silencing of imprinted genes, it seems that in the case of Xist RNA its activity depends on its sequence and/or structural features. It has been demonstrated that the association with X-chromosome chromatin and the recruitment of silencing factors depend on different RNA regions (Wutz et al. 2002). Thus, it is the RNA and not transcriptional activity initiated from Xic which is responsible for the inactivation.

Noncoding RNAs are involved in regulation of the promoter region methylation within the locus encoding rat sphingosine kinase 1 (*Sphk1*). There are several tissue specific isoforms of Sphk1 mRNA initiated from alternative transcription start sites and alternative splicing of untranslated 5′-terminal exons. Expression of Sphk1 isoforms is regulated by a 0.2-kb-long tissue-specific differentially methylated region T-DMR located within a 3.7-kb-long CpG island (Imamura et al. 2001, 2004a). It has been shown that the locus also produces a number of antisense (Khps1) RNAs varying in size from 0.6 to 20 kb. One of the antisense transcripts overlaps the T-DMR and is involved in methylation of non-CpG and demethylation of CpG sites within the T-DMR (Imamura et al. 2004a). The structure of SPHK1 gene is conserved in humans, and the T-DMR region contains a short 38-bp region showing 90% identity between rodents and primates. Also, the tissue specific methylation is evolutionarily conserved and suggests that the mechanism of regulation may be similar (Imamura et al. 2004b). The evolutionarily conserved localization of alternative first exons within CpG islands is a feature of other genes with tissue-specific mRNA isoforms (Turner et al. 2006), and there are numerous CpG islands harboring T-DMRs (Shiota 2004). Thus, the tissue-specific expression of T-DMR-controlled genes may be at least in some cases regulated by overlapping ncRNAs.

3 NcRNAs in Human Diseases

The chromosomal aberrations involving ncRNA-encoding genes and/or their altered expression have been implicated in many human diseases, including cancer, neurobehavioral and developmental disorders. Although the links between ncRNA expression and particular pathological states are in a majority of cases enigmatic, there are clues that point to the regulatory role of RNAs as a key factor in origin of certain diseases.

3.1 Cancer

3.1.1 MicroRNAs

Mammalian ncRNAs often show altered expression patterns in malignantly transformed tissues. In fact, many of the novel ncRNA species have been identified as products of specifically overexpressed genes in cancer cell lines. Some of these RNAs have been shown to represent very sensitive molecular markers that could be used for early diagnosis or prognosis in the treatment of various forms of cancer. The ncRNAs implicated in neoplastic growth belong to various groups and, apart from microRNAs, the functions of the majority of them are unknown. In many cases, it is difficult to determine if altered expression of ncRNA triggers malignant growth or is a secondary effect of the changes in cells' genetic program.

At present, much effort is committed to understanding a role of microRNAs in the origins and development of cancer. The expression patterns of microRNAs are tissue-specific and there is a correlation between their anomalous expression and malignant cells' growth.

A correlation between the expression of microRNAs and the pathogenesis of cancer was first noticed following the observation that the region deleted in over half of human B cell chronic lymphocytic leukemia (B-CLL) cases contains two microRNAs: miR-15a and miR-16a (Calin et al. 2002). These microRNAs target mRNA encoding an antiapoptotic gene BCL2 (B cell lymphoma 2) and induce apoptosis (Cimmino et al. 2005). Such a correlation was further supported by the genome-wide analysis of human microRNAs that often co-localize with chromosomal regions defects which are implicated in cancer (Calin et al. 2004).

In fact, there are numerous cases in which there is a connection between altered expression of microRNAs and cancer. In most cases, however, the link between down- or upregulation of microRNA and malignant growth is not clearly established. A highly aggresive form of brain tumor, glioblastoma multiforme (GBM), shows marked over-expression of miR-21 that is involved in suppression of genes responsible for apoptosis (Chan et al. 2005). The GBM cells also show over-expression of miR-221 and reduced levels of miR-181a, miR-181b and miR-181c (Ciafre et al. 2005). In other cases, certain microRNAs that are present at high levels in normal tissues are significantly downregulated in cancer cells. Prostate and bladder tumors express miR-127 at much lower levels than healthy tissues. In this case, the malignant growth may be due to over-expression of the BCL6 proto-oncogene that is one of the possible targets for miR-127 (Saito et al. 2006). In lung cancer, there is a markedly lower expression of the let-7 microRNA than in normal lung cells. It has been found that target genes for let-7 regulation are members of the RAS oncogene family. All three human RAS genes contain within their 3′-UTRs multiple binding sites for let-7 microRNA, and the downregulation of let-7 is associated with increased levels of RAS protein in lung cancers (Johnson et al. 2005). These results are consistent with computational predictions of potential targets for miRNAs that demonstrated that many of candidate mRNAs originate from genes that are known to be involved in various forms of cancer (John et al. 2004).

The analysis of microRNA expression on a large scale can also be used for diagnostic purposes. It has been demonstrated that the expression profiles of microRNAs provide much more accurate information for classification of cancers than the profiles of protein-coding genes (Lu et al. 2005). A comparison of normal and breast cancer tissues revealed significantly lower levels of expression of miR-125b, miR-145, miR-21, and miR-155 in malignant cells, and the expression profiles for various cell lines differ depending on the tumor stage, proliferation, and responsiveness to hormones (Iorio et al. 2005). Thus, the expression patterns of microRNAs combined with the analysis of the functions of their targets can provide an insight into mechanisms of cancer pathogenesis.

3.1.2 mRNA-Like ncRNAs

Many of the long, polyadenylated ncRNAs were discovered as specific molecular markers over-expressed in cancer cells. The first noncoding transcript associated with malignant growth in humans was H19 RNA that is normally expressed at very high levels in placenta, embryo and in most of fetal tissues. *H19* is transcribed exclusively from the maternal allele. The loss of imprinted expression is often associated with cancer (Ulaner et al. 2003). Interestingly, the expression of H19 can have different effects on the progression of tumors, depending on the type of tissue. In some cell lines, it seems to play a role of tumor suppressor (Hao et al. 1993) while in others it has been proposed to possess oncogenic properties (Lottin et al. 2002, 2005). These differences may depend on the expression of alternatively spliced isoforms (Matouk et al. 2004) or a presence of specific RNA-binding proteins (Ioannidis et al. 2004). Recently, it has been demonstrated that H19 can also play a role of a precursor of microRNA miR-675 (Cai and Cullen 2007).

Another RNA-coding imprinted gene originating from the same cluster, expression of which was shown to be associated with cancer, is PEG8/IGF2AS (paternally expressed gene 8), an antisense transcript from the IGF (insulin-like growth factor 2) locus. Transcription of this RNA is upregulated in several fetal cancers including kidney Wilms' tumor. The increased expression of PEG8 was observed only in malignant cells, but not in normal kidney tissue (Okutsu et al. 2000).

Bladder transitional cell carcinoma (TCC) is the most common genitourinary cancer in China. The search for a molecular marker for this malignancy resulted in identification of a ncRNA gene UCA1 (urothelial carcinoma associated 1) significantly and specifically over-expressed in patients with TCC (Wang et al. 2006). The full-length cDNA sequence of a mature 1.4 kb transcript revealed that UCA1 gene belongs to the HERV-H (human endogenous retroviruses) family. There is no significant open reading which suggests that UCA1 does not code for protein. It is not currently known how the UCA1 RNA's expression contributes to tumorigenesis, but there are known HERVs involved in cancer (Galli et al. 2005; Mangeney et al. 2005) and in the regulation of other oncogenes through their LTR (de Parseval et al. 1999).

In human squamous carcinoma cell lines resistant to drug-induced apoptosis, differential display revealed a novel noncoding transcript, CUDR (cancer up-regulated drug resistant). It has been demonstrated that CUDR expression is in fact associated with resistance to anticancer drugs doxorubicin and etoposide. CUDR shows elevated levels in several human cancer cell lines, including hepatocellular carcinoma, breast carcinoma, HeLa cells, colon carcinoma and lung carcinoma. One of the possible roles of CUDR is the regulation of caspase 3 expression and interference with the apoptotic pathway, as the CUDR-expressing cells show reduced levels of caspase 3 protein (Tsang et al. 2007).

In prostate cancers, there are two ncRNAs that display marked and highly specific over-expression when compared with normal prostatic cells. *PCA3* (prostate cancer antigen 3, DD3) gene produces a series of alternatively spliced and alternatively polyadenylated ncRNAs (Bussemakers et al. 1999). PCA3 shows highly specific expression in prostate tumors, and in over 95% of cases its levels are significantly increased (de Kok et al. 2002). The prostate-specific gene that is upregulated in cancer cells is PCGEM1 expressed exclusively in glandular epithelial cells (Srikantan et al. 2000). PCGEM1 RNA may play a role in regulation of growth and its over-expression in cell cultures results in increased cell proliferation and colony formation (Petrovics et al. 2004). Several other ncRNAs have been also identified as upregulated transcripts associated with various forms of cancer. The NCRMS NCRMS (noncoding RNA in rhabdomyosarcoma) is over-expressed in alveolar rhabdomyosarcoma (RMS), neuroblastoma and synovial sarcoma (Chan et al. 2002). Colon carcinoma cells show significant over-expression of OCC-1 RNA (over-expressed in colon carcinoma 1), but it is absent or expressed at very low levels in normal mucosa (Pibouin et al. 2002). In lung adenocarcinoma, one of the molecular markers of metastasis is ncRNA MALAT-1 (metastasis associated in lung adenocarcinoma transcript 1) (Ji et al. 2003). MALAT-1 was recently identified as one of nuclear-enriched abundant transcripts (NEAT2) originating from chromosome 11. It was shown to be tightly associated with mature SC35 nuclear speckles in human and mouse cells. The role of these structures is not clear, but it seems that they play a role in pre-mRNA processing (Hutchinson et al. 2007).

3.1.3 PolIII Transcripts

The ncRNAs transcribed by RNA polymerase III were also associated with cancers. The brain-specific BC1 and BC200 RNAs are also expressed in certain human and mouse cancers (Chen et al. 1997a,b). In humans, BC200 transcripts were detectable in malignant cells, but not in normal tissues. The expression of otherwise brain-specific BC200 RNA in these cases was specific, and the increased levels of PolIII transcription associated with proliferation was ruled out as an explanation of its activation (Chen et al. 1997b). In breast cancer, BC200 RNA is detectable only in invasive but not benign forms, which indicates a high degree of genomic instability and provides a potential marker for diagnostic purposes (Iacoangeli et al. 2004).

A direct role in the regulation of cancer related genes was proposed for the 7SK RNA involved in expression of *c-myc* protooncogene. Abnormal expression of *c-myc* is observed in many tumors. Transcription of *c-myc* can be initiated from one of four promoters: P0 through P3. The most active P2 is responsible for 75–90% of *c-myc*'s transcription in normal cells. However, SV40-induced transformation results in a significant increase of the utilization of promoters P1 and P3 following an upregulation of 7SK RNA-coding gene. It has been shown that the 7SK RNA can be responsible for increased transcription from P1 and P3 promoters (Luo et al. 1997).

3.2 Developmental and Neurobehavioral Disorders

Several congenital, developmental disorders have been linked to abnormal expression patterns of imprinted ncRNAs. Although the precise role of the majority of these RNAs is still enigmatic, their integrity is crucial for the monoallelic expression of protein-coding genes within imprinted clusters. The defects resulting in aberrant expression of the LIT1 RNA (see Sect. 2.3) and subsequent loss of imprinted expression of several genes within chromosomal region 11p15 has been linked to Romano-Ward, Jervell and Lange-Nielsen syndromes (Neyroud et al. 1999).

Prader–Willi syndrome (PWS) is a neurogenetic disorder characterized by a number of developmental defects and mental retardation. The PWS is caused by genetic defects leading to inactivation of paternally expressed genes within the 15q11-q13 region. There are four protein coding genes (*NDN*, *MKRN3*, *MAGEL2*, and *SNURF-SNRPN*) and one RNA-coding gene (*IC-SNURF-SNRPN/UBE3A-ATS*) expressed exclusively from the paternal chromosome. Human *UBE3A-ATS* is initiated from PWS imprinting control region upstream of *SNURF-SNRPN* and spans over 460 kb. It consists of 148 exons and serves as a host gene for several snoRNAs predominantly located within introns. It also overlaps in antisense orientation maternally expressed *UBE3A* (ubiquitin protein ligase E3A). It has been proposed that the loss of expression of certain snoRNAs processed from the *UBE3A-ATS* transcript may be directly linked to the origin of PWS (Runte et al. 2001; Gallagher et al. 2002). On the other hand, the loss of antisense transcript overlapping *UBE3A* gene may be associated with its abnormal biallelic expression observed in Angelman syndrome (Chamberlain and Brannan 2001; Lalande and Calciano 2007).

DiGeorge syndrome (DGS) correlates with deletions within human chromosome 22q11. Within DGS critical region (DGCR), a gene (*DGCR5*) producing several alternatively spliced and apparently non-protein-coding cDNAs has been identified. These noncoding transcripts are expressed during human and mouse embryogenesis. Interestingly, the first intron of the *DGCR5* contains in a sense orientation a protein-coding *DGCR3* gene (Sutherland et al. 1996). Another ncRNA that may be associated with the origin of DiGeorge syndrome is a product of 22k48 gene transcribed in antisense orientation from the first intron of *HIRA*, a gene deleted in DiGeorge syndrome. 22k48 produces several alternatively spliced RNAs with unknown function expressed in neurons of central and peripheral nervous systems. It has been proposed that 22k48 haploinsufficiency may be one of the causes of DiGeorge syndrome (Pizzuti et al. 1999).

NcRNA-coding genes were also identified within chromosomal regions linked to autism and schizophrenia. Chromosomal defects within the long arm of human chromosome 7 are associated with autism. In one case of an autistic boy, a translocation t(7;13)(q31.2;q21) disrupts a complex *RAY1/ST7* locus (Vincent et al. 2000). Apart from two protein-coding transcripts ST7 and RAY1, there are four major ncRNAs transcribed from sense and antisense strands. It has been proposed that the antisense transcripts overlapping protein-coding genes may play a role of negative regulators, yet their possible role in the pathogenesis of autistic disorder is not known.

The breakpoint region of chromosome 1q43 that is a target of a balanced translocation t(1:11)(q43,q14) found in some schizophrenia patients contains two genes: DISC1 and DISC2 (disrupted in schizophrenia 1 and 2). Alternative *DISC2* transcripts are noncoding RNAs and overlap in antisense orientation the 3′-region of *DISC1* encoding a large protein with similarity to other proteins of the nervous system. It is, therefore, possible that DISC2 RNAs may somehow be involved in the control of DISC1 expression (Millar et al. 2000). On chromosome 11, the same translocation disrupts another *PSZA11q14* gene (putative schizophrenia-associated gene from 11q14). A mouse homolog of that gene exists in mouse genome, but there is no conservation of the putative ORF which strongly suggests that there is no protein product. Expression of PSZA11q14 RNA is reduced in patients with schizophrenia and its localization within the first intron of the *DLG2* suggested that it plays a role of a *cis*-antisense regulator (Polesskaya et al. 2003).

4 Perspectives

The vast majority of noncoding RNAs that have been implicated in human diseases are linked to various forms of cancer and neurobehavioral or developmental defects. It is highly possible that they also play a role in pathogenesis of other diseases. Recently, it has been demonstrated that human cytomegalovirus (HCMV) encodes a microRNA hcmv-mirR-UL112 that is involved in mechanisms that allow a virus to escape host's immune system. The expression of hcmv-mirR-UL112 results in downregulation of major histocompatibility complex class I–related chain B (MICB), a stress-induced ligand of the natural killer cell activating receptor NKG2D. MICB expression is essential for killing of virus-infected cells by natural killer cells (Stern-Ginossar et al. 2007).

Expression of a specific ncRNA was also identified as a factor contributing to psoriasis susceptibility. PRINS RNA (psoriasis susceptibility-related RNA gene induced by stress) is expressed in various normal human tissues, but it shows significant upregulation in the uninvolved epidermis of psoriatic patients and in response to UV-B irradiation or viral infection, and may be involved in protection of cells in under stress conditions (Sonkoly et al. 2005).

The unraveling of mechanisms of gene expression in which RNAs seem to play an essential role will greatly contribute to understanding of many pathological processes leading to human diseases. Specific expression of RNA-coding genes associated with particular pathologies may be exploited to design new diagnostic methods. It has been shown that the microRNA profiling is much more sensitive and accurate in identification of human cancers than other methods taking advantage of protein expression profiles (Lu et al. 2005).

Our today's knowledge of the possible functions of ncRNAs is superficial and there are many unclassified RNAs. Their role in regulation of cellular processes and potential involvement in pathologies still awaits characterization.

References

Adel K, Laurent D, Dominique M (2005) HOPPSIGEN: a database of human and mouse processed pseudogenes. Nucleic Acids Res 33(Database Issue):D59–D66

Allen TA, Von Kaenel S, Goodrich JA et al (2004) The SINE-encoded mouse B2 RNA represses mRNA transcription in response to heat shock. Nat Struct Mol Biol 11:816–821

Bartel DP (2004) MicroRNAs: Genomics, biogenesis, mechanism, and function. Cell 116:281–297

Bussemakers MJ, van Bokhoven A, Verhaegh et al (1999) DD3: A new prostate-specific gene, highly overexpressed in prostate cancer. Cancer Res 59:5975–5979

Cai X, Cullen BR (2007) The imprinted H19 noncoding RNA is a primary microRNA precursor. RNA 13:313–316

Calin GA, Dumitru CD, Shimizu M et al (2002) Frequent deletions and down-regulation of micro-RNA genes miR15 and miR16 at 13q14 in chronic lymphocytic leukemia. Proc Natl Acad Sci USA 99:15524–15529

Calin GA, Sevignani C, Dumitru CD et al (2004) Human microRNA genes are frequently located at fragile sites and genomic regions involved in cancers. Proc Natl Acad Sci USA 101:2999–3004

Caretti G, Schiltz RL, Dilworth FJ et al (2006) The RNA helicases p68/p72 and the noncoding RNA SRA are coregulators of MyoD and skeletal muscle differentiation. Dev Cell 11:547–560

Carninci P, Kasukawa T, Katayama S et al (2005) The transcriptional landscape of the mammalian genome. Science 309:1559–1563

Cawley S, Bekiranov S, Ng HH et al (2004) Unbiased mapping of transcription factor binding sites along human chromosomes 21 and 22 points to widespread regulation of noncoding RNAs. Cell 116:499–509

Chadwick BP, Willard HF (2004) Multiple spatially distinct types of facultative heterochromatin on the human inactive X chromosome. Proc Natl Acad Sci USA 101:17450–17455

Chamberlain SJ, Brannan CI (2001) The Prader-Willi syndrome imprinting center activates the paternally expressed murine Ube3a antisense transcript but represses paternal Ube3a. Genomics 73:316–322

Chan AS, Thorner PS, Squire JA et al (2002) Identification of a novel gene NCRMS on chromosome 12q21 with differential expression between Rhabdomyosarcoma types. Oncogene 21:3029–3037

Chan JA, Krichevsky AM, Kosik KS (2005) MicroRNA-21 is an antiapoptotic factor in human glioblastoma cells Cancer Res 65:6029–6033

Chau YM, Pando S, Taylor HS (2002) HOXA11 silencing and endogenous HOXA11 antisense ribonucleic acid in the uterine endometrium. J Clin Endocrinol Metab 87:2674–2680

Chaumeil J et al (2006) A novel role for Xist RNA in the formation of a repressive nuclear compartment into which genes are recruited when silenced. Genes Dev 20:2223–2231

Chen W, Heierhorst J, Brosius J et al (1997a) Expression of neural BC1 RNA: induction in murine tumours. Eur J Cancer 33:288–292

Chen W, Bocker W, Brosius J et al (1997b) Expression of neural BC200 RNA in human tumours. J Pathol 183:345–351

Chendrimada TP, Finn KJ, Ji X et al (2007) MicroRNA silencing through RISC recruitment of eIF6. Nature 447:823–828

Cheng J, Kapranov P, Drenkow J et al (2005) Transcriptional maps of 10 human chromosomes at 5-nucleotide resolution. Science 308:1149–1154

Ciafre SA, Galardi S, Mangiola A et al (2005) Extensive modulation of a set of microRNAs in primary glioblastoma. Biochem Biophys Res Commun 334:1351–1358

Cimmino A, Calin GA, Fabbri M et al (2005) miR-15 and miR-16 induce apoptosis by targeting BCL2. Proc Natl Acad Sci USA 102:13944–13949

Csankovszki G, Nagy A, Jaenisch R (2001) Synergism of Xist RNA, DNA methylation, and histone hypoacetylation in maintaining X chromosome inactivation. J Cell Biol 153:773–781

DeBaun MR, Niemitz EL, McNeil DE et al (2002) Epigenetic alterations of *H19* and *LIT1* distinguish patients with Beckwith-Wiedemann syndrome with cancer and birth defects. Am J Hum Genet 70:604–611

de Kok JB, Verhaegh GW, Roelofs RW et al (2002) DD3(PCA3), a very sensitive and specific marker to detect prostate tumors. Cancer Res 62:2695–2698

de Napoles M, Mermoud JE, Wakao R et al (2004) Polycomb group proteins Ring1A/B link ubiquitylation of histone H2A to heritable gene silencing and X inactivation. Dev Cell 7:663–676

de Parseval N, Alkabbani H, Heidmann T (1999) The long terminal repeats of the HERV-H human endogenous retrovirus contain binding sites for transcriptional regulation by the Myb protein. J Gen Virol 80:841–845

Emberley E, Huang GJ, Hamedani MK et al (2003) Identification of new human coding steroid receptor RNA activator isoforms. Biochem Biophys Res Commun 301:509–515

Espinoza CA, Allen TA, Hieb AR et al (2004) B2 RNA binds directly to RNA polymerase to repress transcript synthesis. Nat Struct Mol Biol 11:822–829

Faghihi MA, Wahlestedt C (2006) RNA interference is not involved in natural antisense mediated regulation of gene expression in mammals. Genome Biol 7:R38

Fang J, Chen T, Chadwick B et al (2004) Ring1b-mediated H2A ubiquitination associates with inactive X chromosomes and is involved in initiation of X inactivation. J Biol Chem 279: 52812–52815

Gallagher RC, Pils B, Albalwi M et al (2002) Evidence for the role of PWCR1/HBII-85 C/D box small nucleolar RNAs in Prader-Willi syndrome. Am J Hum Genet 71:669–678

Galli UM, Sauter M, Lecher B et al (2005) Human endogenous retrovirus rec interferes with germ cell development in mice and may cause carcinoma in situ, the predecessor lesion of germ cell tumors. Oncogene 24:3223–3228

Haddad F, Bodell PW, Qin AX et al (2003) Role of antisense RNA in coordinating cardiac myosin heavy chain gene switching. J Biol Chem 278:37132–37138

Hao Y, Crenshaw T, Moulton T et al (1993) Tumour-suppressor activity of H19 RNA. Nature 365:764–767

Hastings ML, Ingle HA, Lazar MA et al (2000) Post-transcriptional regulation of thyroid hormone receptor expression by *cis*-acting sequences and a naturally occurring antisense RNA. J Biol Chem 275:11507–11513

Hatchell EC, Colley SM, Beveridge DJ et al (2006) SLIRP, a small SRA binding protein, is a nuclear receptor corepressor. Mol Cell 22:657–668

Hirotsune S, Yoshida N, Chen A et al (2003) An expressed pseudogene regulates the messenger-RNA stability of its homologous coding gene. Nature 423:91–96
Horike S, Mitsuya K, Meguro M et al (2000) Targeted disruption of the human *LIT1* locus defines a putative imprinting control element playing an essential role in Beckwith-Wiedemann syndrome. Hum Mol Genet 9:2075–2083
Hutchinson JN, Ensminger AW, Clemson CM et al (2007) A screen for nuclear transcripts identifies two linked noncoding RNAs associated with SC35 splicing domains. BMC Genomics 8:39
Iacoangeli A, Lin Y, Morley EJ et al (2004) BC200 RNA in invasive and preinvasive breast cancer. Carcinogenesis 25:2125–2133
Imamura T, Ohgane J, Ito S et al (2001) CpG island of rat sphingosine kinase-1 gene: tissue-dependent DNA methylation status and multiple alternative first exons. Genomics 76:113–125
Imamura T, Yamamoto S, Ohgane J et al (2004a) Non-coding RNA directed DNA demethylation of Sphk1 CpG island. Biochem Biophys Res Commun 322:593–600
Imamura T, Miyauchi-Senda N, Tanaka S et al (2004b) Identification of genetic and epigenetic similarities of SPHK1/Sphk1 in mammals. J Vet Med Sci 66:1387–1393
Ioannidis P, Kottaridi C, Dimitriadis E et al (2004) Expression of the RNA-binding protein CRD-BP in brain and non-small cell lung tumors. Cancer Lett 209:245–250
Iorio MV, Ferracin M, Liu CG et al (2005) MicroRNA gene expression deregulation in human breast cancer. Cancer Res 65:7065–7070
Ji P, Diederichs S, Wang W et al (2003) MALAT-1, a novel noncoding RNA, and thymosin beta4 predict metastasis and survival in early-stage non-small cell lung cancer. Oncogene 22:8031–8041
John B, Enright AJ, Aravin A et al (2004) Human microRNA targets. PLoS Biol 2:e363
Johnson SM, Grosshans H, Shingara J et al (2005) RAS is regulated by the let-7 microRNA family. Cell 120:635–647
Kanduri C, Thakur N, Pandey RR (2006) The length of the transcript encoded from the Kcnq1ot1 antisense promoter determines the degree of silencing. EMBO J 25:2096–2102
Kapranov P, Drenkow J, Cheng J et al (2005) Examples of the complex architecture of the human transcriptome revealed by RACE and high-density tiling arrays. Genome Res 5:987–997
Katayama S, Tomaru Y, Kasukawa T et al (2005) Antisense transcription in the mammalian transcriptome. Science 309:1564–1566
Kiernan RE, Vanhulle C, Schiltz L et al (1999) HIV-1 tat transcriptional activity is regulated by acetylation. EMBO J 18:6106–6118
Kiriakidou M, Tan GS, Lamprinaki S et al (2007) An mRNA m^7G cap binding-like motif within human Ago2 represses translation. Cell 129:1141–1151
Lalande M, Calciano MA (2007) Molecular epigenetics of Angelman syndrome. Cell Mol Life Sci 64:947–960
Lander ES et al (2001) Initial sequencing and analysis of the human genome. Nature 409:860–921
Lanz RB, McKenna NJ, Onate SA et al (1999) A steroid receptor coactivator, SRA, functions as an RNA and is present in an SRC-1 complex. Cell 97:17–27
Lanz R, Razani B, Goldberg AD et al (2002) Distinct RNA motifs are important for coactivation of steroid hormone receptors by steroid receptor RNA activator (SRA). Proc Natl Acad Sci USA 99:16081–16086
Lareau LF, Green RE, Bhatnagar RS et al (2004) The evolving roles of alternative splicing. Curr Opin Struct Biol 14:273–282
Lottin S, Adriaenssens E, Dupressoir T et al (2002) Overexpression of an ectopic H19 gene enhances the tumorigenic properties of breast cancer cells. Carcinogenesis 23:1885–1895
Lottin S, Adriaenssens E, Berteaux N et al (2005) The human H19 gene is frequently overexpressed in myometrium and stroma during pathological endometrial proliferative events. Eur J Cancer 41:168–177
Lu J, Getz G, Miska EA et al (2005) MicroRNA expression profiles classify human cancers. Nature 435:834–838

Luo Y, Kurz J, MacAfee N et al (1997) C-myc deregulation during transformation induction: involvement of 7SK RNA. J Cell Biochem 64:313–327
Mancini-Dinardo D, Steele SJ, Levorse JM et al (2006) Elongation of the Kcnq1ot1 transcript is required for genomic imprinting of neighboring genes. Genes Dev 20:1268–1282
Mangeney M, Pothlichet J, Renard M et al (2005) Endogenous retrovirus expression is required for murine melanoma tumor growth in vivo. Cancer Res 65:2588–2591
Martignetti JA, Brosius J (1995) BC1 RNA: transcriptional analysis of a neural cell-specific RNA polymerase III transcript. Mol Cell Biol 15:1642–1650
Matouk I, Ayesh B, Schneider T et al (2004) Oncofetal splice-pattern of the human H19 gene. Biochem Biophys Res Commun 318:916–919
Mattick JS (1994) Introns: evolution and function. Curr Opin Genet Dev 4:823–831
Mattick JS (2001) Non-coding RNAs: the architects of eukaryotic complexity. EMBO Rep 2:986–991
Mattick JS (2003). Introns and noncoding RNAs the hidden layer of eukaryotic complexity. In: Barciszewski J, Erdmann VA (eds) Noncoding RNAs: molecular biology and molecular medicine. Eurekah, Austin, pp 12–33
Millar JK, Wilson-Annan JC, Anderson S et al (2000) Disruption of two novel genes by a translocation co-segregating with schizophrenia. Hum Mol Genet 9:1415–1423
Morison IM, Ramsay JP, Spencer HG (2005) A census of mammalian imprinting. Trends Genet 21:457–463
Newall AE, Duthie S, Formstone E et al (2001) Primary non-random X inactivation associated with disruption of Xist promoter regulation. Hum Mol Genet 10:581–589
Neyroud N, Richard P, Vignier N et al (1999) Genomic organization of the KCNQ1 K1 channel gene and identification of C-terminal mutations in the long-QT syndrome. Circ Res 84:290–297
Nguyen VT, Kiss T, Michels AA et al (2001) 7SK small nuclear RNA binds to and inhibits the activity of CDK9/cyclin T complexes. Nature 414:322–325
Okazaki Y, Furuno M, Kasukawa T et al (2002) Analysis of the mouse transcriptome based on functional annotation of 60,770 full-length cDNAs. Nature 420:563–573
Okutsu T, Kuroiwa Y, Kagitani F et al (2000) Expression and imprinting status of human PEG8/IGF2AS, a paternally expressed antisense transcript from the IGF2 locus, in Wilms' tumors. J Biochem 127:475–483
Paule MR, White RJ (2000) Transcription by RNA polymerases I and III. Nucleic Acids Res 28:1283–1298
Pauler FM, Koerner MV Barlow DP (2007) Silencing by imprinted noncoding RNAs: is transcription the answer? Trends Genet 23:284–292
Petersen CP, Bordeleau ME, Pelletier J et al (2006) Short RNAs repress translation after initiation in mammalian cells. Mol Cell 21:533–542
Petrovics G, Zhang W, Makarem M et al (2004) Elevated expression of PCGEM1, a prostate-specific gene with cell growth-promoting function, is associated with high-risk prostate cancer patients. Oncogene 23:605–611
Pfeffer S, Zavolan M, Grasser FA et al (2004) Identification of virus-encoded microRNAs. Science 304:734–736
Pibouin L, Villaudy J, Ferbus D et al (2002) Cloning of the mRNA of overexpression in colon carcinoma-1: a sequence overexpressed in a subset of colon carcinomas. Cancer Genet Cytogenet 133:55–60
Pillai RS, Bhattacharyya SN, Artus CG et al (2005) Inhibition of translational initiation by Let-7 microRNA in human cells. Science 309:1573–1576
Pizzuti A, Novelli G, Ratti A et al (1999) Isolation and characterization of a novel transcript embedded within HIRA, a gene deleted in DiGeorge syndrome. Mol Genet Metabol 67:227–235
Plath K, Talbot D, Hamer KM et al (2004) Developmentally regulated alterations in Polycomb repressive complex 1 proteins on the inactive X chromosome. J Cell Biol 167:1025–1035

Polesskaya OO, Haroutunian V, Davis KL et al (2003) Novel putative nonprotein-coding RNA gene from 11q14 displays decreased expression in brains of patients with schizophrenia. J Neurosci Res 74:111–122
Pollard KS, Salama SR, Lambert N et al (2006) An RNA gene expressed during cortical development evolved rapidly in humans. Nature 443:167–172
Rodriguez A, Griffiths-Jones S, Ashurst JL et al (2004) Identification of mammalian microRNA host genes and transcription units. Genome Res 14:1902–1910
Rougeulle C, Chaumeil J, Sarma K et al (2004) Differential histone H3 Lys-9 and Lys-27 methylation profiles on the X chromosome. Mol Cell Biol 24:5475–5484
Runte M, Huttenhofer A, Gross S et al (2001) The *IC-SNURF-SNRPN* transcript serves as a host for multiple small nucleolar RNA species and as an antisense RNA for *UBE3A*. Hum Mol Genet 10:2687–2700
Saito Y, Liang G, Egger G et al (2006) Specific activation of microRNA-127 with downregulation of the proto-oncogene BCL6 by chromatin-modifying drugs in human cancer cells. Cancer Cell 9:435–443
Shi Y, Downes M, Xie W et al (2001) SHARP, an inducible cofactor that integrates nuclear receptor repression and activation. Genes Dev 15:1140–1151
Shiota K (2004) DNA methylation profiles of CpG islands for cellular differentiation and development in mammals. Cytogenet Genome Res 105:325–334
Sleutels F, Zwart R, Barlow DP (2002) The non-coding Air RNA is required for silencing autosomal imprinted genes. Nature 415:810–813
Sleutels F, Tjon G, Ludwig T et al (2003) Imprinted silencing of *Slc22a2* and *Slc22a3* does not need transcriptional overlap between *Igf2r* and *Air*. EMBO J 22:3696–3704
Sonkoly E, Bata-Csorgo Z, Pivarcsi A et al (2005) Identification and characterization of a novel, psoriasis susceptibility-related noncoding RNA gene, PRINS. J Biol Chem 280:24159–24167
Southan C (2004) Has the yo-yo stopped? An assessment of human protein-coding gene number. Proteomics 4:1712–1726
Srikantan V, Zou Z, Petrovics G et al (2000) PCGEM1, a prostate-specific gene, is overexpressed in prostate cancer. Proc Natl Acad Sci USA 97:12216–12221
Stern-Ginossar N, Elefant N, Zimmermann A et al (2007) Host immune system gene targeting by a viral miRNA. Science 317:376–381
Storz G, Altuvia S, Wassarman KM (2005) An abundance of RNA regulators. Annu Rev Biochem 74:199–218
Sutherland HF, Wadey R, McKie JM et al (1996) Identification of a novel transcript disrupted by a balanced translocation associated with DiGeorge syndrome. Am J Hum Genet 59:23–31
Teichmann M, Wang Z, Roeder RG (2000) A stable complex of a novel transcription factor IIB-related factor, human TFIIIB50, and associated proteins mediate selective transcription by RNA polymerase III of genes with upstream promoter elements. Proc Natl Acad Sci USA 97:14200–14205
Thakur N, Tiwari VK, Thomassin H et al (2004) An antisense RNA regulates the bidirectional silencing property of the Kcnq1 imprinting control region. Mol Cell Biol 24:7855–7862
Tiedge H, Fremeau RT Jr, Weinstock PH, Arancio O, Brosius J (1991) Dendritic location of neural BC1 RNA. Proc Natl Acad Sci USA 88:2093–2097
Tiedge H, Chen W, Brosius J (1993) Primary structure, neural-specific expression, and dendritic location of human BC200 RNA. J Neurosci 13:2382–2390
Tsang WP, Wong TWL, Cheung AHH, Co CNN, Kwok TT (2007) Induction of drug resistance and transformation in human cancer cells by the noncoding RNA CUDR. RNA 13:890–898
Turner JD, Schote AB, Macedo JA et al (2006) Tissue specific glucocorticoid receptor expression, a role for alternative first exon usage? Biochem Pharmacol 72:1529–1536
Ulaner GA, Vu TH, Li T et al (2003) Loss of imprinting of IGF2 and H19 in osteosarcoma is accompanied by reciprocal methylation changes of a CTCF-binding site. Hum Mol Genet 12:535–549

Umlauf D, Goto Y, Cao R et al (2004) Imprinting along the Kcnq1 domain on mouse chromosome 7 involves repressive histone methylation and recruitment of Polycomb group complexes. Nat Genet 36:1296–1300
Vella MC, Reinert K, Slack FJ (2004) Architecture of a validated microRNA::target interaction. Chem Biol 11:1619–1623
Venter JC, Adams MD, Myers EW et al (2001) The sequence of the human genome. Science 291:1304–1351
Vincent JB, Herbrick JA, Gurling HM et al (2000) Identification of a novel gene on chromosome 7q31 that is interrupted by a translocation breakpoint in an autistic individual. Am J Hum Genet 67:510–514
Wang H, Iacoangeli A, Popp S et al (2002) Dendritic BC1 RNA: Functional role in regulation of translation initiation. J Neurosci 22:10232–10241
Wang XS, Zhang Z, Wang HC et al (2006) Rapid identification of UCA1 as a very sensitive and specific unique marker for human bladder carcinoma. Clin Cancer Res 12:4851–4858
Watanabe M, Yanagisawa J, Kitagawa H et al (2001) A subfamily of RNA-binding DEAD-box proteins acts as an estrogen receptor a coactivator through the N-terminal activation domain (AF-1) with an RNA coactivator, SRA. EMBO J 20:1341–1352
Waterston RH, Lindblad-Toh K, Birney E et al (2002) Initial sequencing and comparative analysis of the mouse genome. Nature 420:520–562
Willingham AT, Orth AP, Batalov S et al (2005) A strategy for probing the function of noncoding RNAs finds a repressor of NFAT. Science 309:1570–1573
Wutz A, Smrzka OW, Schweifer N et al (1997) Imprinted expression of the *Igf2r* gene depends on an intronic CpG island. Nature 389:745–749
Wutz A, Rasmussen TP, Jaenisch R (2002) Chromosomal silencing and localization are mediated by different domains of Xist RNA. Nature Genet 30:167–174
Yang Z, Zhu Q, Luo K et al (2001) The 7SK small nuclear RNA inhibits the CDK9/cyclin T1 kinase to control transcription. Nature 414:317–322
Yano Y, Saito R, Yoshida N et al (2004) A new role for expressed pseudogenes as ncRNA: regulation of mRNA stability of its homologous coding gene. J Mol Med 82:414–422
Yekta S, Shih IH, Bartel DP (2004) MicroRNA-directed cleavage of HOXB8 mRNA. Science 304:594–596
Yik JH, Chen R, Nishimura R et al (2003) Inhibition of P-TEFb (CDK9/Cyclin T) kinase and RNA polymerase II transcription by the coordinated actions of HEXIM1 and 7SK snRNA. Mol Cell 12:971–982
Yik JH, Chen R, Pezda AC et al (2004) A human immunodeficiency virus type 1 Tat-like arginine-rich RNA-binding domain is essential for HEXIM1 to inhibit RNA polymerase II transcription through 7SK snRNA-mediated inactivation of P-TEFb. Mol Cell Biol 24:5094–5105
Zalfa F, Giorgi M, Primerano B et al (2003) The fragile X syndrome protein FMRP associates with BC1 RNA and regulates the translation of specific mRNAs at synapses. Cell 112:317–327

Aptamers and siRNAs in Cardiovascular Disease

C.M. Blake, S. Oney, S.M. Nimjee, and B.A. Sullenger(✉)

Contents

Abstract Cardiovascular disease is the most common cause of death in the United States, and is a leading cause of morbidity. In addition, many drugs used to treat various arenas of cardiovascular disease have untoward side effects and less than desirable safety profiles. To meet the challenge of developing safe and effective treatments, aptamers and siRNAs are being developed that display promising potential. Aptamers are relatively small, synthetic nucleic acid ligands that possess high affinity and specificity for their target proteins. Also, antidote

B.A. Sullenger
Duke Institute for Genome Sciences and Policy, Duke University Medical Center, Durham, North Carolina 27710, USA
e-mail: b.sullenger@cgct.duke.edu

V.A. Erdmann et al. (eds.) *RNA Technologies in Cardiovascular Medicine and Research*, doi: 10.1007/978-3-540-78709-9,

oligonucleotides may be rationally designed against any aptamer, thus introducing an additional level of safety. Similarly, siRNAs may be rationally designed to cleave any messenger RNA through sequence complimentarity, thus silencing gene expression. Also, their effects have been shown to be long lasting, reducing the need for daily treatment. Here, we present background information on both aptamer- and siRNA-based therapeutics, and in particular focus on their applications to many areas of cardiovascular disease, namely anticoagulation, intimal hyperplasia, hypertension, myocardial ischemia, atherosclerosis and heart failure. In addition, potential challenges to both fields are discussed, and ideas on how to overcome these hurdles are offered. Overall, both aptamers and siRNAs represent attractive, novel nucleic acid therapies for the prevention and treatment of cardiovascular disease.

Abbreviations ACS: Acute coronary syndrome; ACT: Activated clotting time; AMD: Age-related macular degeneration; AMI: Acute myocardial infarction; aPTT: Activated partial thromboplastin time; β-AR: Beta adenergic receptor; CPB: Cardiopulmonary bypass; CVB-3: Coxsackievirus B3; dsRNA: Double-stranded RNA; FIX: Factor IX; FVa: Activated Factor V; FVII: Factor VII; FVIIa: Activated Factor VII; FVIIIa: Activated Factor VIIIa; FX: Factor X; FXa: Activated Factor X; HIF-1α: Hypoxia inducible factor-1 alpha; HTN: Hypertension; iNOS: Inducible nitric oxide synthase; LDL: Low-density lipoprotein; LPS: Lipopolysaccharide; miRNA: MicroRNA; PCI: Percutaneous coronary intervention; PCI: Percutaneous coronary intervention; PDGF: Platelet derived growth factor; PEG: Polyethylene glycol; PFA: Platelet function analysis; PKD1: Protein kinase D1; PSMA: Prostate-specific membrane antigen; PT: Prothrombin time; PTGS: Post-transcriptional gene silencing; RIPA: Ristocetin induced platelet aggregation; RISC-RNA: Interfering silencing complex; RNAi: RNA interference; RSV: Respiratory syncytial virus; SELEX: Systematic Evolution of Ligands by Exponential enrichment; shRNA: Short hairpin RNA; siRNA: Short interfering RNA; SMC: Smooth muscle cell; SR: Sarcoplasmic reticulum; SS: Statistically significant; TF: Tissue factor; TLR: Toll-like receptor; VEGF: Vascular endothelial growth factor; VSMC: Vascular smooth muscle cell; vWD: von Willebrand's Disease; VWF: von Willebrand Factor.

1 Introduction

Cardiovascular disease is the most common cause of death in the United States, and is a leading cause of morbidity. In addition, many drugs used to treat various arenas of cardiovascular disease have untoward side effects and less than desirable safety profiles. To meet the challenge of developing safe and effective treatments, nucleic acid therapies, in particular aptamers and siRNAs, are being developed that display promising potential. Here, we present background information on both aptamer- and siRNA-based therapeutics, and in particular focus on their applications to many areas of cardiovascular disease, namely anticoagulation, intimal hyperplasia, hypertension, myocardial ischemia, atherosclerosis and heart failure. In addition, potential challenges to both fields are discussed, and ideas on how to overcome these hurdles are offered. Overall, both aptamers and siRNAs represent attractive, novel nucleic acid therapies for the prevention and treatment of cardiovascular disease.

2 Coagulation and Thrombosis

Coagulation is a complex series of interactions that leads to stable clot formation in an injured blood vessel. A cell-based model characterizes coagulation in three distinct steps: initiation, priming and propagation (Hoffman and Monroe 2005, 2007; Roberts et al. 2006). In the initiation step, tissue factor (TF) comes into contact with activated coagulation Factor VII (FVIIa) and this, in turn, activates Factors IX and X. Factor Xa (FXa) then binds to Factor Va (FVa) on a TF-bearing cell, which produces a sufficient amount of thrombin to activate platelets around the area of injury; this leads to the priming step of coagulation. During priming, thrombin induces platelet activation by cleaving Par1 and Par4, and also activates Factors V, VIII and XI. Factor VIIIa (FVIIa) then assembles on the platelet surface to form the IXa/VIIIa complex (tenase complex) which leads to further production of FXa and thrombin, and the beginning of the propagation phase. The propagation phase generates large amounts of thrombin through the Xa/Va complex (prothrombinase complex), resulting in a stabilized fibrin and platelet-rich clot.

Hemostasis is the product of these well-regulated steps that prevents blood loss following injury to blood vessels; it is therefore a vital process. However, undesirable clot formation can occur in these vessels due to a number of reasons, leading to thrombus formation and occlusion in an otherwise patent blood vessel. This acute condition, known as pathologic thrombosis, can lead to coronary, cerebral and peripheral vascular stenosis, and represents the most common cause of morbidity and mortality in the United States today (Bhatt and Topol 2003; Jackson and Schoenwaelder 2003; Nimjee et al. 2005a; Thom et al. 2006). Furthermore, there are a number of procedures (i.e., percutaneous coronary intervention (PCI), cardiopulmonary bypass surgery (CPB), hemodialysis, cardiac valve replacement, etc.) where controlled inhibition of coagulation is necessary. Because thrombosis is associated with such clinical problems, a number of pharmacological agents have been developed to inhibit coagulation and thrombus formation by targeting clotting factors and platelets. Despite the progress that has been made in the development of these therapies, their use is associated with an increased risk of bleeding and adverse immune reactions (i.e., heparin-induced thrombocytopenia), which limits their clinical utility. Thus, development of new anticoagulant agents against targets with improved safety and efficacy is a medical priority. Here, we summarize the present knowledge of aptamers and siRNAs in preventing and treating cardiovascular disease.

3 Aptamers as Anticoagulants

Aptamers are single-stranded nucleic acid molecules that fold into three-dimensional structures and can directly inhibit protein function by binding to their targets with high affinity and specificity (Nimjee et al. 2005b). Sullenger et al. were the first to show the potential therapeutic utility of aptamers by inhibiting HIV-replication with a TAR aptamer (Sullenger et al. 1990). In building on this principle, a number of aptamers have been developed in the last decade that have substantial therapeutic value (Nimjee et al. 2005b). Macugen, an anti-VEGF RNA aptamer, received FDA approval for the treatment of age-related macular degeneration in December 2004, setting the precedent for further aptamer-based therapeutics (EyetechStudyGroup 2002, 2003). Aptamers have been generated to numerous other targets, a number of which are in pre-clinical development.

The in vitro selection technique called SELEX (Systematic Evolution of Ligands by EXponential enrichment), which allows for the isolation of high affinity, specific aptamers against a wide range of proteins, was developed simultaneously by Tuerk and Gold, and Ellington and Szostak in 1990 (Ellington and Szostak 1990, 1992; Tuerk and Gold 1990). The SELEX process begins with a combinatorial library of more than 10^{14} random RNA (or DNA) sequences that fold into various three-dimensional structures. This library of nucleic acid molecules is incubated with a target protein, and the molecules that bind to the target protein are separated from those that do not bind. When using RNA, the molecules which bind to the target are amplified by RT-PCR, and are in vitro transcribed to generate an enriched pool of RNA molecules with increased affinity for the target. This selection process is repeated (usually 8–12 rounds) under increasingly stringent conditions until the RNA molecules with the highest affinity for the target protein are isolated. In order to produce nucleic acids that are nuclease-resistant, all RNA molecules are transcribed with 2′-modified nucleotides such as 2′-fluoro-pyrimidines. Over the past decade, SELEX technology has yielded numerous aptamers that antagonize a broad spectrum of proteins, ranging from reverse transcriptases to cell adhesion molecules to growth factors (Nimjee et al. 2005b). In theory, using this procedure, aptamers can be isolated to any protein target. For example, this technology has been employed successfully to generate potent antagonists of coagulation factors, including factors VIIa (Rusconi et al. 2000), IXa (Rusconi et al. 2002), Xa (Layzer and Sullenger 2007) and thrombin (White et al. 2001).

Aptamers isolated using SELEX exhibit a high degree of specificity for their target protein. Although advantageous in inhibiting a particular protein, this property of aptamer molecules can potentially cause significant problems in pre-clinical studies. For example, an aptamer that binds only to the human ortholog of a protein has a limited future for becoming a clinically relevant compound. However, a strategy has been developed called "toggle" SELEX (White et al. 2001) to overcome this issue. Briefly, the RNA library is exposed to the human form of a given protein, and in the subsequent round of selection, the enriched group of ligands is bound to the animal ortholog from the species in which preclinical studies will be conducted. Furthermore, development of tightly-controlled therapeutic agents can occur through the use of rationally-designed antidotes which control the activity of the aptamer (Rusconi et al. 2002, 2004).

Aptamers represent an attractive class of therapeutic compounds for many reasons. They are relatively small (8–15 kDa), synthetic compounds that possess high affinity and specificity for their target proteins (equilibrium dissociation constants typically range from 0.05 to 10 nM). Thus, they embody the affinity properties of monoclonal antibodies with the chemical production properties of small peptides. In addition, pre-clinical and clinical studies to date have shown that aptamers and compounds of similar composition are well tolerated, exhibit low or no immunogenicity, and are suitable for repeated administration as therapeutic compounds. Moreover, the bioavailability and clearance mechanisms of aptamers can be rationally altered by molecular modification of the ligand (Nimjee et al. 2005a,b). It has been shown that conjugation of an aptamer to a high molecular weight inert carrier molecule such as polyethylene glycol can substantially increase its circulating half-life. Dyke et al. demonstrated that the circulating half-life of an aptamer can be increased from ~10 min to 6–12 h by a PEG modification to the 5′ end of the molecule (Rusconi et al. 2004). Coupled with antidote-mediated control, these properties of aptamers make them ideal for use as therapeutic agents.

3.1 Antidote Control

It has been shown that antidotes can be rationally designed to inhibit an aptamer's activity (Rusconi et al. 2002, 2004). The theory of antidote control is based on the hypothesis that any molecule which can critically change the aptamer's three-dimensional structure will inhibit its function. By exploiting the properties of Watson-Crick base pairing between the aptamer and its complimentary oligonucleotide, the three-dimensional shape of the aptamer is altered and becomes functionally inactive. The outcome of this research generated the first antidote-controlled regulation of a molecule and produced the first rationally designed reversible anticoagulant targeting coagulation factor IXa. Versions of this aptamer-antidote pair were tested in vivo in a porcine systemic anticoagulation model as well as a murine carotid artery thrombosis and tail transection model (Rusconi et al. 2004), compared with heparin and protamine in a porcine model of cardiopulmonary bypass (Nimjee et al. 2006), and successfully went through phase I clinical trials (Dyke et al. 2006).

A number of drugs exist that are routinely used in clinics to prevent thrombosis. These drugs have shown remarkable clinical efficiency in reducing the morbidity and mortality associated with thrombosis. However, these agents also have a number of drawbacks, most significant of which is hemorrhage, the largest limitation in their clinical utility. Therefore, a pressing need exists for anticoagulant drugs with improved safety profiles. Antidote development represents a key strategy to overcome the obstacle of hemorrhage and opens the door for a novel class of anticoagulants.

3.2 Factor VIIa Aptamers

Coagulation is initiated by the binding of tissue factor (TF), an integral membrane protein, to coagulation factor VIIa (FVIIa). This TF/FVIIa complex formation represents an important step in hemostasis and thrombosis, and inhibition of this step represents an attractive target for the development of anticoagulants.

Previously, a 2′-amino modified RNA aptamer that binds to Factor VIIa with high affinity and specificity and inhibits Factor IX (FIX) and FX activation was isolated (Rusconi et al. 2000). This aptamer significantly prolonged the prothrombin time (PT) in in vitro clotting assays in a dose dependent manner. While this aptamer was not tested in in vivo models, it was the first aptamer generated against a coagulation protein involved in the generation of thrombin and suggested that aptamers targeting such upstream coagulation factors may represent potential therapeutics. Recently, a 2′-fluoropyrimidine modified aptamer has been developed against Factor VII (FVII) that has potent anticoagulant activity as determined by PT clotting times in human plasma (Layzer and Sullenger 2007).

3.3 Factor IXa Aptamer-Antidote Pair

In the cell-based model of coagulation, thrombin generation occurs on the surface of platelets during the propagation phase. The FIXa/VIIIa (tenase) complex plays a vital role in activating Factor X (FX) on the platelet surface, leading to FXa/Va (prothrombinase) complex formation. This complex then generates a burst of thrombin, leading to stable clot formation (Hoffman and Monroe 2005, 2007; Roberts et al. 2006). A number of studies to date have demonstrated that activation of FX by the FIXa/VIIIa complex is ~50 times more efficient than its activation by the Factor VIIa/TF pathway (Lawson and Mann 1991). In this model, Factor IXa (FIXa) is a very potent initiator of thrombin generation and represents a very exciting target for anticoagulant therapy.

Rusconi et al. developed a 2′-fluoropyrimidine modified RNA aptamer of 31-nucleotides (9.3t) that binds to FIXa and inhibits the activity of the FIXa/FVIIIa complex. In vitro studies have demonstrated that this molecule can effectively inhibit FX activation, thereby prolonging clotting times in activated partial thromboplastin time (aPTT) assays using plasma from humans and several relevant animal models (Rusconi et al. 2002). Furthermore, the circulating half-life and stability of this compound was increased by attaching a cholesterol or polyethylene glycol (PEG) molecule to the 5′ end of the RNA. In vivo studies have shown that this molecule can systemically anticoagulate animals and inhibit thrombus formation in an animal vascular injury model (Rusconi et al. 2004).

To improve the safety profile of this molecule, an antidote molecule was designed using properties inherent to nucleic acids. Binding of the antidote to the aptamer by Watson-Crick base pairing prevents the aptamer from folding into its appropriate conformation, which is necessary for binding to its target protein (i.e.,

FIXa); therefore, the activity of the aptamer is diminished, and aptamer inhibition of the target is thus reversed. The antidote oligonucleotide 5-2C was able to reverse Ch-9.3t-induced anticoagulation of human plasma rapidly and durably in a dose-dependent manner (Rusconi et al. 2002). This aptamer-antidote pair was further tested and validated in a number of animal models. Administration of the Ch-9.3t aptamer resulted in increased bleeding when mice were surgically challenged (Rusconi et al. 2004). This increased bleeding was prevented, and the effects of the aptamer were completely reversed, by administration of the antidote immediately following tail injury (Rusconi et al. 2004). Moreover, Nimjee et al. tested this aptamer-antidote pair in a porcine cardiopulmonary bypass model as a replacement therapy for unfractionated heparin and protamine (Nimjee et al. 2006). The aptamer was able to anticoagulate the animal while keeping the bypass circuit patent throughout the entire procedure. Upon antidote administration, the activity of the aptamer was rapidly reversed for a prolonged period of time. This study also suggested that the FIXa aptamer-antidote pair may have advantages over using heparin and protamine (which is currently used during CPB), including better inflammatory and cardiac profiles and lower thrombin generation. These seminal studies paved the way for this aptamer-antidote pair to move into clinical trials.

Dyke et al. demonstrated that the aptamer-antidote pair approach is both safe and effective in a phase 1a blinded, dose-escalation, placebo-controlled clinical trial with 85 healthy volunteers. An optimized version of the FIXa aptamer (RB006) demonstrated a clear anticoagulant effect at 30, 60 and 90 mg doses with mean aPTT values 1.3-, 2.1- and 2.9-fold higher than pretreatment times, respectively. Administration of its complementary antidote (RB007) 3 h later through an intravenous bolus resulted in a rapid and durable return of the aPTT value to baseline. During this study, no major hemorrhagic events were observed, and adverse effects were similar to placebo (Dyke et al. 2006). This study is the first of its kind in that an RNA aptamer and its complementary oligonucleotide antidote were evaluated in humans. The results of this study validate the potential utility of this emerging technology. Moreover it is the first published report of the systemic administration of any aptamer to a patient.

3.4 Thrombin Aptamers

Thrombin plays a key role in coagulation by converting fibrinogen to fibrin, which is the building block of a thrombus. In addition, thrombin also activates upstream procoagulant factors such as Factors IX and XI. Because of these functions, thrombin has been considered an attractive target for anticoagulation therapy.

To date, a number of groups, starting as early as 1992, have developed aptamers that bind to thrombin. Bock and colleagues isolated a family of single-stranded DNA aptamers with binding affinities to thrombin ranging from 25 to 200 nM. The most promising of these aptamers, termed GS-522 (later renamed ARC 183 by Archemix Corp.), a 15-nucleotide-long ssDNA molecule of ~5,000 Da, was studied further in in vitro and in vivo experiments (Bock et al. 1992; Griffin et al. 1993). Bock et al. demonstrated that this aptamer could inhibit the thrombin-catalyzed conversion of fibrinogen to fibrin. Furthermore, they illustrated aptamer prolongation of clotting times in pure fibrinogen and plasma assays.

Griffin et al. investigated the anticoagulant effects of ARC 183 in a series of experiments using cynomolgus monkeys and sheep undergoing hemofiltration with an extracorporeal circuit. This aptamer prolonged the prothrombin time in both of these models and also inhibited thrombin-induced platelet aggregation (Bock et al. 1992; Griffin et al. 1993). This aptamer was also tested in a canine cardiopulmonary bypass model to determine its efficiency and pharmacokinetics, and subsequently move it closer to clinical trials. During this experiment, at an infusion rate of 0.5 mg kg^{-1} min^{-1}, the aptamer prolonged PT, aPTT and activated clotting time (ACT) assays, which rapidly returned to baseline following discontinuation of the drug. The calculated plasma half life of the aptamer was 1.9 min pre-CPB and 7.7 min during CPB. This change in plasma half-life could have been due to altered renal clearance or the effects of the bypass circuitry. There were no significant adverse effects of this aptamer compared to the heparin control group during or after the operatation (DeAnda et al. 1994). This data suggested the potential clinical use of a thrombin aptamer. Nearly a decade later, ARC 183 was taken to phase I clinical trials by Archemix Corporation in late 2005 for use in on-pump coronary artery bypass graft surgery. Rather than using an antidote-based strategy of regulation, they focused on the short half-life of the drug, such that when the infusion is turned off, the drug would clear quickly and the patient's clotting parameters would quickly return to normal. Unfortunately, results from this study have not been published, so it is currently unclear how effective and practical this approach is in humans.

In recent years, a number of new aptamers (both DNA and modified RNA) that bind to the different exosites of thrombin have been selected by different groups. These aptamers are currently being tested in in vivo and in vitro models.

4 VWF Aptamers as Antiplatelet Agents

In the last 20 years, a number of antiplatelet agents have emerged as more information about platelet biology revealed novel targets for therapeutic inhibition. Although these agents have had a major impact on reducing the complications of cardiovascular disease, none of the current antiplatelet therapies prevent the initiating step for thrombus formation, platelet adhesion, which is principally mediated by circulating von Willebrand Factor (VWF) and subendothelial matrix proteins (Bhatt and Topol 2003; Jackson and Schoenwaelder 2003; Oney et al. 2007). Furthermore, many patients on aspirin and clopidorel are advised to stop taking their medication 7–10 days prior to surgery in order to prevent bleeding complications. This places the patient at risk for thrombosis and creates a therapeutic dilemma for surgeons. Moreover, the use of antiplatelet agents in the setting of cardiopulmonary bypass surgery is counter-indicated since lack of control of such agents places patients at a high risk for developing uncontrollable bleeding.

VWF is a multimeric plasma glycoprotein that is critical for primary hemostasis. It plays a direct role in allowing platelet attachment to sites of vascular injury through their GP Ib-IX-V receptors, and aids in platelet plug formation by enabling platelet–platelet interactions through GP IIb-IIIa (Sadler 2005). Qualitative or quantitative deficiencies of VWF can lead to von Willebrand's Disease (vWD), which is characterized by a mild to severe bleeding phenotype (Sadler 2005). The fact that vWD results in patients with bleeding disorders that may require medical therapy validates VWF as an attractive target for platelet inhibition. However, its inhibition would require stringent control so as to provide clinical benefit without the risk of hemorrhage.

In attempts to generate a safer antiplatelet agent, a modified version of SELEX against VWF was performed (Oney et al. 2007). First, an RNA library containing 2′-flouropyrimidines was incubated with total plasma proteins; the RNA ligands that bound to this proteome were then recovered. Four additional rounds of SELEX were performed against the plasma proteome to generate a focused library that was highly enriched for RNA ligands that bound plasma proteins. Next, convergent SELEX was performed to isolate those RNA aptamers from the focused library that specifically bound VWF. This method yielded aptamers that bound to VWF with high affinity. Two of these molecules were selected and tested in in vitro and ex vivo assays. Both of these molecules completely inhibited VWF-mediated platelet aggregation in ristocetin induced platelet aggregation (RIPA) and a platelet function analysis (PFA-100) assays using human plasma and human whole blood, respectively (Oney et al. 2007).

Because the aforementioned VWF aptamers inhibit platelet function to the same degree as seen in the vWD population, specific antidotes for the VWF aptamers were designed using properties inherent to nucleic acid molecules. The most potent antidote, antidote oligonucleotide 6 (AO6), rapidly and completely reversed the activity of the VWF aptamer in a PFA-100 assay (>2 min) for a prolonged period of time (>4 h) (Oney et al. 2007). These findings have significant clinical implications. For the first time, a potent antiplatelet agent (VWF aptamer R9.14) and an antidote molecule that can quickly and effectively reverse its activity in the setting of high shear stress has been developed. Once further developed, this aptamer-antidote pair can potentially provide physicians and surgeons a quick, effective and continual way to regulate antiplatelet therapy and thus allow for effective and safe treatment of their patients. Currently, this aptamer-antidote pair is being tested in in vivo models.

Recently, Archemix Corporation has announced the development of an aptamer molecule targeted towards VWF's function in platelet aggregation. This aptamer molecule (ARC1779) is currently being developed to treat patients diagnosed with Acute Coronary Syndrome (ACS) who undergo a percutaneous coronary intervention (PCI). According to the Archemix website, phase I testing for ARC1779 was initiated in December 2006, and a phase 2 study for patients with ACS is planned to begin by the end of 2007 (http://www.archemix.com).

5 Aptamers and Intimal Hyperplasia

As a result of atherosclerosis leading to myocardial ischemia, many patients have to undergo coronary artery bypass surgery and angioplasty in order to restore blood flow to the heart and prevent further infarction. These operations involve engraftment of a blood vessel from another location in the body to positions proximal and distal to the site of occlusion, or the placement of a mechanical stent to keep the vessel patent, respectively. Intimal hyperplasia is a common cause of long-term bypass graft and stent failure due to pathological migration, proliferation, and accumulation of vascular smooth muscle cells (VSMCs) in the intima of the treated vessel. This accumulation of cells narrows the lumen, thus decreasing blood flow and concomitantly causing turbulent flow; this increases the propensity of platelets to become activated and thus causes thrombus formation, which may again occlude the vessel.

There are, of note, two aptamers that have been shown to prevent intimal hyperplasia in animal models of vascular injury. Giangrande et al. selected an RNA aptamer against E2F3, a transcriptional activator whose accumulation correlates with increased cell proliferation in most cell types. In a murine model of venous bypass grafting, they were able to achieve 50–70% uptake of the aptamer into the graft, which then produced a 52% decrease in the intimal-to-medial ratio (Giangrande et al. 2007). In addition to targeting the E2F family of transcription factors, one can look to inhibit growth factors to prevent intimal hyperplasia. Leppänen et al. developed an aptamer against the B-chain of platelet derived growth factor (PDGF), which is a growth factor and chemoattractant for all cells of mesenchymal origin (which includes fibroblasts and macrophages, in addition to the platelet precursor, megakaryocytes). They showed that daily intraperitoneal administration of their aptamer, NX1975, for 2 weeks following balloon injury of the common carotid artery reduced the intimal lesion size by 50% (Leppanen et al. 2000). However, they also noted a transient benefit, as administration for 2 weeks followed by 6 weeks of rest resulted in a lesion size statistically indistinguishable from the control group and significantly worse than that noted in the 2-week treatment and immediate observation group, indicating that their PDGF aptamer would require chronic administration. For information concerning siRNA treatment of intimal hyperplasia, see Sect. 10.4 and Table 1.

6 Potential Challenges Facing Aptamers in Cardiovascular Disease

Aptamers as a class of therapeutic agents have traversed large distances during the last 17 years. With pegaptanib (Macugen) available to treat macular degeneration, and a number of other molecules in the clinical pipeline, their usefulness is incontrovertible. However, there are a number of challenges that will need to be addressed

Table 1 In vitro and in vivo effects of siRNA inhibition on cardiovascular disease

Target	Description	siRNA	Functional inhibition	In vitro outcome	In vivo outcome	References
Intimal hyperplasia						
E2F3	Transcription factor that regulates the cell cycle	21-mer siRNA	Qualitative decrease (↓) in E2F3 protein levels in VSMCs	~90% ↓ in VSMC proliferation	42% ↓ in protein expression in venous grafts; ~50% ↓ in intima-media ratio in a murine venous bypass graft model	Giangrande et al. (2007)
CDH11	Cell surface adhesion molecule; also involved in cell signaling	Cocktail of siRNAs; structures not given in paper	69% ↓ in CDH11 mRNA levels in smooth muscle cells (SMCs)	33% ↓ in VSMC migration; 1.3-fold ↓ in proliferation	–	Monahan et al. (2007)
Midkine	Heparin-binding growth factor involved in inflammation, cancer, and the nervous system	19-mer modified siRNA with 3′ dTdT overhangs; used siSTABLE (Dharmacon) to improve stability and silencing longevity	91% ↓ in midkine protein expression in a rabbit kidney cell line	–	90% ↓ in intima-media ratio and intima thickness in rabbit venous bypass graft; ↓ in proliferating cells and leukocyte count in veins	Banno et al. (2006)
MMP-2, MMP-9	Proteases that degrade basement membrane	19-mer with 5′ phosphorylation and 3′ dTdT overhangs	90% ↓ in PDGF/IL-1 induced protein levels in saphenous vein SMC supernatant (for both MMP-2 and MMP-9)	65% and >80% ↓ in PDGF/IL-1 induced cell invasion for siRNA inhibition of MMP-2 and MMP-9, respectively	–	Turner et al. (2007)
ClC-2	Volume regulated chloride channel implicated in cellular proliferation	19-mer modified siRNA with 3′ dTdT overhangs	Qualitative ↓ in ClC-2 mRNA and protein expression levels	Statistically significant (SS) ↓ in IGF-1 induced VSMC DNA synthesis	–	Cheng et al. (2007)

(continued)

Table 1 (continued)

Target	Description	siRNA	Functional inhibition	In vitro outcome	In vivo outcome	References
CTGF	Profibrotic factor implicated in vascular remodeling and atherosclerosis	21-mer sequence; used a plasmid expressing short hairpin siRNA	79% ↓ in high glucose-induced CTGF upregulation	SS ↓ in extracellular membrane protein deposition, as well as high-glucose induced VSMC proliferation, and migration	–	Liu et al. (2007)
CyclinD1	Protein involved in the progression of cells through the cell cycle	21-mer siRNA	Qualitative ↓ in CyclinD1 protein levels	SS ↓ in IL-6 induced VSMC motility	–	Wang et al. (2007)
Intimal hyperplasia, con't						
EGR-1	Transcriptional regulator that mediates gene expression after vascular injury	19-mer modified siRNA with 3′ dTdT overhangs	SS ↓ in EGR-1 protein expression levels	Virtually complete inhibition of serum-induced SMC proliferation; blocked SMC regrowth following in vitro injury; ↓ G-CSF and FGF-2 expression	–	Fahmy and Khachigian (2007)
rK(Ca)3.1	Gene that produces an intermediate-conductance calcium activated K^+ channel which regulates proliferation	19-mer with two-3′ tt overhangs	Complete abolishment of functional IK_{Ca} channels	Complete abolishment of VSMC mitogenesis	–	Si et al. (2006)
Skp2	Responsible for polyubiquitylation of cell cycle regulators	19-mer siRNA	SS ↓ in Skp2 mRNA and protein expression levels	SS ↓ in VSMC proliferation	–	Wu et al. (2006)

PLD2, cPLA2	Phospholipases; after activation by ANG II, cPLA2 activates PLD2, which then hydrolyzes mainly phosphatidylcholine into phosphatidic acid and choline	Used a plasmid containing a retroviral vector expressing the retro-siRNA insert	Immunohistochemical decrease in PLD2 and cPLA2 protein levels noted in the arteries	–	↓ protein expression levels in injured arteries; ↓ ANG II-induced, Akt mediated neointmal thickening post rat carotid artery balloon injury model	Li et al. (2005)
FoxO4	Transcription factor that is a member of the Forkhead Box O family; regulates many pathways and other transcription factors	19-mer with two-3′ tt overhangs	↓ in FoxO4 mRNA expression	SS ↓ in SMC migration	↓ intima- media ratio; ↓ MMP-9 expression	Li et al. (2007)
Atherosclerosis						
HIF-1α	Transcription factor present only during hypoxia (undetectable during normoxia)	19-mer; delivered in an expression vector	↓ to 27% of normal HIF-1α mRNA level; ↓ to 30% of normal protein level	Downregulated 57 of 96 key genes known to be involved in atherosclerosis; qualitative ↓ in foam cell formation	–	Jiang et al. (2007)
PTOV1	Androgen-induced mitogenic factor shown to enhance cell proliferation	19-mer with 3′ dTdT overhangs	SS ↓ in mRNA expression level	SS ↓ in testosterone-induced human aorta VSMC proliferation	–	Nakamura et al. (2006)
AIF-1	Inflammation-responsive scaffolding protein; considered a marker for macrophage activation	19-mer; delivered in an expression vector	79% ↓ in AIF-1 protein expression	52% ↓ in macrophage proliferation; 60% ↓ in macrophage migration	–	Tian et al. (2006)

(continued)

Table 1 (continued)

Target	Description	siRNA	Functional inhibition	In vitro outcome	In vivo outcome	References
ApoB48R	Protein involved in the uptake of triglyceride rich-lipoproteins	19-mer with 3′ TT overhangs	Qualitative ↓ in ApoB48R protein levels	SS ↓ in triglyceride accumulation in macrophages	–	Kawakami et al. (2005)
SphK1	Kinase involved in the cell signaling cascade; forms S1P; both proteins are important in endothelial cell function	21-mer with 3′-fluorescein modification	Qualitative↓ in SphK1 mRNA levels; SS ↓ in SphK1 activity levels	SS ↓ in adhesion of monocytes to endothelial cells	–	Wang et al. (2005)
ATF2	Transcription factor that is a component of AP-1, which induces inflammatory gene expression	Purchased from Ambion; structure not given in paper	80% ↓ in ATF2 mRNA and protein levels	SS suppression of basal proinflammatory gene expression under no-flow conditions	–	Fledderus et al. (2007)
MK2	Mitogen-activated kinase that leads to production of proinflammatory mediators	Purchased from Qiagen; structure not given in paper	Qualitative ↓ in MK2 protein levels	SS ↓ in IL-1 induced expression of VCAM-1/MCP-1	–	Jagavelu et al. (2007)
CHOP	Protein which triggers apoptosis	19-mer with 3′ TT overhangs	Qualitative ↓ in CHOP protein levels	SS ↓ in endoplasmic reticulum-stress dependant death of coronary artery SMCs	–	Myoishi et al. (2007)
Atherosclerosis, con't						
Nox4	NADPH oxidase that co-localizes with CRP in VSMCs	pSUPER vector containing siRNA sequences	Unable to detect Nox4 mRNA after siRNA administration	SS ↓ in CRP-induced ROS generation by VSMCs	–	Ryu et al. (2007)

Furin	Proteinase that cleavage-activates MT-MMPs	19-mer siRNA	60% ↓ in furin protein levels	SS ↓ in pro-MMP-2 levels	–	Stawowy et al. (2005)
Bfl-1, Mcl-1	Antiapoptotic genes	19-mers siRNA with 3′ dTdT overhangs	60% and 80% ↓ in mRNA and protein levels for Mcl-1 (Bcl-1 data lacking)	~50% ↓ in IL-10 mediated lipid accumulation in macrophages in the presence of both siRNAs	–	Halvorsen et al. (2005)
REDD2	Gene highly upregulated due to LDL treatment of macrophages	Purchased from Ambion; structure not given in paper	Reduction of REDD2 protein to basal levels when stimulated by oxLDL	Protected monocytes against oxLDL induced necrosis	–	Cuaz-Perolin et al. (2004)
GADD153	Transcription factor upregulated after DNA damage/oxidative stress	19-mer with two-3′ TT overhangs	47% ↓ in CRP-induced mRNA expression	SS ↓ in CRP-induced apoptosis of VSMCs	–	Blaschke et al. (2004)
Adipophilin	A gene upregulated by modified LDL; considered to be a marker of lipid accumulation	29-mer siRNA	↓ in adipophilin protein levels to 13% of its normal level	SS ↓ in accumulation of triglycerides and cholesteryl esters in macrophages	–	Larigauderie et al. (2004)
Myocarditis						
CAR	Common receptor of Coxsackievirus B3 (CVB-3) and adenovirus	shRNA delivered in an expression cassette with a 19-mer stem sequence	85–90% ↓ in CAR mRNA levels; ablation of protein expression	SS ↓ in CVB-3 cellular infection; ↓ (but not SS) in adenovirus infection	–	Fechner et al. (2007)
3D RNA-dependent RNA polymerase of CVB-3	Viral enzyme that mediates RNA synthesis to aid in viral replication	siRNA purchased from MWG Biotech and Dharmacon; structure not given in paper	Qualitative ↓ in protein expression levels	80–90% ↓ in CVB-3 viral propagation; ↓ in viral titre	–	Werk et al. (2005)

(continued)

Table 1 (continued)

Target	Description	siRNA	Functional inhibition	In vitro outcome	In vivo outcome	References
CVB-3 viral protease 2A	Viral enzyme that mediates protein synthesis to aid in viral replication	19-mer siRNA	93% ↓ in CVB-3 replication in cardiomyocytes	↓ (though not SS) in viral protein expression levels; protection against CVB-3 induced cytopathic effects	–	Yuan et al. (2005)
Myocardial ischemia						
PDH2	Prolyl hydroxylase that catalyzes the hydroxylation of HIF-1, targeting it for degradation under normoxic conditions	19-mer sequences; used SMARTpool and siSTABLE (Dharmacon)	89% ↓ in PDH2 mRNA expression	300% increase (↑) in HIF-1 transcriptional activity (under normoxia)	61% ↓ in mRNA levels in mice; >60% ↓ in infarct size in murine ischemia/reperfusion model	Natarajan et al. (2006, 2007)
SHP-1	Phosphatase that plays a negative regulatory role in signal transduction pathways; inhibits VEGF receptor-2 (KDR)	21-mer; delivered in an expression vector	Qualitative ↓ in SHP-1 mRNA expression level	↑ endothelial cell growth	SS ↓ in SHP-1 mRNA and protein levels as well as infarct size in the heart in a rat model of acute MI; SS ↑ in capillary density in rat ischemic limb model	Sugano et al. (2005, 2007)
Caveolin	Protein required for caveolar structure and thus endocytosis of key receptors and their targets	Structure not mentioned in paper	Caveolin protein expression ↓ to ~15% of normal level	↑ in attachment of endothelial progenitor cells to endothelial cells	SS ↓ in ischemic tissue damage and ambulatory impairment in murine model of hindlimb ischemia	Sbaa et al. (2006)

FoxO1, FoxO3a	Transcription factors that are members of the Forkhead Box O family; regulates many pathways and other transcription factors	19-mer siRNA	Qualitative ↓ in FoxO1 and FoxO3a protein expression levels	SS ↑ in eNOS levels (required for postnatal neovascularization); SS ↑ in migration of endothelial cells and vessel sprout length	–	Potente et al. (2005)
Omi/HtrA2	Mitochondrial serine protease involved in cell death	19-mer with 3′ dTdT overhangs	↓ in Omi/HtrA2 protein levels to nearly undetectable levels	SS ↓ in ischemia/reperfusion (I/R) induced cardiomyocyte apoptosis	–	Liu et al. (2005)
Anticoagulation						
Versican G3 domain	C-terminus domain of versican, a member of the large aggregating chondroitin sulfate proteoglycan family; shown here to inhibit TFPI	19-mer siRNA inserted into a plasmid	SS ↓ in versican protein levels	SS ↑ in time for blood to coagulate as compared to G3-induced coagulation times	–	Zheng et al. (2006)
Hypertension						
preproTRH	Precursor gene to TRH, which releases thyroid hormone	Structure not given in paper	Statistically significant ↓ in preproTSH mRNA expression level and TRH protein levels	–	SS ↓ systolic BP in obesity-induced hypertensive rats	Landa et al. (2007)
β1-AR	Beta adrenergic receptor that mediates the sympathetic response to endogenous catecholamines	19-mer siRNA with two-3′ TT overhangs; complexed with liposomes	70% ↓ in mRNA levels, with no overlap to β2-AR	Specific ↓ in β1-AR protein levels without overlap to β2-AR	Demonstrated in vivo silencing of β1-AR; SS ↓ in systemic blood pressure in spontaneously hypertensive rats; ↓ cardiac hypertrophy	Arnold et al. (2007)

(continued)

Table 1 (continued)

Target	Description	siRNA	Functional inhibition	In vitro outcome	In vivo outcome	References
MR	Mineralocorticoid receptor that upon activation leads to sodium retention	shRNA delivered in an adeno-associated virus vector	SS ↓ in MR protein expression levels in kidney	–	Sustained SS prevention of progression to cold induced hypertension in rats	Wang et al. (2006)
AT1	Angiotensin-II receptor that mediates angiotensin-induced vasoconstriction and increases in blood pressure	shRNA delivered in a vector (22-mer stem with a loop)	AT1 mRNA and protein levels reduced to 21 and 37% their normal levels, respectively	–	–	Xiao et al. (2006)
$AT_{1a}R$	Angiotensin-II receptor that mediates angiotensin-induced vasoconstriction and increases in blood pressure	21-mer siRNA	80% decrease in binding to AT1aR without decrease in $AT_{1b}R$ orAT_2R subtypes; SS ↓ in $AT_{1a}R$ mRNA expression levels	–	–	Vazquez et al. (2005)
RhoA	Kinase involved in regulating cellular calcium sensitivity	23-mer siRNA	SS ↓ in RhoA protein levels	SS ↓ in contractility of coronary artery SMCs	–	Bi et al. (2005)
Heart failure						
Phospholamban	Regulatory protein that inhibits sarcoplasmic reticulum calcium intake (by inhibiting the SERCA2a pump)	19-mer siRNA with 3′ dTdT overhangs (Watanabe)/ Short hairpin RNA (shRNA) delivered in an adeno-associated virus vector (Fechner)	↓ phospholamban mRNA to 6%; protein to 12% (Watanabe)/ ↓ mRNA to 0.01%; protein to 7% (Fechner)	SS ↑ in calcium uptake affinity after exogenously inhibiting SERCA2a (Watanabe)/ SS ↑ in calcium uptake (Fechner); both used neonatal cardiac myocyte cultures	–	Watanabe et al. (2004), Fechner et al. (2007)

PKD1	Kinase which can phosphorylate and hence lead to the nuclear export of HDAC5, a transcriptional repressor of cardiac remodeling	21-mer siRNA	Qualitative ↓ in PDK1 protein levels	SS ↓ in nuclear export of HDAC5; SS ↓ phenylephrine-induced cardiomyocyte growth	–	Harrison et al. (2006)
Aldose Reductase	Cytosolic reductase that regulates postreceptor events that lead to NF-κB activation	21-mer siRNA	↓ AR protein to <5% of normal levels	SS ↓ in NF-κB activation and cytokine production in LPS-stimulated macrophages	–	Ramana et al. (2006)
AMPKα1	Kinase that regulates enzymes involved in fat and carbohydrate metabolism and thus ATP levels	siRNA purchased from Santa Cruz; structure not given in paper	Qualitiative ↓ in AMPKα1 protein expression	SS ↓ in Ang-II induced cell proliferation	–	Hattori et al. (2006)
Vascular disease						
MRP1	Exports oxidized GSH (GSSG) which affects other redox couples (e.g., NAD^+/NADH) and can lead to apoptosis	21-mer siRNA	70 and 64% ↓ in mRNA and protein levels, respectively	SS ↓ in endothelial cell markers of apoptosis; 40% ↑ in GSH reductase activity under oscillatory sheer stress	–	Mueller et al. (2005)
RhoGDIα	Protein that negatively regulates (dissociation inhibitor) rho-family GTPases (like Rac1)	19-mer siRNA with two-3′ tt overhangs	SS ↓ in RhoGDIα protein production	SS ↓ in Ang-II induced superoxide production and lipid peroxidation	–	Custodis et al. (2006)

in order to realize the potential of these compounds, including toxicity and immune reactions, off-target effects, oral delivery and cost.

Pre-clinical data exists that indicates that no toxic effects are elicited by aptamers as a class of agents when they are administered systemically. In contrast, very limited clinical data exists to support this conclusion in patients. Pegaptanib benefits from being a therapeutic agent that is locally delivered. Compounds like the FIXa and VWF aptamers will face the challenge of proving that they have limited systemic toxicity. However, in doing so, should the data be favorable, they would set the precedent for systemically administered therapeutic RNA agents. With respect to reversible drugs, the potential of immune reactions is a realistic consideration in aptamer-antidote development. Data have shown that double-stranded RNA molecules can induce innate immune responses via the toll-like receptor pathway (for more information, see Sect. 11). Aptamer-antidote pairs theoretically could induce such a response. Also, as with all drugs, there is the potential for off-target effects. Aptamer binding is believed to be mediated by their three-dimensional conformation, and they tend to bind to positively charged surfaces on proteins. Therefore, aptamers could potentially bind to positively charged moieties on other proteins and elicit undesirable effects. In vitro binding studies suggest that aptamers are highly specific, but in vivo and clinical specificity data remain limited.

The current inability to orally deliver aptamers is perhaps the greatest limitation of aptamers as therapeutic agents. This is especially true in the field of cardiovascular disease. As patients often require chronic therapy, oral medication would provide the most convenient delivery method. It seems reasonable that, like low molecular weight heparin, aptamers could be delivered subcutaneously. Oral delivery would, however, expand the role of aptamers in cardiovascular disease. Finally, cost will be another consideration in moving an aptamer compound from the bench-top to the bedside. This challenge is being helped by the demand for siRNA synthesis for research and clinical development; as a result, the cost of synthetic RNA production has decreased dramatically in even the last 10 years. As large scale production of synthetic RNA continues to improve, we believe the cost of aptamers will compare favorably to that of antibody and small peptide production.

7 siRNA Background

The Nobel Prize winning discovery of RNA interference (RNAi) occurred in 1998 when Craig Mello and Andrew Fire demonstrated that double-stranded RNA (dsRNA) was able to inhibit complimentary gene expression in *C. elegans* through degradation of its cytoplasmic mRNA (Fire et al. 1998). RNA interference pathways are mediated by short interfering RNA (siRNA), as well as microRNA (miRNA), which are derived from endogenously transcribed, non-coding short hairpin RNAs that demonstrate imperfect pairing between their two strands. For a more detailed review of microRNA, please see the chapter entitled *MicroRNA Systems Biology*. siRNA may be introduced to the cell as long (>38 bp), double-stranded RNA or by a vector as short hairpin RNA (shRNA). Both long dsRNA and shRNA undergo processing by Dicer, an RNAse III enzyme, into smaller, usually 21–23 base pair fragments with a 5′ monophosphate and two nucleotide single-stranded overhangs on the 3′ ends. However, siRNAs which mimic the Dicer cleavage product may also be introduced directly into the cytoplasm. This method of introducing synthetic siRNAs is most commonly undertaken in developing siRNA as therapeutic agents. Following delivery, the siRNA is incorporated into the Argonaute 2/RNA-interfering silencing complex (RISC). The passenger (or sense) strand is then cleaved, leaving the guide (or antisense) strand to base pair with its complimentary mRNA sequence, which is then cleaved by the catalytic domain of AGO2. This knockdown of mRNA thus effectively silences the gene and results in a decrease in its protein product.

Following the initial linkage of post-transcriptional gene silencing (PTGS) with dsRNA in nematodes in 1998, dsRNA quickly became utilized for uncovering gene function in plants, fungi, and other eukaryotes (reviewed in Hammond et al. 2001). Although use of siRNA to mediate gene-specific silencing was growing in many model systems, it was difficult to translate this work into mammalian cells. This challenge was met by Elbashir et al. in 2001, when they demonstrated the ability to silence several genes in multiple mammalian cell lines (Elbashir et al. 2001). The field then expanded exponentially, as several gene functions and pathways were now able to be studied in a much more specific manner. However, the realization that siRNA could be used for therapeutic application occurred in 2003, when in vivo evidence demonstrated the ability of siRNA directed against Fas to protect mice against the development of liver failure by preventing hepatocyte cytotoxicity (Song et al. 2003).

8 siRNA as Therapeutic Agents

Since 2003, siRNA has been viewed as a viable means to potentially protect against and treat a multitude of diseases and disorders. Several encouraging in vivo studies have been published, and siRNA-based therapeutic agents for the treatment of wet age-related macular degeneration (AMD) and respiratory syncytial virus (RSV) have successfully completed phase I clinical trials and are now in (or have successfully completed) phase II clinical trials. This opens several avenues for siRNA-based treatment of cardiovascular disease, as many current drugs for treating such diseases require frequent dosing and are associated with untoward side effects such as toxicity and development of resistance. In contrast, siRNA has been shown to have effects lasting up to several weeks, decreasing the need for inconvenient multiple daily dosing. Also, because the targeted mRNA undergoes rapid degradation following cleavage, the RISC complex becomes free to target another mRNA, rendering treatment extremely potent; effects have even been demonstrated at the nanomolar level (Hutvagner and Zamore 2002). Furthermore, recent clinical trials have shown siRNA-based therapeutics to be generally well tolerated.

In addition to the aforementioned advantages to using siRNA-based therapeutics, the properties inherent to siRNA make it an attractive class of treatment agents. As basic science research continues to elucidate pathways mediating the pathophysiology of cardiovascular disease, new targets for inhibition are generated. However, many of these targets are unable to be inhibited using current technology. Because siRNA can be rationally designed to target any mRNA, the possibilities of viable gene targets are limited only by the current level of knowledge. Moreover, because siRNA can be used to elucidate these pathways, a positive feedback loop is created which ultimately serves to increase the odds of successful treatment of all diseases. In this section we present the current clinical applications of siRNA technology and take a detailed look at its use in the treatment of cardiovascular disease.

9 Current Clinical Uses of siRNA

As mentioned above, there are three ongoing clinical trials involving siRNA-based therapeutics for the treatment of age-related macular degeneration (AMD) and respiratory syncytial virus (RSV). Age-related macular degeneration is the leading cause of blindness in adults over the age of 50 in the US. The wet form of this disease is characterized by abnormal blood vessel growth into the eye which leads to irreversible damage to photoreceptors and ultimately loss of focused vision. As vascular endothelial growth factor (VEGF) stimulates angiogenesis, it was made the target of both siRNA therapeutic agents for AMD. Cand5 from Acuity Pharmaceuticals targets all VEGF-A spliced isoforms and has successfully completed phase II clinical trials (Acuity Pharmaceuticals, press release). Sirna-021 is siRNA directed against the VEGF receptor-1, and is currently in phase II clinical trials (Sirna Pharmaceuticals, press release).

According to the CDC, respiratory syncytial virus is the most common cause of pneumonia and bronchiolitis in infants and children under 1 year of age. Treatment is largely supportive, as the efficacy of the only antiviral agent previously used, Ribavirin, was called into question. Alnylam Pharmaceuticals has thus developed siRNA against mRNA used to encode the nucleocapsid protein, which is necessary for viral replication. ALN-RSV01 successfully completed phase I clinical trials in May of 2007 and has currently moved into phase II studies (Alnylam Pharmaceuticals, press release).

In addition to these current studies, other siRNA-based therapeutic agents are in the pipeline to enter clinical trails. Calando Pharmaceuticals announced in May of 2007 that they would enter their lead molecule, CALAA-01, into phase I studies for nanoparticle-delivered, targeted treatment of cancer. Along those lines, Silence Technologies, a UK-based company, plans on filing an investigational new drug application in 2007 for its siRNA-based cancer treatment. In addition, there are plans to bring about clinical trials for the siRNA-based treatment of HIV/AIDS, Hepatitis B, and Hepatitis C.

10 siRNA for the Treatment of Cardiovascular Disease

As shown above, there are many potential siRNA-based therapeutic agents in clinical trials for a multitude of diseases and disorders. However, as the term "cardiovascular disease" may refer to any condition or disorder affecting the heart or blood vessels, the range of ailments in which this umbrella diagnosis covers is wide. This allows for great opportunities in siRNA research, and many laboratories across the world have answered the call. Presented here are selected siRNA targets which fall under the heading of five common cardiovascular diseases: hypertension, myocardial ischemia, atherosclerosis, intimal hyperplasia, and heart failure. For a more comprehensive list of targets and their applications which apply to cardiovascular disease, see Table 1.

10.1 Hypertension

Hypertension (HTN) is the most common cardiovascular disease worldwide, and according to the CDC, 29% of US adults aged 20 or older have been diagnosed with this malady. It is frequently called the "silent killer", as symptoms are not readily apparent until the disease has progressed significantly. Because HTN is a common disease with often unknown incidence, and is a risk factor for a multitude of diseases such as myocardial infarction and stroke, it is important that once diagnosed, hypertension be treated efficiently and with as few side effects as possible. There are many classes of antihypertensive agents, and each has their advantages and disadvantages. For example, agents which inhibit the binding to beta adenergic receptors (β-ARs) are commonly used to treat HTN; these drugs are called beta-blockers (also written as β-blockers). However, in patients with asthma, their general use is contraindicated because blockade of β2-ARs may cause bronchoconstriction, triggering an asthma attack. Therefore, specificity in receptor inhibition may be desired. Recently, Arnold et al. developed an siRNA that specifically inhibits the expression of β1-ARs. They showed that in addition to reducing the level of β1-AR mRNA by 70% without overlap to the β2-AR, the siRNA significantly lowered systolic blood pressure in spontaneously hypertensive rats and improved overall cardiac function (Arnold et al. 2007).

Increases in intravascular volume lead to increases in systolic blood pressure. One of the methods by which this can occur is through retention of sodium by the kidneys, which then pulls water into the vessels in order to balance this change in osmotic pressure. Diuretic agents tend to reduce blood pressure by utilizing this balance between sodium and water; by decreasing the reabsorption of sodium, the excretion of water is increased and ultimately intravascular volume is reduced. One agent that increases sodium retention is aldosterone, a mineralocorticoid. As a new mechanism to treat HTN, Wang et al. demonstrated that shRNA directed against the mineralocorticoid receptor (MR) prevented the progression of cold-induced hypertension in rats (Wang et al. 2006). For more information on siRNA therapeutics directed towards hypertension, see Table 1.

10.2 Myocardial Ischemia

One of the most recognizable cardiovascular diseases is myocardial ischemia, which may progress to irreversible infarction or what is commonly known as heart attack. Ischemia occurs when oxygen deprivation of the cells is paired with inadequate removal of metabolites due to decreased blood flow or perfusion; if not reversed, the end result is cell death. siRNAs are currently being developed to impair this progression at various stages. For example, siRNA directed against HIF-1 alpha-prolyl-4 hydroxylase-2 (PDH2) has been shown to reduce myocardial infarct size by more than 60% in murine and rat models of acute myocardial infarction (AMI) through increasing hypoxia inducible factor-1 alpha (HIF-1α) levels; this, in turn, decreases the inflammatory response to hypoxia and increases inducible nitric oxide synthase (iNOS) levels, which presumably leads to vasodilation through nitric oxide and hence enhanced blood flow (Natarajan et al. 2006, 2007).

In addition to improving perfusion through structures already present, siRNAs have been developed that can stimulate angiogenesis and even postnatal vasculogenesis. siRNA-mediated reductions in caveolin expression have been shown to increase the engraftment of endothelial progenitor cells and reduce ischemic tissue damage following a murine model of hindlimb ischemia (Sbaa et al. 2006), and knockdown of both FoxO1 and FoxO3a lead to increased migration of endothelial cells and longer vessel sprout lengths in a VEGF-stimulated model of in vitro angiogenesis (Potente et al. 2005). Also, targeting of SHP-1 demonstrated enhanced in vitro endothelial cell growth and increases in capillary density in an ischemic rat hindlimb. These effects were shown to be mediated through disinhibition of VEGF, as SHP-1 removes the phosphate required for VEGF receptor-2 function (Sugano et al. 2007).

Lastly, if cardiomyocytes are unable to be rescued through vessel dilation or formation, it is possible that inhibition of apoptosis can serve as last resort to prevent cell death. In addition to inhibition of SHP-1 allowing for angiogenesis, Sugano et al. have also shown that siRNA-mediated SHP-1 downregulation decreases apoptosis through disinhibition of Akt, a kinase involved in inhibiting apoptosis. To further the evidence that SHP-1 suppression by siRNA may serve as a viable target for cardiovascular disease, a rat model of acute myocardial infarction was performed, and SHP-1 siRNA was shown to significantly reduce infarct size (Sugano et al. 2005). For additional information on factors involved in myocardial ischemia that have been targeted by siRNAs, see Table 1.

10.3 Atherosclerosis

Atherosclerosis is a common chronic inflammatory disease that affects arteries and is the most common cause of myocardial infarction, due to the thrombogenic nature of its atheromatous plaques. Although the exact mechanics of atherosclerosis have yet to be uncovered, the most widely accepted view of its pathogenesis is the injury model. This states that endothelial injury, for example, due to oxidized low-density lipoprotein (LDL), high glucose, or toxins in cigarette smoke, causes vascular inflammation and a fibroproliferative response; monocytes then adhere to the vessel and migrate into its intimal layer. Once out of the vascular environment, macrophages (the former monocytes) then take up triglycerides to form pathologic foam cells. These now activated macrophages produce numerous factors that contribute to further endothelial injury, and along with smooth muscle cells, the formation of an atheromatous plaque. The instability of this plaque may then lead to rupture and subsequent thrombo-occlusion of the artery.

To prevent the formation and progression of atherosclerosis, siRNAs have been created to inhibit multiple factors involved in its pathogenesis. Wang et al. have developed an siRNA against SphK1, a kinase upregulated under hyperglycemic conditions; this siRNA was shown to significantly reduce monocyte adhesion to endothelial cells (Wang et al. 2005). To counter the next steps in atherosclerosis, siRNA against allograft inflammatory factor-1 (AIF-1) demonstrated a 62% decrease in macrophage migration and a 52% decrease in their proliferation (Tian et al. 2006).

Gene products play diverse roles in multiple processes and pathways. For example, upregulation in one situation may be beneficial, but under different pathological conditions, its inhibition may be more desirable. Such is the case for the abovementioned HIF-1α. It was shown that its upregulation through PDH2 inhibition protects against ischemia. However, its suppression by direct siRNA inhibition leads to a decrease in pathologic foam cell formation (Jiang et al. 2007). In addition to foam cells, vascular smooth muscle cells (VSMCs) are another component of atheromas, and siRNA-induced PTOV1 suppression significantly decreased testosterone-induced VSMC proliferation, which may help to offset some of the gender differences noted in the development of atherosclerosis (Nakamura et al. 2006). For a more comprehensive list of targets involved in atherosclerosis, see Table 1.

10.4 Intimal Hyperplasia

As explained in Sect. 5, intimal hyperplasia is a common cause of long-term graft failure following angioplasty and stent placement, and coronary artery bypass surgery. In attempts to prevent this stenosis of the vessel due to thickening of the intima, many groups have designed siRNAs against various molecules involved in this process. For example, following gene chip analysis of vein grafts, Monahan et al. noted the upregulation of CDH11 mRNA, a cadherin involved in cell surface adhesion. They then designed siRNA against the molecule and noted a 33% decrease in vascular smooth muscle cell migration (Monahan et al. 2007). In a rabbit venous bypass graft model, siRNA designed to inhibit midkine, a heparin-binding growth factor, decreased the leukocyte count and amount of proliferating cells found within the graft (Banno et al. 2006). Moreover, similar to the E2F3 aptamer described in Sect. 5, siRNAs designed against E2F1 and E2F3 by Giangrande et al. were able to decrease VSMC proliferation by as much as 90% and demonstrated in vivo efficacy in a murine venous bypass graft model, where a 50% reduction in the intima-media ratio was achieved (Giangrande et al. 2007). For more siRNA-directed targets involved in intimal hyperplasia, see Table 1.

10.5 Heart Failure

Heart failure is the leading cause of hospitalization in people over the age of 65. By definition, heart failure is any structural or functional heart disorder that decreases the heart's ability to fill with or pump a sufficient amount of blood through the body. In order for cardiomyocytes to contract and thus pump blood throughout the body, there must be a release of calcium from the sarcoplasmic reticulum (SR) within the cell. In fact, decreases in SR calcium levels have been linked to excitation-contraction uncoupling and failure of canine hearts (Hobai and O'Rourke 2001). Therefore, increasing SR calcium may be seen as an attractive treatment for heart failure. One method of achieving this increase is through releasing any inhibition of the SERCA2 SR calcium pump. Watanabe et al. and Fletcher et al. both used siRNA technology to inhibit phospholamban, a protein known to regulate SERCA2. These groups were able to demonstrate statistically significant increases in calcium uptake, which represents a novel possibility for the treatment of heart failure (Watanabe et al. 2004; Fechner et al. 2007).

There are many factors which may lead to heart failure; for example, as an adaptive response to hypertension, cardiomyocytes may enlarge in response to the increased arterial resistance. However, prolonged hypertrophy of the heart can lead to chamber dilation, a decreased ability to pump, and ultimately heart failure. There are many molecular pathways that lead to cardiac hypertrophy, and one such pathway involves PDK1. Protein Kinase D1 (PDK1) is able to initiate hypertrophy through its phosphorylation of HDAC5, a transcriptional repressor of pathogenic cardiac remodeling. Once HDAC5 is phosphorylated, it undergoes nuclear export and is no longer able to repress cardiac growth. Harrison et al. was able to take advantage of this knowledge and demonstrated that through siRNA inhibition of PDK1, the levels of HDAC5 in the nucleus increased, and subsequently, myocytes underwent significantly less growth when stressed with phenylephrine (Harrison et al. 2006).

Endotoxic shock which leads to sepsis is another potential cause of heart failure. When the blood is infected with Gram negative bacteria (such as *E. coli* or *Neisseria meningitidis*), lipopolysaccharide (LPS; an endotoxin) is released. This then stimulates the immune response through TLR4 or soluble CD14 immune receptors, which ultimately leads to NF-κB activation and an upregulation of various cytokines such as TNF-α and IL-1. When the body can no longer regulate this immune reaction, cardiomyopathy, organ failure, and shock ensue, leading to sepsis. As an attempt to halt this progression, Ramana et al. have developed an siRNA against aldose reductase, which regulates the postreceptor stimulation and nuclear translocation of NF-κB. Through siRNA-mediated reduction of aldose reductase protein levels, they were able to demonstrate a statistically significant decrease in NF-κB activation and cytokine production in LPS-stimulated macrophages (Ramana et al. 2006).

The aforementioned targets represent novel potential treatments for heart failure. For an expanded list of siRNAs targeting heart failure, as well as additional targets relevant to cardiovascular disease, see Table 1.

11 Potential Challenges Facing siRNA in Cardiovascular Disease

There are three major areas of concern facing the use of siRNA as therapeutic agents in general: avoiding "off-targeting" of the siRNA, circumventing immune reactions, and improving delivery of the agent to its intended site of action. Ongoing research is currently being conducted to overcome these challenges, and much headway has already been made. For example, off-targeting occurs when genes with partial homology to the selected target undergo untoward silencing by the siRNA (Jackson et al. 2003). Moreover, off-targeting may be toxic to the cell, in addition to causing the improper downregulation of unintended genes (Fedorov et al. 2006). It has been shown that off-targeting generally results from complimentarity between the 3′ UTR of a gene and a specific location within the siRNA called the "seed region," which is located in positions 2 through 7 or 8 from the 5′ end of the guide strand (Lin et al. 2005; Birmingham et al. 2006; Jackson et al. 2006). Through this enhanced knowledge, strategies to predict and hence avoid off-targeting can be made (Nielsen et al. 2007). Furthermore, off-targeting can be reduced by a 2′-*O*-methyl modification at position 2 in the guide strand of the siRNA, without compromising the silencing of perfectly matched targets (Jackson et al. 2006). Taken together, off-targeting is becoming less of a hurdle for application of siRNA-based therapeutics.

siRNA-instigated activation of the immune system is another area of concern, as double-stranded RNAs are common intermediates in the life cycle of many viruses. Given this resemblance, siRNAs may trigger the innate immune responses designed to combat viral infections. For dsRNA with less than 30 base pairs (as is the case for siRNA), the RNA-sensing immunoreceptors belonging to the toll-like receptor (TLR) family, particularly TLR7, mediate this response. Activation of TLR7 then leads to the induction of type I interferon and proinflammatory cytokines (Schlee et al. 2006). There are currently two strategies in place to avoid this reaction. One is to avoid the TLR response by avoiding immune cells that can mount a TLR response (such as plasmacytoid dendritic cells) or by bypassing the endosomal compartment where TLRs are present by delivering the siRNA directly to the cytoplasm (de Fougerolles et al. 2007). However, an alternative approach is to incorporate modifications in the siRNA sequence. It has been shown that 2′-*O*-Methyl modifications and avoidance of GU-rich sequences decrease TLR7 recognition and hence abrogate the innate immune response (Schlee et al. 2006; Sioud et al. 2007). Therefore, immune responses are now seen as less overwhelming and are now deemed as something to avoid.

Although much progress has been made to avoid off-targeting and activation of the immune system, the largest challenge hindering widespread use of siRNA as a therapeutic agent is delivery. As a small nucleic acid, siRNAs have a half-life in the body lasting only several minutes. This short length of time is a product of its degradation by nucleases and its rapid clearance by the kidneys. These obstacles need to be avoided or overcome. The siRNA therapeutic agents currently in clinical trials avoid both high concentrations of nucleases and rapid renal clearance through local delivery at the siRNA's site of action. Both the eyes and the lungs have much lower concentrations of nucleases than the intravascular environment, and their local delivery thus allows for lower concentrations of the drugs to be used, with decreased opportunities for the development of systemic side effects.

However, as most targets in cardiovascular disease are intravascular, avoidance of nucleases is not an option; because of this, the siRNA must be rendered nuclease resistant in order to achieve efficacious results. This can be accomplished through 2′ base sugar modifications (2′OMe, 2′F) to confer endonuclease resistance, along with phosphorothioate (P=S) backbone linkages to render the siRNA exonuclease resistant. However, these internal modifications alone may not be enough to extend the bioavailability of the siRNA to the level required for sustained activity. In this case, there may be a need to package the siRNA in a delivery system that allows for an additional degree of nuclease resistance (e.g., conjugation to cholesterol, which may then incorporate into circulating lipoprotein particles).

Once the siRNA has been made resistant to nucleases and can thus remain in the intravascular environment without immediate degradation, the matters of getting to the correct location where its effects are to be exerted, and next entering into the cellular compartment must be addressed. This is an area of intense research and discovery, as delivery is the last major obstacle facing siRNA therapeutics. McNamara et al. have merged two forms of RNA therapeutics to create an aptamer-conjugated siRNA chimera molecule (McNamara et al. 2006). The aptamer portion was directed against prostate-specific membrane antigen (PSMA), a cell surface receptor over-expressed on prostate cancer cells. This allowed for cell-type specific delivery of the siRNA, as the chimera did not bind to cells which did not express PSMA. Following binding, the chimera was subsequently internalized and processed by Dicer, and the siRNA portion was able to mediate death of the cancerous cell. This was the first example of using aptamer-siRNA chimeras for cell-type specific delivery of siRNA, and its application is sure to expand, as many targets in cardiovascular medicine, for example, are directed specifically against vascular smooth muscle cells. In addition to aptamer-mediated cell-type specific delivery, a similar approach has been used with antibodies (Song et al. 2005). Also, cholesterol conjugation and liposome complexes have improved cellular uptake, and nanoparticles can be conjugated with molecules to improve targeting to their natural receptors.

As research in this area continues, additional methods to enhance delivery of siRNAs to their targets will emerge. Given the compelling advantages of siRNA-based therapeutics and its growing number of viable targets in cardiovascular medicine, these challenges are sure to be met in time.

12 Conclusion

Overall, both aptamers and siRNAs demonstrate extreme potential for the treatment and prevention of cardiovascular disease. Both classes of nucleic acids have their individual strengths and challenges, and both present novel methods to better patient health and improve drug safety. As new targets are uncovered through continued basic science research, additional aptamer and siRNA therapeutic agents will emerge, which may ultimately bring about new safe and effective medicines for patients with cardiovascular disease.

References

Arnold AS, Tang YL, Qian K et al (2007) Specific beta1-adrenergic receptor silencing with small interfering RNA lowers high blood pressure and improves cardiac function in myocardial ischemia. J Hypertens 25:197–205

Banno H, Takei Y, Muramatsu T et al (2006) Controlled release of small interfering RNA targeting midkine attenuates intimal hyperplasia in vein grafts. J Vasc Surg 44:633–641

Bhatt DL, Topol EJ (2003) Scientific and therapeutic advances in antiplatelet therapy. Nat Rev Drug Discov 2:15–28

Bi D, Nishimura J, Niiro N et al (2005) Contractile properties of the cultured vascular smooth muscle cells: the crucial role played by RhoA in the regulation of contractility. Circ Res 96:890–897

Birmingham A, Anderson EM, Reynolds A et al (2006) 3′ UTR seed matches, but not overall identity, are associated with RNAi off-targets. Nat Methods 3:199–204

Blaschke F, Bruemmer D, Yin F et al (2004) C-reactive protein induces apoptosis in human coronary vascular smooth muscle cells. Circulation 110:579–587

Bock LC, Griffin LC, Latham JA et al (1992) Selection of single-stranded DNA molecules that bind and inhibit human thrombin. Nature 355:564–566

Cheng G, Kim MJ, Jia G et al (2007) Involvement of chloride channels in IGF-I-induced proliferation of porcine arterial smooth muscle cells. Cardiovasc Res 73:198–207

Cuaz-Perolin C, Furman C, Larigauderie G et al (2004) REDD2 gene is upregulated by modified LDL or hypoxia and mediates human macrophage cell death. Arterioscler Thromb Vasc Biol 24:1830–1835

Custodis F, Eberl M, Kilter H et al (2006) Association of RhoGDIalpha with Rac1 GTPase mediates free radical production during myocardial hypertrophy. Cardiovasc Res 71:342–351

DeAnda A, Jr., Coutre SE, Moon MR et al (1994) Pilot study of the efficacy of a thrombin inhibitor for use during cardiopulmonary bypass. Ann Thorac Surg 58:344–350

de Fougerolles A, Vornlocher HP, Maraganore J et al (2007) Interfering with disease: a progress report on siRNA-based therapeutics. Nat Rev Drug Discov 6:443–453

Dyke CK, Steinhubl SR, Kleiman NS et al (2006) First-in-human experience of an antidote-controlled anticoagulant using RNA aptamer technology: a phase 1a pharmacodynamic evaluation of a drug-antidote pair for the controlled regulation of factor IXa activity. Circulation 114:2490–2497

Elbashir SM, Harborth J, Lendeckel W et al (2001) Duplexes of 21-nucleotide RNAs mediate RNA interference in cultured mammalian cells. Nature 411:494–498

Ellington AD, Szostak JW (1990) In vitro selection of RNA molecules that bind specific ligands. Nature 346:818–822

Ellington AD, Szostak JW (1992) Selection in vitro of single-stranded DNA molecules that fold into specific ligand-binding structures. Nature 355:850–852

EyetechStudyGroup (2002) Preclinical and phase 1A clinical evaluation of an anti-VEGF pegylated aptamer (EYE001) for the treatment of exudative age-related macular degeneration. Retina 22:143–152

EyetechStudyGroup (2003) Anti-vascular endothelial growth factor therapy for subfoveal choroidal neovascularization secondary to age-related macular degeneration: phase II study results. Ophthalmology 110:979–986

Fahmy RG, Khachigian LM (2007) Suppression of growth factor expression and human vascular smooth muscle cell growth by small interfering RNA targeting EGR-1. J Cell Biochem 100:1526–1535

Fechner H, Pinkert S, Wang X et al (2007a) Coxsackievirus B3 and adenovirus infections of cardiac cells are efficiently inhibited by vector-mediated RNA interference targeting their common receptor. Gene Ther 14:960–971

Fechner H, Suckau L, Kurreck J et al (2007b) Highly efficient and specific modulation of cardiac calcium homeostasis by adenovector-derived short hairpin RNA targeting phospholamban. Gene Ther 14:211–218

Fedorov Y, Anderson EM, Birmingham A et al (2006) Off-target effects by siRNA can induce toxic phenotype. RNA 12:1188–1196

Fire A, Xu S, Montgomery MK et al (1998) Potent and specific genetic interference by double-stranded RNA in *Caenorhabditis elegans*. Nature 391:806–811

Fledderus JO, van Thienen JV, Boon RA et al (2007) Prolonged shear stress and KLF2 suppress constitutive proinflammatory transcription through inhibition of ATF2. Blood 109:4249–4257

Giangrande PH, Zhang J, Tanner A et al (2007) Distinct roles of E2F proteins in vascular smooth muscle cell proliferation and intimal hyperplasia. Proc Natl Acad Sci USA 104:12988–12993

Griffin LC, Tidmarsh GF, Bock LC et al (1993) In vivo anticoagulant properties of a novel nucleotide-based thrombin inhibitor and demonstration of regional anticoagulation in extracorporeal circuits. Blood 81:3271–3276

Halvorsen B, Waehre T, Scholz H et al (2005) Interleukin-10 enhances the oxidized LDL-induced foam cell formation of macrophages by antiapoptotic mechanisms. J Lipid Res 46:211–219

Hammond SM, Caudy AA, Hannon GJ (2001) Post-transcriptional gene silencing by double-stranded RNA. Nat Rev Genet 2:110–119

Harrison BC, Kim MS, van Rooij E et al (2006) Regulation of cardiac stress signaling by protein kinase d1. Mol Cell Biol 26:3875–3888

Hattori Y, Akimoto K, Nishikimi T et al (2006) Activation of AMP-activated protein kinase enhances angiotensin ii-induced proliferation in cardiac fibroblasts. Hypertension 47:265–270

Hobai IA, O'Rourke B (2001) Decreased sarcoplasmic reticulum calcium content is responsible for defective excitation-contraction coupling in canine heart failure. Circulation 103:1577–1584

Hoffman MM, Monroe DM (2005) Rethinking the coagulation cascade. Curr Hematol Rep 4:391–396

Hoffman M, Monroe DM (2007) Coagulation 2006: a modern view of hemostasis. Hematol Oncol Clin North Am 21:1–11

Hutvagner G, Zamore PD (2002) A microRNA in a multiple-turnover RNAi enzyme complex. Science 297:2056–2060

Jackson SP, Schoenwaelder SM (2003) Antiplatelet therapy: in search of the 'magic bullet'. Nat Rev Drug Discov 2:775–789
Jackson AL, Bartz SR, Schelter J et al (2003) Expression profiling reveals off-target gene regulation by RNAi. Nat Biotechnol 21:635–637
Jackson AL, Burchard J, Leake D et al (2006a) Position-specific chemical modification of siRNAs reduces "off-target" transcript silencing. RNA 12:1197–1205
Jackson AL, Burchard J, Schelter J et al (2006b) Widespread siRNA "off-target" transcript silencing mediated by seed region sequence complementarity. RNA 12:1179–1187
Jagavelu K, Tietge UJ, Gaestel M et al (2007) Systemic Deficiency of the MAP Kinase Activated Protein Kinase 2 Reduces Atherosclerosis in Hypercholesterolemic Mice. Circ Res [ePub ahead of print]
Jiang G, Li T, Qiu Y et al (2007) RNA interference for HIF-1alpha inhibits foam cells formation in vitro. Eur J Pharmacol 562:183–190
Kawakami A, Tani M, Chiba T et al (2005) Pitavastatin inhibits remnant lipoprotein-induced macrophage foam cell formation through ApoB48 receptor-dependent mechanism. Arterioscler Thromb Vasc Biol 25:424–429
Landa MS, Garcia SI, Schuman ML et al (2007) Knocking down the diencephalic thyrotropin-releasing hormone precursor gene normalizes obesity-induced hypertension in the rat. Am J Physiol Endocrinol Metab 292:E1388–E1394
Larigauderie G, Furman C, Jaye M et al (2004) Adipophilin enhances lipid accumulation and prevents lipid efflux from THP-1 macrophages: potential role in atherogenesis. Arterioscler Thromb Vasc Biol 24:504–510
Lawson JH, Mann KG (1991) Cooperative activation of human factor IX by the human extrinsic pathway of blood coagulation. J Biol Chem 266:11317–11327
Layzer JM, Sullenger BA (2007) Simultaneous generation of aptamers to multiple gamma-carboxyglutamic acid proteins from a focused aptamer library using DeSELEX and convergent selection. Oligonucleotides 17:1–11
Leppanen O, Janjic N, Carlsson MA et al (2000) Intimal hyperplasia recurs after removal of PDGF-AB and -BB inhibition in the rat carotid artery injury model. Arterioscler Thromb Vasc Biol 20:E89–E95
Li F, Zhang C, Schaefer S et al (2005) ANG II-induced neointimal growth is mediated via cPLA2- and PLD2-activated Akt in balloon-injured rat carotid artery. Am J Physiol Heart Circ Physiol 289:H2592–H2601
Li H, Liang J, Castrillon DH et al (2007) FoxO4 regulates tumor necrosis factor alpha-directed smooth muscle cell migration by activating matrix metalloproteinase 9 gene transcription. Mol Cell Biol 27:2676–2686
Lin X, Ruan X, Anderson MG et al (2005) siRNA-mediated off-target gene silencing triggered by a 7 nt complementation. Nucleic Acids Res 33:4527–4535
Liu HR, Gao E, Hu A et al (2005) Role of Omi/HtrA2 in apoptotic cell death after myocardial ischemia and reperfusion. Circulation 111:90–96
Liu X, Luo F, Pan K et al (2007) High glucose upregulates connective tissue growth factor expression in human vascular smooth muscle cells. BMC Cell Biol [ePub ahead of print]
McNamara JO, 2nd, Andrechek ER, Wang Y et al (2006) Cell type-specific delivery of siRNAs with aptamer-siRNA chimeras. Nat Biotechnol 24:1005–1015
Monahan TS, Andersen ND, Panossian H et al (2007) A novel function for cadherin 11/osteoblast-cadherin in vascular smooth muscle cells: modulation of cell migration and proliferation. J Vasc Surg 45:581–589
Mueller CF, Widder JD, McNally JS et al (2005) The role of the multidrug resistance protein-1 in modulation of endothelial cell oxidative stress. Circ Res 97:637–644
Myoishi M, Hao H, Minamino T et al (2007) Increased endoplasmic reticulum stress in atherosclerotic plaques associated with acute coronary syndrome. Circulation 116:1226–1233
Nakamura Y, Suzuki T, Igarashi K et al (2006) PTOV1: a novel testosterone-induced atherogenic gene in human aorta. J Pathol 209:522–531

Natarajan R, Salloum FN, Fisher BJ et al (2006) Hypoxia inducible factor-1 activation by prolyl 4-hydroxylase-2 gene silencing attenuates myocardial ischemia reperfusion injury. Circ Res 98:133–140
Natarajan R, Salloum FN, Fisher BJ et al (2007) Activation of hypoxia-inducible factor-1 via prolyl-4 hydoxylase-2 gene silencing attenuates acute inflammatory responses in postischemic myocardium. Am J Physiol Heart Circ Physiol 293:H1571–H1580
Nielsen CB, Shomron N, Sandberg R et al (2007) Determinants of targeting by endogenous and exogenous microRNAs and siRNAs. RNA 13:1894–1910
Nimjee SM, Rusconi CP, Harrington RA et al (2005a) The potential of aptamers as anticoagulants. Trends Cardiovasc Med 15:41–45
Nimjee SM, Rusconi CP, Sullenger BA (2005b) Aptamers: an emerging class of therapeutics. Annu Rev Med 56:555–583
Nimjee SM, Keys JR, Pitoc GA et al (2006) A novel antidote-controlled anticoagulant reduces thrombin generation and inflammation and improves cardiac function in cardiopulmonary bypass surgery. Mol Ther 14:408–415
Oney S, Nimjee SM, Layzer J et al (2007) Antidote-controlled platelet inhibition targeting von Willebrand factor with aptamers. Oligonucleotides 17:265–274
Potente M, Urbich C, Sasaki K et al (2005) Involvement of Foxo transcription factors in angiogenesis and postnatal neovascularization. J Clin Invest 115:2382–2392
Ramana KV, Willis MS, White MD et al (2006) Endotoxin-induced cardiomyopathy and systemic inflammation in mice is prevented by aldose reductase inhibition. Circulation 114:1838–1846
Roberts HR, Hoffman M, Monroe DM (2006) A cell-based model of thrombin generation. Semin Thromb Hemost 32 Suppl 1:32–38
Rusconi CP, Yeh A, Lyerly HK et al (2000) Blocking the initiation of coagulation by RNA aptamers to factor VIIa. Thromb Haemost 84:841–848
Rusconi CP, Scardino E, Layzer J et al (2002) RNA aptamers as reversible antagonists of coagulation factor IXa. Nature 419:90–94
Rusconi CP, Roberts JD, Pitoc GA et al (2004) Antidote-mediated control of an anticoagulant aptamer in vivo. Nat Biotechnol 22:1423–1428
Ryu J, Lee CW, Shin JA et al (2007) FcgammaRIIa mediates C-reactive protein-induced inflammatory responses of human vascular smooth muscle cells by activating NADPH oxidase 4. Cardiovasc Res 75:555–565
Sadler JE (2005a) New concepts in von Willebrand disease. Annu Rev Med 56:173–191
Sadler JE (2005b) von Willebrand factor: two sides of a coin. J Thromb Haemost 3:1702–1709
Sbaa E, Dewever J, Martinive P et al (2006) Caveolin plays a central role in endothelial progenitor cell mobilization and homing in SDF-1-driven postischemic vasculogenesis. Circ Res 98:1219–1227
Schlee M, Hornung V, Hartmann G (2006) siRNA and isRNA: two edges of one sword. Mol Ther 14:463–470
Si H, Grgic I, Heyken WT et al (2006) Mitogenic modulation of Ca2+ -activated K+ channels in proliferating A7r5 vascular smooth muscle cells. Br J Pharmacol 148:909–917
Sioud M, Furset G, Cekaite L (2007) Suppression of immunostimulatory siRNA-driven innate immune activation by 2′-modified RNAs. Biochem Biophys Res Commun 361:122–126
Song E, Lee SK, Wang J et al (2003) RNA interference targeting Fas protects mice from fulminant hepatitis. Nat Med 9:347–351
Song E, Zhu P, Lee SK et al (2005) Antibody mediated in vivo delivery of small interfering RNAs via cell-surface receptors. Nat Biotechnol 23:709–717
Stawowy P, Meyborg H, Stibenz D et al (2005) Furin-like proprotein convertases are central regulators of the membrane type matrix metalloproteinase-pro-matrix metalloproteinase-2 proteolytic cascade in atherosclerosis. Circulation 111:2820–2827
Sugano M, Tsuchida K, Hata T et al (2005) RNA interference targeting SHP-1 attenuates myocardial infarction in rats. Faseb J 19:2054–2056

Sugano M, Tsuchida K, Maeda T et al (2007) SiRNA targeting SHP-1 accelerates angiogenesis in a rat model of hindlimb ischemia. Atherosclerosis 191:33–39

Sullenger BA, Gallardo HF, Ungers GE et al (1990) Overexpression of TAR sequences renders cells resistant to human immunodeficiency virus replication. Cell 63:601–608

Thom T, Haase N, Rosamond W et al (2006) Heart disease and stroke statistics – 2006 update: a report from the American Heart Association Statistics Committee and Stroke Statistics Subcommittee. Circulation 113:e85–e151

Tian Y, Kelemen SE, Autieri MV (2006) Inhibition of AIF-1 expression by constitutive siRNA expression reduces macrophage migration, proliferation, and signal transduction initiated by atherogenic stimuli. Am J Physiol Cell Physiol 290:C1083–C1091

Tuerk C, Gold L (1990) Systematic evolution of ligands by exponential enrichment: RNA ligands to bacteriophage T4 DNA polymerase. Science 249:505–510

Turner NA, Hall KT, Ball SG et al (2007) Selective gene silencing of either MMP-2 or MMP-9 inhibits invasion of human saphenous vein smooth muscle cells. Atherosclerosis 193:36–43

Vazquez J, Correa de Adjounian MF, Sumners C et al (2005) Selective silencing of angiotensin receptor subtype 1a (AT1aR) by RNA interference. Hypertension 45:115–119

Wang L, Xing XP, Holmes A et al (2005) Activation of the sphingosine kinase-signaling pathway by high glucose mediates the proinflammatory phenotype of endothelial cells. Circ Res 97:891–899

Wang X, Skelley L, Cade R et al (2006) AAV delivery of mineralocorticoid receptor shRNA prevents progression of cold-induced hypertension and attenuates renal damage. Gene Ther 13:1097–1103

Wang D, Liu Z, Li Q et al (2007) An essential role for gp130 in neointima formation following arterial injury. Circ Res 100:807–816

Watanabe A, Arai M, Yamazaki M et al (2004) Phospholamban ablation by RNA interference increases Ca2+ uptake into rat cardiac myocyte sarcoplasmic reticulum. J Mol Cell Cardiol 37:691–698

Werk D, Schubert S, Lindig V et al (2005) Developing an effective RNA interference strategy against a plus-strand RNA virus: silencing of coxsackievirus B3 and its cognate coxsackievirus-adenovirus receptor. Biol Chem 386:857–863

White R, Rusconi C, Scardino E et al (2001) Generation of species cross-reactive aptamers using "toggle" SELEX. Mol Ther 4:567–573

Wu YJ, Bond M, Sala-Newby GB et al (2006) Altered S-phase kinase-associated protein-2 levels are a major mediator of cyclic nucleotide-induced inhibition of vascular smooth muscle cell proliferation. Circ Res 98:1141–1150

Xiao C, Qiu L, Zen Q (2006) Construction of shRNA targeted to the rat angiotensin II type 1 receptors and its RNAi in cytoplasma. J Huazhong Univ Sci Technolog Med Sci 26:4–8

Yuan J, Cheung PK, Zhang HM et al (2005) Inhibition of coxsackievirus B3 replication by small interfering RNAs requires perfect sequence match in the central region of the viral positive strand. J Virol 79:2151–2159

Zheng PS, Reis M, Sparling C et al (2006) Versican G3 domain promotes blood coagulation through suppressing the activity of tissue factor pathway inhibitor-1. J Biol Chem 281:8175–8182

Nucleic Acid Aptamers for Cardiovascular Therapeutics

P.S. Pendergrast(✉), K.M. Thompson, and R.G. Schaub

Contents

Abstract Nucleic acid aptamers can fold into a multitude three-dimensional structures allowing them to bind to almost any small molecule, nucleic acid or protein target with high specificity and affinity. Aptamers can be generated in a controlled and entirely in vitro process known as SELEX. Advances in aptamer technology have made possible the application of this therapeutic modality to many therapeutic areas including cardiovascular indications. To that end, aptamers have been generated to coagulants, adhesion, and angiogenic targets leading to the pre-clinical and clinical development of numerous aptamer drugs.

1 Nucleic Acid Aptamers

An aptamer is a structured nucleic acid molecule that binds tightly to a specific molecular target. The term "aptamer" is derived from the Latin word "aptus" that means "to fit." Aptamers fold into unique, stable tertiary structures that allow for molecular recognition through van der Waals, hydrogen bonding, and electrostatic interactions. From a mode of action point of view, aptamers can be thought of as nucleic acid or "chemical" antibodies. They bind with high affinity, K_D = pM–nM, and have the ability to disrupt protein–protein interactions. Aptamers are typically 15–40 nucleotides in length or 5–15 KDa and can be produced by a readily scalable

P.S. Pendergrast
Archemix Corp. 300 Third St. 2nd Floor, Cambridge, MA 02139, USA
e-mail: pendergrast@archemix.com

V.A. Erdmann et al. (eds.) *RNA Technologies in Cardiovascular Medicine and Research*,
doi: 10.1007/978-3-540-78709-9,

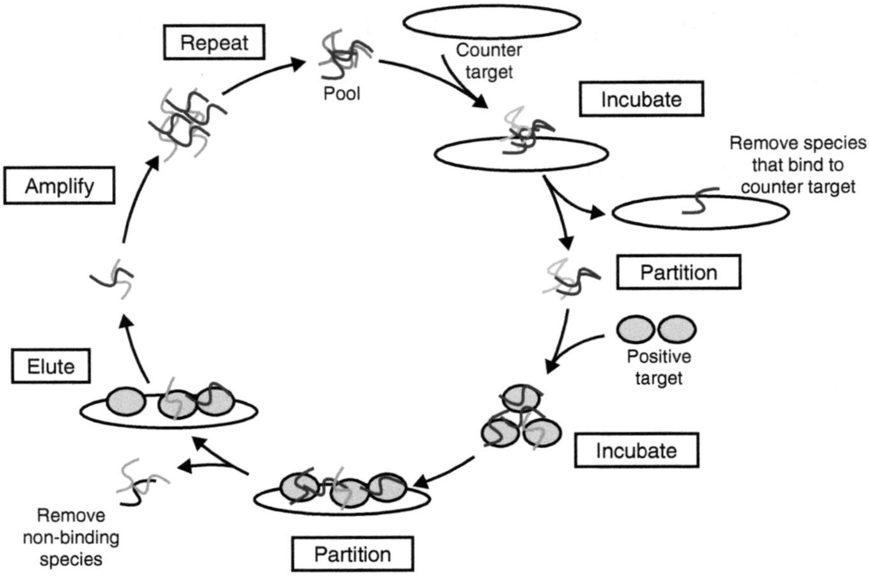

Fig. 1 General SELEX scheme. Pool RNA is mixed with the counter target (ex: selection matrix or counter-SELEX target). Molecules of the pool that do not bind to the counter target are partitioned from the bound molecules. These molecules are then incubated with the positive selection target. Target:RNA complexes are partitioned from non-binding molecules and bound molecules are eluted and then amplified. Multiple cycles of SELEX result in RNA aptamers that have high affinity and specificity toward the target

chemical manufacturing process. Aptamers are discovered through the application of the SELEX (Systematic Evolution of Ligands through EXponential enrichment) process (Fig. 1) (Tuerk and Gold 1990). The initial SELEX experiment was designed to study the key determinants for the binding interaction between T4 DNA polymerase and the RBS of its mRNA, yet Tuerk and Gold proposed that this process would allow for the ability to select nucleic acid-based binders to any molecule and these nucleic acid-based binders could modulate protein function. Since the introduction of the SELEX process in 1990, researchers have identified high affinity aptamers targeting a broad cross-section of protein families including cytokines, proteases, kinases, cell-surface receptors and cell adhesion molecules (Fig. 2).

2 Therapeutic Potential of Aptamers

In addition to target validation and research applications, aptamer technology has progressed to the point that aptamers can be developed as therapeutic agents. Although natural RNAs/DNAs are subject to nuclease degradation when dosed into animals, chemical modifications can be introduced to protect aptamers from serum endonucleases and exonucleases. For instance, endonucleases can be blocked by

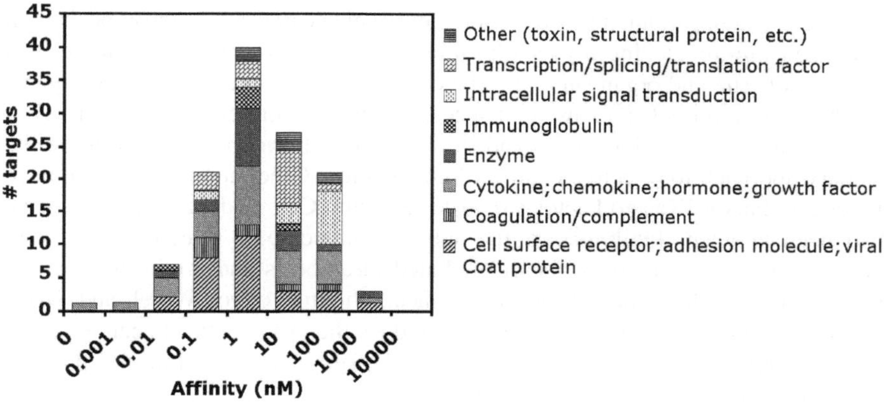

Fig. 2 Affinities for aptamers for different target types. Affinity versus number of aptamers selected plotted for a broad cross-section of targeted protein families

modifications to the 2′-hydroxyl position (Green et al. 1995). Nucleic acid-derived pools with chemical modifications at the 2′ position on the ribose can be used in the SELEX process to modulate the nuclease stability of output aptamers. These modified bases are chemically stable, nuclease resistant, and can be efficiently incorporated into the SELEX process through the use of T7 polymerase mutants (Burmeister et al. 2006). Exonucleases can be blocked by modification of the 5′- and 3′-ends of an aptamer by using a 3′-3′-linked thymidine cap for the 3′ end (Dougan et al. 2000) or by attaching a high molecular weight polyethylene glycol (PEG) group to either end. High molecular weight PEG has the added benefit of dramatically decreasing the clearance via the kidney of aptamer (Healy et al. 2004) and allowing increases in pharmacokinetic half-life by more than ten-fold (Ostendorf et al. 2001; Eyetech Study Group 2003).

Aside from their great specificity, affinity, efficacy and ability to modulate pharmacokinetic properties, aptamers have other competitive advantages over protein biologics. For instance, unlike antibodies, aptamers are produced by an entirely in vitro process. This allows for a more rapid generation of initial therapeutic leads and gives researchers more control over determining the specificity and affinity of said leads. Aptamers have also demonstrated little or no toxicity or immunogenicity and because of their superior solubility can be administered at high doses by subcutaneous injection as well as by intravenous injection (Cload et al. 2006).

3 Anticoagulant and Antithrombotic Aptamers

The primary control of bleeding is via the biochemical cascade that results in the formation of the hemostatic plug through fibrin polymerization and platelet activation. In the healthy blood vessel this cascade is critical to control bleeding. In the atherosclerotic or inflamed blood vessel, these events can result in thrombosis with associated life-threatening myocardial infarction, stroke or other ischemic events that effect tissue and organ survival. The development of safe and effective agents to inhibit thrombosis has focused on specific enzymatic or cellular components of the clotting pathways. Factor Xa and thrombin have been and continue to be the primary approach to effective control of coagulation. Other approaches have been directed to Factor IXa and Factor VIIa/Tissue Factor. Cell mediated pathways have also been targeted. Inhibitors of the selectin molecule, P-selectin, have been shown to reduce both tissue factor mediated thrombosis and prothrombotic cell accumulation on the damaged vascular surface. Inhibitors of von Willebrand Factor (VWF) inhibitors have also been shown to reduce the initial platelet accumulation. Inhibitory aptamers have been developed to all of these targets. Multiple aptamers have entered into clinical development.

Thrombin specific inhibitory aptamers were first reported by Toole and colleagues (Bock et al. 1995). They reported on a single-stranded DNA aptamer containing a highly conserved 14–17 base region with binding affinities for thrombin of 25–200 nM. One aptamer, with binding specificity for alpha-thrombin (GGTTGGTGTGGTTGG), inhibited fibrin formation at nM concentrations. A role for the anion binding exosite in aptamer binding was suggested by the fact that hirudin competed with the aptamer for binding. The specific binding site of the aptamer was identified as Lys-21 and Lys-65 on the B chain located within or in proximity to the anion binding site, suggesting that the aptamer competes with fibrinogen and platelet thrombin receptor binding and inhibits thrombin-catalyzed cleavage of PAR-1 in a dose-dependent manner (Paborsky et al. 1993; Boncler et al. 2001). The anticoagulant activity of the aptamer was reported in two studies. The aptamer inhibited thrombin induced platelet aggregation with an IC50 of 70–80 nM and, in an ex vivo whole artery angioplasty model, the aptamer inhibited fibrinopeptide A production and platelet adhesion at both low and high shear conditions. Clot bound thrombin was also inhibited by approximately 80% (Li et al. 1994). Non-human primate studies in cynomolgus monkeys demonstrated a rapid prolongation of the prothrombin time during infusion and a rapid reversal once infusion was stopped. The half-life was estimated at 108 ± 14 s. Thrombin-induced platelet aggregation was also inhibited, while collagen aggregation was unchanged. The aptamer was evaluated in a sheep model of extracorporeal hemofiltration. The prothrombin time was doubled in the circuit during treatment. The authors concluded that the rapid onset and the short half-life could make such an aptamer a useful anticoagulant for extracorporeal circuits with some distinct advantages over current anticoagulant therapy (Griffin et al. 1993). As of 2007, the only thrombin inhibitory aptamer to reach a phase I clinical trials is ARC183. The administration of ARC183 resulted in a rapid onset of anticoagulation and demonstrated a stable, dose-associated anticoagulation. The effects of ARC183 rapidly reversed after drug infusion was stopped. The efficacy of ARC183 was limited by the amount of drug needed to achieve the desired anticoagulation. This resulted in an unacceptable dosing profile for use in coronary artery bypass graft surgery (CABG). A second generation molecule with greater potency is currently under development (Hutabarat et al. 2007). This molecule has a very high affinity ($\underline{K}_D$ ~0.1 nM) for thrombin and, when administered by IV bolus + infusion to pigs and monkeys, it achieves a significant anticoagulation effect. In these studies, an ACT ≥400 s and a rapid reversal of the activity was noted within 25 min of stopping infusion. In a pig bypass pump model, this second generation molecule was administered at doses sufficient to maintain ACTs ≥400 s throughout a 3-h bypass procedure. It prevented clot formation while clots formed in the circuits of the saline control animals (Wagner-Whyte et al. 2007).

A primary stimulus to physiologic and pathologic clot formation is the Tissue factor/Factor VIIa complex. A specific stable RNA aptamer inhibitor of Factor VII/VIIa has been developed to inhibit this pathway. The aptamer was shown to inhibit tissue factor activation of Xa by VIIa with a prolonged half-life of approximately 15 h (Rusconi et al. 2000)

The Factor IXa inhibitory apatmer RB006 (Rusconi et al. 2004a,b; Nimjes et al. 2006) has resulted in a significant advancement in the development of anticoagulant aptamers. This aptamer did not use the strategy of unstable constructs to promote rapid inactivation by endo- and exonucleases that was part of the thrombin inhibition strategy. The high affinity and specific Factor IXa inhibitory RNA-based aptamer RB006 has been formulated to have a prolonged duration of action, using both chemical stabilization against nucleases and conjugation with 40-kD PEG to improve pharmacokinetic parameters. The mechanism of action for RB006 results from the inhibition of factor IXa's role in the generation of factor Xa. Its better stability and PK characteristics result in a more durable and controlled systemic anticoagulant and antithrombotic activity and a much slower reversal of Factor IXa inhibition compared to the thrombin apatmers previously discussed. In order to accelerate reversal, a modified-RNA oligonucleotide antidote complementary to RB006 and designated RB007 was developed. RB007 retains the rapid clearance of the less stable aptamers resulting in both neutralization and clearance of the RB006/RB007 complex. The results of the Phase I clinical trial was recently reported by Dyke et al. (2006). A predictable dose–response as measured by activated partial thromboplastin time was observed with escalating doses. Volunteers treated with the antidote RB007 had a rapid (1–5 min) and sustained return to baseline.

4 Antiadhesion Aptamers

Platelets and leukocytes have a central role in the thrombotic process. At the same time, the antithrombotic efficacies of inhibitors of platelet-leukocyte adhesion have been reported (Furie and Furie 2007; Hennan et al. 2006). P-selectin and von Willebrand Factor are two important mediators of platelet/leukocyte and platelet adhesion that have been targets for anti-adhesion apatmer development. Anti-P-selectin aptamers with an affinity of 16–710 pM (10^5–10^6 higher than the native minimal carbohydrate ligand sialyl Lewis X) were reported by Jenison et al. (1998). The aptamers were found to bind with subnanomolar affinity to P-selectin expressed on the surface of thrombin-activated platelets and blocked the binding of activated platelets to neutrophils. In vitro studies on sickle cell disease were recently reported, with one of these aptamer clones (PF377), for its ability to prevent cellular adhesion of SS-RBCs to endothelial cells. The aptamer binds P-selectin with a Kd of 6 nM and a Bmax of 87%. The aptamer at 60 nM had anti-adhesion activity similar to heparin and an inhibitory antibody to P-selectin in inhibiting sickle rbc adhesion (Nishimura et al. 2007). No in vivo data have been reported.

von Willebrand Factor (VWF) multimers adhere to exposed collagen via the vWF A3 domain and, under conditions of high shear force common to diseased arteries with luminal atherosclerotic lesions, platelets in transit through these vessels become immobilized through interactions between the adherent VWF A1 domain and the platelet glycoprotein Ib (GPIb) receptors. This interaction stimulates platelet activation, recruitment and thrombus formation. Aptamers to VWF that inhibit these interactions have been developed. A high affinity VWF aptamer that can inhibit VWF mediated platelet aggregation and can be reversed by an antidote similar to that described above for Factor IXa has been reported (Oney et al. 2007). A more complete characterization of a second VWF aptamer with specificity for the A1 domain, has been completed (Lagassé et al. 2007a; Rottman et al. 2007). The aptamer has high affinity for human VWF (K_D = 1.4 nM, or ~0.02 μg ml^{-1}) and minimal binding to other closely related A1 domain-containing proteins CD49c, ICAM, collagen type VI and Factor B, as well as complement factors C3 and C5, prothrombin, thrombin, Factor Xa, fibrinogen and HSA. Assessment by ELISA for inhibition of VWF activity demonstrated that the aptamer was a potent inhibitor of VWF activity. The IC_{90} value for inhibition of VWF activity in human plasma was 3.72 μg ml^{-1} (~283 nM). Human platelet aggregation was also assessed with the PFA-100®. The IC_{90} and maximal prolongation of closure times (≥300 s) values were 1.15 μg ml^{-1} (0.088 μM). Despite potent inhibition of VWF-mediated platelet aggregation, concentrations up to 130 μg ml^{-1} or 10 μM had no effect on the VWF-independent platelet aggregation induced by platelet agonists such as epinephrine, arachidonic acid, ADP, collagen and thrombin. The aptamer demonstrated a significant anti-thrombotic activity when it was administered to 15 monkeys by a bolus + continuous iv infusion at doses of 100 μg kg^{-1} bolus + a 1 μg kg^{-1} min^{-1} infusion up to a 600 μg kg^{-1} bolus + a 3.7 μg kg^{-1} min^{-1} infusion. In this study, it was compared to saline and the GPIIb/IIIa inhibitor abciximab (250 μg kg^{-1} bolus + a 0.125 μg kg^{-1} min^{-1} infusion. Bleeding times were minimally elevated over saline except at the highest dose of aptamer tested. This aptamer has also been evaluated in a Phase I clinical trial (Gilbert et al. 2007). This was a randomized, double-blind, placebo-controlled study in 47 healthy volunteers at doses of 0.05–1.0 mg kg^{-1}. PD effects were measured by an ELISA for free VWF A1 binding sites and by a platelet function analyzer. The concentration-time profile appeared monophasic. C_{max} and AUC were dose-proportional. The mean apparent elimination half-life ($t_{1/2\beta}$) was ~2 h and mean residence time (MRT) was ~3 h. The mean apparent volumes of distribution (V_z and V_{ss}) were ~1/2 the blood volume, suggesting that distribution is in the central compartment. The mean clearance (CL) ranged from ~10 to 21% of GFR, suggesting that renal filtration may not be a major mechanism of clearance. Inhibition of VWF A1 binding activity was achieved with an EC_{90} value of 2.0 μg ml^{-1} (151 nM) and of platelet function with an EC_{90} value of 2.6 μg ml^{-1} (196 nM). The aptamer was generally well tolerated and no bleeding was observed. Adverse events tended to be minor and not dose related.

5 Antiangiogenic Aptamers and Potential for Treatment of Atherosclerosis

Inhibition of angiogenesis has been one of the early target areas of interest for the development of therapeutic aptamers. In fact, the first approved aptamer (Macugen, penaptamib sodium) inhibits vascular endothelial cell growth factor (VEGF) and the abnormal vasculogenesis associated with wet age-related macular degeneration (AMD) (Ng and Adamis 2006). An aptamer inhibiting the angiogenic factor platelet derived growth factor (PDGF-B) and an antiangiopoietin-2 (Ang-2) aptamer are also under development for treatment of AMD. Antiangiogenic aptamers have also been a focus for the treatment of cancer by reducing vascular growth, and thus the nutrient supply necessary for tumor growth and vascular channels that carry metastatic cells. This effort has included studies of aptamers inhibiting PDGF-B, Ang-2, and VEGF (Sennino et al. 2007; Sarraf-Yazdi et al. 2007; Huang et al. 2001). Angiogenesis and the inhibition of angiogenesis by aptamers may also have an important therapeutic benefit in the treatment of atherosclerosis. The outer layer of blood vessels, the adventitia, is characterized by a vascular network known as the vasa vasorum. The normal function of this network is the supply of nutrient flow to the blood vessel. However, in the developing atherosclerotic lesion, there is significant vascular proliferation in the vasa vasorum, and this proliferation can extend into the media (smooth muscle rich portion of the vessel wall) and the atherosclerotic lesion. Like the abnormal, leaky vessels observed in AMD and within tumors, these vessels are also leaky. In addition, it is thought that they may also be the source of infiltrating inflammatory cells that can both increase the lesion growth as well as further destabilize the lesion resulting in intralesion hemorrhage and thrombosis (Doyle and Caplice 2007; Kahlon et al. 1992; Michel et al. 2007; Chyu and Shah 2007; Petrovan et al. 2007; Kolodgie et al. 2007). The source of the angiogenic stimulus may come from one of the key cells known to be involved in the development of the proliferative atherosclerotic lesion, the platelet. Angiogenic growth factors have been reported to be stored and released from platelet granules (Italiano et al. 2008). Thus, localization of platelets in an area of lesion formation could also be expected to release factors that would also promote local angiogenesis. It would be expected that aptamers inhibiting one or more of these angiogenic factors could have therapeutic benefit in the treatment of lesions, especially the highly vascular unstable lesion or early lesions created after dilation by stent placement. PDGF inhibitory aptamers have been evaluated their effect on smooth muscle proliferation following balloon injury of the rat carotid artery, a model of post angioplasty re-occlusion. Treatment was shown to be effective during a 2-week period.

However, once treatment was stopped, the beneficial effect was lost. This suggested to the authors that a longer treatment would be required for full therapeutic benefit (Leppanen et al. 2000). No other studies with antiangiogenic aptamers in atherosclerosis have been reported.

References

Bock LC, Griffin LC, Latham JA et al (1995) Selection of single-stranded DNA molecules that bind and inhibit human thrombin. Nature 355:564–566

Burmeister PE, Wang C, Killough J R et al (2006) 2′-Deoxy purine, 2′-*O*-methyl pyrimidine (dRmY) aptamers as candidate therapeutics. Oligonucleotides 16(4):337–351

Boncler MA, Koziolkiewicz M, Watala C (2001) Aptamer inhibits degradation of platelet proteolytically activatable receptor, PAR-1, by thrombin. Thromb Res 104:215–222

Chyu K-Y, Shah PK (2007) Choking off plaque neovascularity: a promising atheroprotective strategy or a double-edged sword? Arterioscl Thromb Vasc Biol 27:993–995

Cload ST, McCauley TG, Keefe AD et al (2006) Properties of therapeutic aptamers. In: Klussman S (ed) The aptamer handbook. Wiley, Weinheim, pp 363–416

Dougan H, Lyster DM, Vo CV et al (2000). Extending the lifetime of anticoagulant oligodeoxynucleotide aptamers in blood. Nucl Med Biol 27:289–297

Doyle B, Caplice N (2007) Plaque neovascularization and antiangiogenic therapy for atherosclerosis. J Am Coll Cardiol 49:2102–2104

Dyke CK, Steinhubl SR, Kleiman NS et al (2006) First-in-human experience of an antidote-controlled anticoagulant using RNA aptamer technology. A Phase Ia pharmacodynamic evaluation of a drug-antidote paior for the controlled regulation of factor IXa activity. Circulation 114:2490–2497

Eyetech Study Group (2003). Anti-vascular endothelial growth factor therapy for subfoveal choroidal neovascularization secondary to age-related macular degeneration: phase II study results. Ophthalmology 110:979–986

Furie B, Furie B (2007) In vivo thrombosis. J Thromb Haemost Suppl 1:12–17

Gilbert JC, DeFeo-Fraulini T, Hutabarat RM et al (2007) Pharmacokinetis, pharmacodynamics, and safety of an anti-von Willebrand factor therapeutic aptamer, ARC1779, in healthy volunteers. Circulation 116:2678–2686

Green LS, Jellinek D, Bell C et al (1995) Nuclease-resistant nucleic acid ligands to vascular permeability factor/vascular endothelial growth factor. Chem Biol 2:683–695

Griffin LC, Tidmarsh GF, Bock LC et al (1993) In vivo anticoagulant properties of a novel nucleotide-based thrombin inhibitor and demonstration of regional anticoagulation in extracorporeal circuits. Blood 81:3271–3276

Healy JM, Lewis SD, Kurz M et al (2004) Pharmacokinetics and biodistribution of novel aptamer compositions. Pharm Res 21:2234–2246

Hennan JK, Swillo RE, Morgan GA et al (2006) Pharmacologic inhibition of platelet vWF-GPIb alpha interaction prevents coronary artery thrombosis. Thromb Haemost 95:469–475

Huang J, Moore J, Soffer S et al (2001) Highly specific antiangiogenic therapy is effective in suppressing growth of experimental Wilms tumors. J Pediatr Surg 36:357–361

Hutabarat R, Horvath C, Funk WD (2007) Discovery of a potent, direct thrombin inhibiting aptamer, J Thromb Haemost 5(Suppl 2):P-S-067

Italiano JE, Richardson JL, Patel-Hett S Et al (2008) Angiogenesis is regulated by a novel mechanism: Pro- and anti-angiogenic proteins are organized into separate platelet a-granules and differentially released. Blood 111(3):1227–1233 (published ahead of print On-Line October 25, 2007)

Jenison RD, Jennings SD, Walker DW et al (1998) Oligonucleotide inhibitors of P-selectin-dependent neutrophil-platelet adhesion. Antisense Nucleic Acid Drug Dev 8:265–279
Kahlon R, Shapero J, Gotlieb AI (1992) Angiogenesis in atherosclerosis. Can J Cardiol 8:60–64
Kolodgie FD, Narulo J, Yuan C et al (2007) Elimination of neoangiogenesis for plaque stabilization: Is there a role for local drug therapy? J Am Coll Cardiol 49:2093–2101
Lagassé HAD, Merlino PG, Marsh HN et al (2007) Discovery and characterization of a potent neutralizing anti-VWF A1 domain specific aptamer. J Thromb Haemost 5(Suppl 2):P-S-665
Leppanen O, Janjic N, Carlsson M-A et al (2000) Intimal hyperpasia recurs after removal of PDGF-AB and BB inhibition by the rat carotid artery injury model. Arterioscl Thromb Vasc Biol 20:89–95
Li W-X, Kaplan AV, Grant GW et al (1994) A novel nucleotide-based thrombin inhibitor inhibits clot-bound thrombin and reduces arterial platelet thrombus formation. Blood 83:677–682
Michel JB, Thaunat O, Houard X, et al. (2007) Topological determinants and consequences of adventitial responses to arterial wall injury. Arterioscler Thromb Vasc Biol 27:1259–1268.
Ng EWM, Adamis AP (2006) Anti-VEGF aptamer (pegaptanib) yherapy for ocular vascular diseases. Oligonucleotide thrapeutics: First Annual Meeting of the Oligonucleotide Therapeutics Society. Ann NY Acad Sci 1082:151–171
Nimjes SM, Keys JR, Pitoc GA et al (2006) A novel antidote-controlled anticoagulant reduces thrombin generation and inflammation and improves cardiac function in cardiopulmonary bypass surgery. Mol Ther 24:408–415
Nishimura JI, Burnette AD, Oney S et al (2007) Blood 110(11): Abstract 147
Oney S, Nimjee SM, Layzer J et al (2007) Antidote-Controlled Platelet inhibition targeting von Willebrand factor with aptamers. Oligonucleotides 17:265–274
Ostendorf T, Kunter U, Grone HJ et al (2001). Specific antagonism of PDGF prevents renal scarring in experimental glomerulonephritis. J Am Soc Nephrol 12:909–918
Paborsky LR, McCurdy SN, Griffin LC et al (1993) The single stranded DNA aptamer-binding site of human thrombin. J Biol Chem 268:20808–20811
Petrovan PJ, Kaplan CD, Reisfeld RA et al (2007) DNA vaccination against VEGF receptor 2 reduces atherosclerosis in LDL receptor-deficient mice. Arterioscl Thromb Vasc Biol 27:1095–1100
Rottman RB, Gilbert M, Marsh HN et al (2007) An anti-Von Willebrand's factor A1 domain aptamer inhibits arterial thrombosis induced by electrical injury in cynomolgus macaques. J Thromb Haemost 5(Suppl 2):P-S-664
Rusconi CP, Yeh A, Lyerly HK et al (2000) Blocking the initiation of coagulation by RNA aptamers to Factor VIIa. Thromb Haemost 84:841–848
Rusconi CP, Roberts JD, Pitoc GA et al (2004a) Antidote-mediated control of an anticoagulant aptamer in vivo. Nat Biotechnol 22:1423–1428
Rusconi CP, Scardino E, Layzer J et al (2004b) RNA aptamers as reversible antagonists of coagulation factor IXa. Nature 419:90–94
Sarraf-Yazdi S, Mi J, Moeller BJ et al (2007) Inhibition of in vivo tumor angiogenesis and growth via systemic delivery of an angiopoietin-2 specific RNA aptamer. J Surg Res (published on-line Oct 22 2007)
Sennino B, Falcon BL, McCauley D et al (2007) Sequential loss of tumor vessel pericytes and endothelial cells after inhibition of platelet-derived growth factor B by selective aptamer AX102. Cancer Res 67:7358–7367
Tuerk C, Gold L (1990) Systematic evolution of ligands by exponential enrichment: RNA ligands to bacteriophage T4 DNA polymerase. Science 249:505–510
Wagner-Whyte J, Khuri SF, Preiss JR et al (2007) Discovery of a potent, direct thrombin inhibiting aptamer. J Thromb Haemost 5(Suppl 2):P-S-067

NFκB Decoy Oligodeoxynucleotide-Based Therapy in Cardiovascular Diseases

H. Nakagami(✉), M.K. Osako, N. Tomita, and R. Morishita

Contents

Abstract Gene therapy based on oligodeoxynucleotides (ODNs) offers a novel approach for the prevention and treatment of cardiovascular diseases. We focused on the regulation of powerful transcriptional factors which could be mainly involved in the process of atherosclerosis, myocardial infarction, and vascular remodeling, etc. The transcription factor nuclear factor-kappa B (NF-κB) plays a pivotal role in the coordinated transactivation of cytokine and adhesion molecule genes such as tumor necrosis factor (TNF)-α, inlerleukin-6, and ICAM-1, and we utilized ODNs as "decoy" *cis*-elements that block the binding of nuclear factors to promoter regions of targeted genes, resulting in the inhibition of gene transactivation. Transfection of NF-κB decoy ODNs into coronary artery effectively prevented transactivation of essential cytokine and adhesion molecule protein expression. In the myocardial infarction model of rat, transfection of NF-κB decoy decreased the infarction size induced by reperfusion injury. Transfection of NF-κB decoy ODNs into balloon-injured carotid artery or porcine coronary artery also markedly reduced neointimal formation. For the next generation of OSN-based gene therapy, we designed chimeric decoy ODNs, containing the consensus sequences

H. Nakagami
Division of Clinical Gene Therapy, Osaka University Graduate School of Medicine, Suita 565-0871, Japan and Division of Gene Therapy Science, Osaka University Graduate School of Medicine, Suita 565-0871, Japan
e-mail: nakagami@gts.med.osaka-u.ac.jp

V.A. Erdmann et al. (eds.) *RNA Technologies in Cardiovascular Medicine and Research*,
doi: 10.1007/978-3-540-78709-9,

of both the NF-κB- and ets-binding sites, to inhibit both the transcription factors simultaneously. The covalently modified ODNs were developed by enzymatically ligating two identical molecules, thereby preventing their degradation by exonucleases. Although there are still unresolved issues, decoy ODN drugs should become a reality in the near future.

1 Introduction

Gene therapy based on oligodeoxynucleotides (ODNs) offers a novel approach for the prevention and treatment of cardiovascular diseases. We focused on the regulation of powerful transcriptional factors which could be mainly involved in the process of atherosclerosis, myocardial infarction, and vascular remodeling, etc. Nuclear factor-kappa B (NF-κB) is a transcriptional factor of the Rel family, and the vertebrate family includes five cellular proteins: p50, p52, p65, RelB and c-Rel, which can form various homo- and hetero-dimeric complexes with diverse DNA-binding and transcriptional activating properties. The most common of them consists of a p65:p50 heterodimer. NF-κB is activated by the release from cytoplasmic IκB proteins and subsequently translocation into the nucleus (Baeuerle and Baltimore 1996; Baldwin 1996). The activation is triggered by signal-induced phosphorylation of Inhibitor of kappa B (IκB), which targets the inhibitor for rapid degradation by proteasome (Verma et al. 1995). In the nucleus, NF-κB binds to specific sequences in the promoter region, called kappa B-sites, regulating the expression of genes that leads to cell growth and differentiation (Guttridge et al. 1999), inflammatory responses (Baeuerle 1998), the regulation of apoptosis (Wang et al. 1998), and neoplastic transformation (Pahl 1999). Although selective inhibition of NF-κB has been studied previously (Berman et al. 2002; Kopp and Ghosh 1994; Wahl et al. 1998), we utilized synthetic double-stranded oligodeoxynucleotides (ODN) as "decoy" *cis* element that block the binding of nuclear factor to promoter regions of targeted gene, resulting in the inhibition of gene transactivation, and have reported the potency of decoy ODN against NF-κB in several models (Kawamura et al. 2001; Miyake et al. 2006; Nakashima et al. 2004; Ono et al. 1998; Tomita et al. 2001; Yamasaki et al. 2003; Yoshimura et al. 2001). The principle of the transcription factor decoy ODN approach is simply the reduction of promoter activity as a result of the inhibition of binding of a transcription factor to a specific sequence in the promoter region (Bielinska et al. 1990). The decoy ODN strategy is particularly attractive for several reasons: potential drug targets (transcription factors) are plentiful and readily identifiable; knowledge of the exact molecular structure of the target transcription factor is unnecessary; and the synthesis of a sequence-specific decoy is relatively simple and can be targeted to specific tissues. In this paper, we will describe the therapeutic effect of NF-κB decoy for cardiovascular diseases and recent progress in modification of NF-κB decoy.

2 Application of NF-κB Decoy ODN for Cardiovascular Diseases

2.1 *Cardiac Diseases*

Myocardial reperfusion injury develops, to a large degree, as a result of severe damage to myocytes and endothelial cells, probably induced by the complex interaction of multiple cytokines and adhesion molecules activated by reperfusion. The process of ischemic reperfusion might be dependent upon the coordinated activation of a series of cytokine and adhesion molecule genes that results in the attachment of leukocytes and release of cytotoxic molecules. Importantly, increased NF-κB binding activity has been confirmed in hearts with myocardial infarction (Morishita et al. 1995). Our previous study provided the first evidence of the feasibility of a decoy strategy against NF-κB in treating myocardial reperfusion injury (Morishita et al. 1997; Sawa et al. 1997). Transfection of NF-κB decoy ODNs into rat coronary artery before occlusion of the left ascending coronary artery markedly reduced the damaged area of myocytes at 24 h after reperfusion. The therapeutic efficacy of this strategy by intra-coronary administration immediately after reperfusion, similar to the clinical situation, was also examined. NF-κB decoy ODNs reduced the damage to myocytes resulting from reperfusion. Inhibition of vascular smooth muscle cells (VSMC) replication was confirmed by the observation that transfection of NF-κB decoy ODNs inhibited the progression of vasculopathy in cardiac transplantation models (Yokoseki et al. 2001; Suzuki et al. 2000).

2.2 *Restenosis After Angioplasty or Stenting*

Balloon angioplasty is one of the major therapeutic approaches to coronary artery stenosis. However, restenosis after angioplasty still remains an issue in the field of cardiovascular disease, as the long-term effectiveness of this procedure is limited by the development of restenosis (Gibbons and Dzau 1994). Thus, the possibility of ODN-based therapy for restenosis was expected, and its effect has been evaluated in many animal models. Most strategies targeted VSMC proliferation after balloon injury, which is one of the major mechanisms of chronic restenosis. Intimal hyperplasia develops largely as a result of VSMC proliferation and migration induced by a complex interaction of multiple growth factors that are activated by vascular "injury" (Gibbons and Dzau 1994). Initially, the effectiveness of antisense ODNs against a proto-oncogene, *c-myb*, was reported for the treatment of restenosis (Simons et al. 1992). Accordingly, inhibition of other proto-oncogenes, such as *c-myc*, by antisense ODNs was also reported to inhibit neointimal formation in several animal models (Shi et al. 1994). Recently, the results from a phase II trial using antisense *c-myc* to treat restenosis have been reported (Kutryk et al. 2002).

Treatment with 10 mg phosphorothioate-modified ODNs directed against *c-myc* did not reduce neointimal volume obstruction or the angiographic restenosis rate (Kutryk et al. 2002). However, that trial utilized intra-coronary infusion of antisense *c-myc* ODNs without any vectors, and several factors such as low transfection efficiency might limit the efficacy of this strategy.

NF-κB also plays a pivotal role in the coordinated transactivation of cytokine and adhesion molecule genes whose activation has been postulated to be involved in numerous diseases including restenosis. Accordingly, we hypothesized that restenosis could be prevented by the blockade of genes regulating cell inflammation, the final common pathway induced by NF-κB binding. The necessity to block cytokine and adhesion molecule genes at more than one site to achieve maximum inhibitory effects might occur because of the redundancy and complexity of interactions of these genes. Importantly, increased NF-κB binding activity has been confirmed in balloon-injured blood vessels (Yoshimura et al. 2001). Transfection of NF-κB decoy ODNs into balloon-injured carotid artery (Yoshimura et al. 2001) or porcine coronary artery (Yamasaki et al. 2003) markedly reduced neointimal formation, whereas no difference was observed between scrambled decoy ODNs and sham operation. In addition to VSMC proliferation, endothelial damage is one of the major mechanisms of restenosis, as endothelial cells play an important role as a biological barrier in the suppression of VSMC growth, maintenance of vascular tonus and protection from monocyte and platelet adhesion. Severe damage to endothelial cells by balloon injury induces vascular inflammation and has been assumed to be responsible for restenosis. From this viewpoint, inhibition of NF-κB might be ideal. Importantly, transfection of NF-κB decoy ODNs inhibited endothelial cell death (Matsushita et al. 2000). Thus, the protective effects of NF-κB decoy ODNs against restenosis can be mediated by several mechanisms: (1) prevention of endothelial damage induced by inflammation, (2) prevention of macrophage or leukocyte recruitment through suppression of adhesion molecules, (3) inhibition of VSMC growth by improvement of endothelial function, and (4) inhibition of VSMC growth by induction of VSMC apoptosis. Based on the data, the novel therapy using NF-κB decoy ODNs can be conducted for restenosis after angioplasty by NF-κB decoy eluting balloon or prevention of in-stent restenosis in humans by NF-κB decoy eluting stent. For the clinical application, the examination of potential adverse effects or toxicity of NF-κB decoy may be important. In histopathological analysis, no adverse reactions such as incomplete healing or impaired endothelial regeneration were noted. Measurements of serum blood markers (glucose, aspartate aminotransferase, alanine aminotransferase, creatine kinase, γ-GTP, and C-reactive protein) showed no systemic adverse effects in our study.

2.3 Abdominal Aortic Aneurysm by Chimera Decoy

Abdominal aortic aneurysm (AAA) is a common degenerative condition associated with aging and atherosclerosis (Baxter et al. 1992). Although elective surgical repair is an effective approach to prevent deaths from AAA rupture, there is a conspicuous absence of alternative therapeutic strategies for this disease (MacSweeney et al. 1994). Because human aneurysmal tissues are characterized by destructive remodeling of the elastic media and outer aortic wall, recent investigations have emphasized disease mechanisms involving chronic aortic wall inflammation and the progressive degradation of fibrillar matrix proteins (Anidjar and Kieffer 1992; Brophy et al. 1991; Freestone et al. 1995). The dissolution of elastic fibers requires the presence of specific proteinases, and several elastolytic matrix metalloproteinases (MMP) are thought to contribute to aneurysm development, including MMP-2 and MMP-9 (Longo et al. 2002; Thompson et al. 1995; Yamashita et al. 2001). MMP-9 expression also correlates with increasing aneurysm diameter (Petersen et al. 2000; Satta et al. 1998), and it is elevated in the circulating plasma of patients with AAA (Lindholt et al. 2000). To consider the pathophysiology of AAA, we focused on the transcription factors nuclear factor NF-κB and ets-1. In addition to mediating inflammatory changes, NF-κB regulates the transcription of MMP-1, MMP-2, MMP-3 and MMP-9 (Bond et al. 1999; Kim and Koh 2000; Takeshita et al. 1999). In contrast, the ets family activates the transcription of genes encoding MMP-1, MMP-3, MMP-9 and urokinase plasminogen activator, which are proteases involved in extracellular matrix degradation (Campbell et al. 2001; Chase et al. 2002). We hypothesized that inhibition of both NF-κB and ets-1 activation would prevent aneurysmal development by a reduction of MMP expression. To test this hypothesis, we used a novel strategy to inhibit the activation of multiple transcription factors by using chimeric decoy ODN in this study (Fig. 1).

AAA was induced in rats by transient aortic perfusion with elastase, whereas transfection of decoy ODN was performed by wrapping a delivery sheet containing decoy ODN around the aorta. Gel-mobility shift assay after treatment demonstrated that both NF-κB and ets binding activities were simultaneously inhibited by chimeric decoy ODN. Transfection of chimeric decoy ODN resulted in significant inhibition of the progression of AAA such as aneurysmal dilation at 4 weeks after treatment as compared with control, accompanied by a reduction of MMP expression. Moreover, the destruction of elastin fibers was inhibited in the aorta transfected with chimeric decoy ODN. Importantly, transfection of chimeric decoy ODN demonstrated potent inhibition of aneurysmal dilatation compared with NF-κB decoy ODN alone, whereas scrambled decoy ODN had no effects. Interestingly, the migration of macrophages was significantly inhibited by chimeric decoy ODN. We demonstrated successful treatment of AAA by using chimeric decoy ODN against NF-κB and ets-1 in a rat model (Nakashima et al. 2004).

The specificity of the inhibitory effect of chimeric decoy ODN on the progression of AAA is supported by several lines of evidence: (1) in vivo administration of chimeric decoy ODN but not scrambled ODN markedly inhibited dilation of the aorta, accompanied by inhibition of NF-κB and ets-1 binding activity, (2) reduction of the aortic diameter by chimeric decoy ODN delivered by a peripheral wrapping sheet was more potent than a single transfection of NF-κB decoy ODN, and (3) the decrease in matrix degradation activity in adventitia transfected with chimeric decoy ODN was associated with decreased aortic diameter. As NF-κB also regulates intracellular adhesion molecule (ICAM)-1 or vascular endothelial adhesion molecule (VCAM)-1, which are well-known adhesion molecules, transfection of

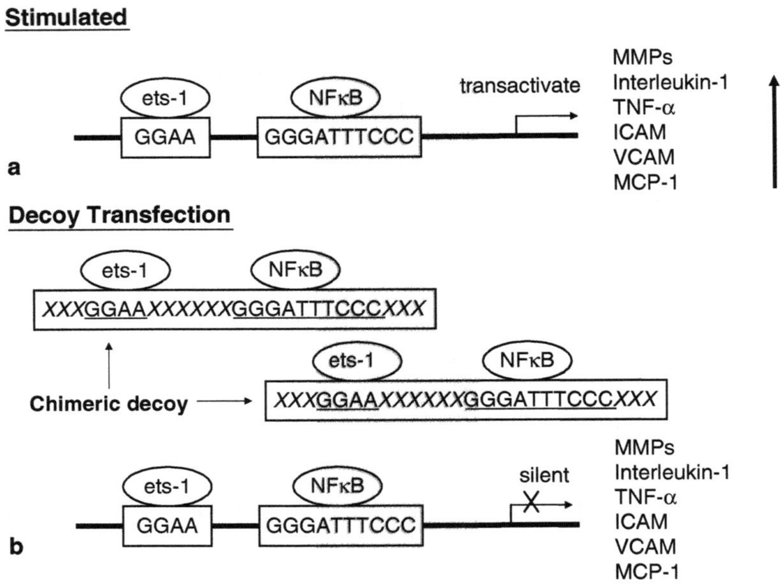

Fig. 1 Scheme of chimeric decoy strategy. (**a**) In the *Stimulated* state, transcription factor (i.e., NF-κB or ets-1) is bound to *cis* element (GGAA for ets-1 and GGGATTTCCC for NF-κB, resulting in continuous activation of target gene expression of each transcriptional factor (i.e., MMPs, interleukin-1, TNF-α). The *black arrow* shows the increase of the cytokines. (**b**) Under the *Decoy Transfection* state, chimeric decoy which includes ets-1 and NF-κB decoy *cis*-element ODNs bind to ets-1 and/or NF-κB, resulting in the prevention of ets-1 or NF-κB interaction and transactivation of ets-1 or NF-κB-promoting target gene expression (i.e., MMPs, interleukin-1, TNF-α)

chimeric decoy ODN would result in a beneficial effect on macrophage migration. Overall, the suppression of AAA by chimeric decoy ODN could be mediated by three pathways: (1) direct inhibition of MMP gene expression driven by either the NF-κB or ets-1 binding site, (2) indirect inhibition of MMP secretion by migrating macrophages and VSMC, and (3) inhibition of migration of macrophages that secrete MMP. In this study, we demonstrated that inhibition of the progression of experimental AAA in rats was achieved by using a new tool: the chimeric decoy strategy against both NF-κB and ets. The present data suggest an important role of NF-κB and ets-1 in the pathogenesis of AAA (Nakashima et al. 2004).

3 Recent Progress in Modification of NF-κB Decoy

The therapeutic effectiveness of synthetic double-stranded ODNs in modulating specific gene expression largely depends on several factors, including stability, specificity, and efficient cellular and tissue uptake of ODN. One of the obstacles to use of the decoy ODN strategy as a pharmaceutical drug is related to the stability of ODN in cells and blood. Since phosphodiester ODN, non-modified (non-phosphorothioate) ODN (N-ODN), are precluded because of their instability under physiological conditions, chemical modifications such as phosphorothioation and methylphosphonation of ODN has been employed in order to decrease their susceptibility to degradation by exo- and endonucleases (Kume et al. 2002; Miller et al. 1981). Although the efficacy in inhibiting a large variety of transcription factors has been reported (Morishita et al. 1995, 1996), use of the phosphorothioate ODN (S-ODN) has brought other problems such as safety and the cost of production resulting from that chemical modifications (Brown et al. 1994; Burgess et al. 1995; Gao et al. 1992). To overcome that, we developed a non-chemical modified decoy ODN, ribbon-type decoy ODN, targeting NF-κB by ligation of the extremities of two single phosphodiester strands that resulted in a dumbbell shape structure. Such a construction significantly increased the stability of the phosphodiester backbone against nucleases compared to N-ODN, and showed to be as efficient as S-ODN in inhibiting the transcription factor NF-κB in vascular smooth muscle cells. Further development of ribbon-type decoy ODN (R-ODN) against NF-κB would provide a useful tool for basic and clinical research (Fig. 2).

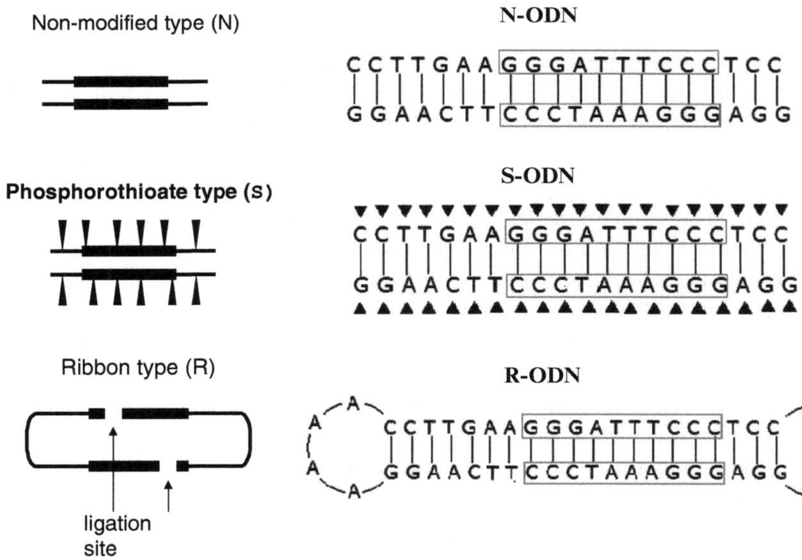

Fig. 2 Scheme of non-modified phosphorothioate ODN (*N-ODN*), phosphorothioate ODN (*S-ODN*) and ribbon-type ODN (*R-ODN*). The NF-κB binding sequence is marked in *rectangles* (GGGATTTCCC). The symbol *filled triangle* in the S-ODN sequence (both *left* and *right panel*) represents the replacement of one non-bridging oxygen by a sulfur in the phosphate group which indicates that all bases are phosphorothioate in S-ODN. R-ODN has been made by the ligation of two ODN within the NF-κB binding sequence. The *circles* in the R-ODN sequence indicate the bases involved in the ligation reaction

Regarding the ODN-based strategy as molecular therapy, one of the major concerns is non-specific effects, particularly those attributed to the polyanionic nature of phosphorothioate-substituted ODN. Non-sequence-specific inhibition may operate through blockade of cell surface receptor activity or interference with other proteins (Gibson 1996). In addition, the toxicity of phosphorothioate ODN may also be relevant. Although low dosage administration does not seem to cause any toxicity, bolus infusions may be dangerous. Higher doses over prolonged periods of time may cause kidney damage, as evidenced by proteinuria and leukocytes in urine in animals (Henry et al. 1997a). Liver enzymes may also be increased in animals treated with moderate to high doses. Several phosphorothioate ODN have been shown to cause acute hypotensive events in monkeys (Iversen et al. 1999; Srinivasan and Iversen 1995), probably due to complement activation (Henry et al. 1997b). More recently, prolongation of prothrombin, partial thromboplastin, and bleeding times has been reported in monkeys (Crooke 1995). Although these effects are transient and relatively uncommon, this toxicity might be avoided by using a construction that resembles the natural DNA oligomer, therefore that does not have chemical modifications, like the ribbon-type decoy ODN.

ODN-based gene therapy still shows many unsolved problems, such as a short half-life, low uptake efficiency, and degradation by endocytosis and nucleases. Therefore, many groups are currently focusing on modifications of the gel approach using a catheter delivery system. Further modification of ODN pharmacokinetics will facilitate the potential clinical utility of the agents by: (1) allowing a shorter intraluminal incubation time to preserve organ perfusion, (2) prolonging the duration of biological action, and (3) enhancing efficacy in a such way that the non-specific effects of high doses of ODNs can be avoided. Although direct transfer of "naked" decoy ODNs can be achieved through passive uptake, the transfection efficiency seems to be lower than that with single-stranded antisense ODNs. To enhance the transfection efficiency of decoy ODNs, the cationic liposome (hemagglutinating virus of Japan–liposome) method or other vector systems are generally used. The majority of ODNs is sequestered and degraded in lysosomes and never reaches the nucleus. Because the site of decoy effects is in the nucleus, bypassing the endocytotic pathway and translocation of decoy ODNs from the cytoplasm are extremely important in the practical application of therapeutics.

Transfection of *cis*-element double-stranded oligodeoxynucleotides has been reported as a powerful tool in a new class of anti-gene strategies for molecular therapy. Although there are still many unresolved issues, decoy ODN drugs should become a reality.

References

Anidjar S, Kieffer E (1992) Pathogenesis of acquired aneurysms of the abdominal aorta. Ann Vasc Surg 6:298–305

Baeuerle PA (1998) IkappaB-NF-kappaB structures: at the interface of inflammation control. Cell 95:729–731

Baeuerle PA, Baltimore D (1996) NF-kappa B: ten years after. Cell 87:13–20
Baldwin AS Jr (1996) The NF-kappa B and I kappa B proteins: new discoveries and insights. Annu Rev Immunol 14:649–683
Baxter BT, McGee GS, Shively VP et al (1992) Elastin content, cross-links, and mRNA in normal and aneurysmal human aorta. J Vasc Surg 16:192–200
Berman KS, Verma UN, Harburg G et al (2002) Sulindac enhances tumor necrosis factor-alpha-mediated apoptosis of lung cancer cell lines by inhibition of nuclear factor-kappaB. Clin Cancer Res 8:354–360
Bielinska A, Shivdasani RA, Zhang LQ et al (1990) Regulation of gene expression with double-stranded phosphorothioate oligonucleotides. Science 250:997–1000
Bond M, Baker AH, Newby AC (1999) Nuclear factor kappaB activity is essential for matrix metalloproteinase-1 and -3 upregulation in rabbit dermal fibroblasts. Biochem Biophys Res Commun 264:561–567
Brophy CM, Reilly JM, Smith GJ et al (1991) The role of inflammation in nonspecific abdominal aortic aneurysm disease. Ann Vasc Surg 5:229–233
Brown DA, Kang SH, Gryaznov SM et al (1994) Effect of phosphorothioate modification of oligodeoxynucleotides on specific protein binding. J Biol Chem 269:26801–26805
Burgess TL, Fisher EF, Ross SL et al (1995) The antiproliferative activity of c-myb and c-myc antisense oligonucleotides in smooth muscle cells is caused by a nonantisense mechanism. Proc Natl Acad Sci USA 92:4051–4055
Campbell SE, Nasir L, Argyle DJ et al (2001) Molecular cloning and characterization of canine metalloproteinase-9 gene promoter. Gene 273:81–87
Chase AJ, Bond M, Crook MF et al (2002) Role of nuclear factor-kappa B activation in metalloproteinase-1, -3, and -9 secretion by human macrophages in vitro and rabbit foam cells produced in vivo. Arterioscler Thromb Vasc Biol 22:765–771
Crooke ST (1995) Progress in antisense therapeutics. Hematol Pathol 9:59–72
Freestone T, Turner RJ, Coady A et al (1995) Inflammation and matrix metalloproteinases in the enlarging abdominal aortic aneurysm. Arterioscler Thromb Vasc Biol 15:1145–1151
Gao WY, Han FS, Storm C et al (1992) Phosphorothioate oligonucleotides are inhibitors of human DNA polymerases and RNase H: implications for antisense technology. Mol Pharmacol 41:223–229
Gibbons GH, Dzau VJ (1994) The emerging concept of vascular remodeling. N Engl J Med 330:1431–1438
Gibson I. (1996) Antisense approaches to the gene therapy of cancer-'Recnac'. Cancer Metastasis Rev 15:287–299
Guttridge DC, Albanese C, Reuther JY et al (1999) NF-kappaB controls cell growth and differentiation through transcriptional regulation of cyclin D1. Mol Cell Biol 19:5785–5799
Henry SP, Bolte H, Auletta C et al (1997a) Evaluation of the toxicity of ISIS 2302, a phosphorothioate oligonucleotide, in a four-week study in cynomolgus monkeys. Toxicology 120:145–155
Henry SP, Giclas PC, Leeds J et al (1997b) Activation of the alternative pathway of complement by a phosphorothioate oligonucleotide: potential mechanism of action. J Pharmacol Exp Ther 281:810–816
Iversen PL, Cornish KG, Iversen LJ et al (1999) Bolus intravenous injection of phosphorothioate oligonucleotides causes hypotension by acting as alpha(1)-adrenergic receptor antagonists. Toxicol Appl Pharmacol 160:289–296
Kawamura I, Morishita R, Tsujimoto S et al (2001) Intravenous injection of oligodeoxynucleotides to the NF-kappaB binding site inhibits hepatic metastasis of M5076 reticulosarcoma in mice. Gene Ther 8:905–912
Kim H, Koh G (2000) Lipopolysaccharide activates matrix metalloproteinase-2 in endothelial cells through an NF-kappaB-dependent pathway. Biochem Biophys Res Commun 269:401–405
Kopp E, Ghosh S (1994) Inhibition of NF-kappa B by sodium salicylate and aspirin. Science 265:956–959

Kume M, Komori K, Matsumoto T et al (2002) Administration of a decoy against the activator protein-1 binding site suppresses neointimal thickening in rabbit balloon-injured arteries. Circulation 105:1226–1232

Kutryk MJ, Foley DP, van den Brand M et al (2002) Local intracoronary administration of antisense oligonucleotide against c-myc for the prevention of in-stent restenosis: results of the randomized investigation by the Thoraxcenter of antisense DNA using local delivery and IVUS after coronary stenting (*ITALICS*) trial. J Am Coll Cardiol 39:281–287

Lindholt JS, Vammen S, Fasting H et al (2000) The plasma level of matrix metalloproteinase 9 may predict the natural history of small abdominal aortic aneurysms: a preliminary study. Eur J Vasc Endovasc Surg 20:281–285

Longo GM, Xiong W, Greiner TC et al (2002) Matrix metalloproteinases 2 and 9 work in concert to produce aortic aneurysms. J Clin Invest 110:625–632

MacSweeney ST, Powell JT, Greenhalgh RM (1994) Pathogenesis of abdominal aortic aneurysm. Br J Surg 81:935–941

Matsushita H, Morishita R, Nata T et al (2000) Hypoxia induced endothelial apoptosis through NFkB-mediated bcl-2 suppression: in vivo evidence of importance of NFκB in endothelial cell regulation. Circ Res 86:974–981

Miller PS, McParland KB, Jayaraman K et al (1981) Biochemical and biological effects of nonionic nucleic acid methylphosphonates. Biochemistry 20:1874–1880

Miyake T, Aoki M, Nakashima H et al (2006) Prevention of abdominal aortic aneurysms by simultaneous inhibition of NFkappaB and ets using chimeric decoy oligonucleotides in a rabbit model. Gene Ther 13:695–704

Morishita R, Gibbons GH, Horiuchi M et al (1995) A gene therapy strategy using a transcription factor decoy of the E2F binding site inhibits smooth muscle proliferation in vivo. Proc Natl Acad Sci USA 92:5855–5859

Morishita R, Sugimoto T, Aoki M et al (1997) In vivo transfection of *cis* element "decoy" against NF-κB binding site prevented myocardial infarction as gene therapy. Nat Med. 3:894–899

Nakashima H, Aoki M, Miyake T et al (2004) Inhibition of experimental abdominal aortic aneurysm in the rat by use of decoy oligodeoxynucleotides suppressing activity of nuclear factor kappaB and ets transcription factors. Circulation 109:132–138

Ono S, Date I, Onoda K et al (1998) Decoy administration of NF-kappaB into the subarachnoid space for cerebral angiopathy. Hum Gene Ther 9:1003–1011

Pahl HL (1999) Activators and target genes of Rel/NF-kappaB transcription factors. Oncogene 18:6853–6866

Petersen E, Gineitis A, Wagberg F et al (2000) Activity of matrix metalloproteinase-2 and -9 in abdominal aortic aneurysms: relation to size and rupture. Eur J Vasc Endovasc Surg 20:457–461

Satta J, Laurila A, Paakko P et al (1998) Chronic inflammation and elastin degradation in abdominal aortic aneurysm disease: an immunohistochemical and electron microscopic study. Eur J Vasc Endovasc Surg 15:313–319

Sawa Y, Morishita R, Suzuki K et al (1997) A novel strategy for myocardial protection using in vivo transfection of cis element 'decoy' against NFκB binding site: evidence for a role of NFκB in ischemic-reperfusion injury. Circulation 965:280–285

Shi Y, Fard A, Galeo A et al (1994) Transcatheter delivery of c-myc antisense oligomers reduces neointimal formation in a porcine model of coronary artery balloon injury. Circulation 90:944–951

Simons M, Edelman ER, DeKeyser JL et al (1992) Antisense c-myb oligonucleotides inhibit intimal arterial smooth muscle cell accumulation in vivo. Nature 359:67–80

Srinivasan SK, Iversen P (1995) Review of in vivo pharmacokinetics and toxicology of phosphorothioate oligonucleotides. J Clin Lab Anal 9:129–137

Suzuki J, Morishita R, Amano J et al (2000) Decoy against nuclear factor-kappa B attenuates myocardial cell infiltration and arterial neointimal formation in murine cardiac allografts. Gene Ther 7:1847–1852

Takeshita H, Yoshizaki T, Miller WE et al (1999) Matrix metalloproteinase 9 expression is induced by Epstein-Barr virus latent membrane protein 1 C-terminal activation regions 1 and 2. J Virol 73:5548–5555
Tomita N, Morishita R, Tomita S, et al (2001) Inhibition of TNF-alpha, induced cytokine and adhesion molecule. Expression in glomerular cells in vitro and in vivo by transcription factor decoy for NFkappaB. Exp Nephrol 9:181–190
Thompson RW, Holmes DR, Mertens RA et al (1995) Production and localization of 92-kilodalton gelatinase in abdominal aortic aneurysms: an elastolytic metalloproteinase expressed by aneurysm-infiltrating macrophages. J Clin Invest 96:318–326
Verma IM, Stevenson JK, Schwarz EM et al (1995) Rel/NF-kappa B/I kappa B family: intimate tales of association and dissociation. Genes Dev 9:2723–2735
Wahl C, Liptay S, Adler G et al (1998) Sulfasalazine: a potent and specific inhibitor of nuclear factor kappa B. J Clin Invest 101:1163–1174
Wang CY, Mayo MW, Korneluk RG et al (1998) NF-kappaB antiapoptosis: induction of TRAF1 and TRAF2 and c-IAP1 and c-IAP2 to suppress caspase-8 activation. Science 281:1680–1683
Yamasaki K, Asai T, Shimizu M et al (2003) Inhibition of NFkappaB activation using *cis*-element 'decoy' of NFkappaB binding site reduces neointimal formation in porcine balloon-injured coronary artery model. Gene Ther 10:356–364
Yamashita A, Noma T, Nakazawa A et al (2001) Enhanced expression of matrix metalloproteinase-9 in abdominal aortic aneurysms. World J Surg 25:259–265
Yokoseki O, Suzuki J, Kitabayashi H et al (2001) *Cis* element decoy against nuclear factor-kappaB attenuates development of experimental autoimmune myocarditis in rats. Circ Res 89:899–906
Yoshimura S, Morishita R, Hayashi K et al (2001) Inhibition of intimal hyperplasia after balloon injury in rat carotid artery model using *cis*-element 'decoy' of nuclear factor-kappaB binding site as a novel molecular strategy. Gene Ther 8:1635–1642

Antisense Therapy for Restenosis Following Percutaneous Coronary Interventions

N. Kipshidze

Contents

Abstract Antisense refers to a gene inactivating technology which blocks the "sense" of the genetic code (hence antisense) and prevents the normal wound healing responses that can lead to vessel obstruction or restenosis following injury.

Recent studies have focused on the use of antisense compounds to prevent recurrent vessel obstruction or restenosis following PTCA. Some antisense compounds can inhibit the cell cycle response to injury in the G-1 by blocking *c-myc,* a regulatory gene that is the key factor in the cascade of effects that leads to restenosis in many angioplasty patients.

N. Kipshidze
Lenox Hill Hospital, Department of Interventional Cardiac & Vascular Services, 30 East 77th Street, New York, NY 10021, USA
e-mail: NKipshidze@Lenoxhill.net

V.A. Erdmann et al. (eds.) *RNA Technologies in Cardiovascular Medicine and Research,* 311
doi: 10.1007/978-3-540-78709-9,

Until recently, the clinical applicability of antisense technology to the problem of restenosis has been limited due to a relative lack of target specificity, slow uptake across the cell membranes, and rapid intracellular degradation of the antisense oligonucleotides. The only randomized study in humans with *c-myc* antisense demonstrated no reduction in restenosis after stent implantation when arteries were pretreated with the drug.

However, the recently introduced AVI-4126 (Resten-NG) belongs to a new family of molecules known as the phosphorodiamidate morpholino oligomers (PMO). These oligomers are comprised of dimethylamino phosphinylideneoxy-linked morpholino subunits. The morpholino subunits contain a heterocyclic base recognition moiety of DNA (A,C,G,T) attached to a substituted morpholine ring system. In general, PMO are capable of binding to ribonucleic acid (RNA) in a sequence-specific fashion with sufficient avidity to be useful for the inhibition of the translation of mRNA into protein in vivo, a result commonly referred to as an antisense effect.

The most robust of the observations to date include the fact that AVI-4126 is safe and effective in reduction of restenosis in multiple species and conducted by multiple investigators. Three different methods for local and systemic delivery have been described, each with advantages and limitations. Efficacy in animal models is encouraging. Further, clinical trials with AVI-4126 indicate that the agent is very safe. The last remaining question: Will AVI-4126 find a place in the future therapeutic regimen for the prevention of restenosis remains unanswered.

1 Introduction

Antisense refers to a gene inactivating technology which blocks the "sense" of the genetic code (hence antisense) and prevents the normal wound healing responses that can lead to vessel obstruction or restenosis following injury.

Percutaneous transluminal coronary angioplasty (PTCA) was introduced in 1977 (Gruntzig et al. 1977) and has become an effective treatment for limited coronary artery disease (Parisi et al. 1992). Since then, PTCA treatment has become more extensive and gained favor as an alternative treatment for coronary artery bypass grafting (CABG) (King et al. 1994). Frequently, however, the artery is injured at the site of PTCA leading to wound-healing responses that include thrombosis, smooth muscle proliferation and migration, elastic recoil and vascular remodeling. Each of these responses may contribute to recurrent obstruction or vessel narrowing referred to as restenosis.

Recent studies have focused on the use of antisense compounds to prevent recurrent vessel obstruction or restenosis following PTCA. Some antisense compounds can inhibit the cell cycle response to injury in the G-1 by blocking *c-myc*, a regulatory gene that is the key factor in the cascade of effects that leads to restenosis in many angioplasty patients. Compounds that inhibit the cell cycle in the early phase are often less toxic, a description that fits both rapamycin and AVI-4126 (Resten-NG).

Until recently, the clinical applicability of antisense technology to the problem of restenosis has been limited due to a relative lack of target specificity, slow uptake across the cell membranes, and rapid intracellular degradation of the antisense oligonucleotides (Zalewski et al. 1998). The only randomized study in humans with *c-myc* antisense demonstrated no reduction in restenosis after stent implantation when arteries were pretreated with the drug (Serruys et al. 1998). However, the recently introduced AVI-4126 (Resten-NG) belongs to a new family of molecules known as the phosphorodiamidate morpholino oligomers (PMO).

These oligomers are comprised of *dimethylamino phosphinylideneoxy*-linked morpholino subunits. The morpholino subunits contain a heterocyclic base recognition moiety of DNA (A,C,G,T) attached to a substituted morpholine ring system. In general, PMO are capable of binding to ribonucleic acid (RNA) in a sequence-specific fashion with sufficient avidity to be useful for the inhibition of the translation of mRNA into protein in vivo, a result commonly referred to as an antisense effect.

Although the PMO share many similarities with other substances capable of producing antisense effects, such as deoxyribonucleic acid (DNA), RNA, and their analogous oligonucleotide analogs such as the phosphorothioates (PSO), there are specific differences. Most importantly, the PMO are uncharged and resistant to degradation under biological conditions. The combination of efficacy, potency and lack of nonspecific activities of the PMO chemistry has compelled us to reexamine the approach to antisense *c-myc* for the prevention of restenosis following balloon angioplasty.

AVI-4126 is an antisense phosphorodiamidate morpholino oligomer (PMO) with sequence complementary to the translation initiation start site of the *c-myc* mRNA. The mechanism of action of AVI-4126 involves the interference with ribosomal assembly thus preventing translation of *c-myc* and the interference with intron 1-exon 2 splicing of the *c-myc* pre-mRNA preventing appropriate translation of the *c-myc* mRNA. The IC_{50} for inhibition of *c-myc* by AVI-4126 is 0.3 μM in cell culture (Hudziak et al. 2000). The cellular response to AVI-4126 is diminished cell growth associated with arrest of cells in the G_0/G_1 phase of the cell cycle. Inhibition of *c-myc* would also interfere with expression of downstream genes such as those associated with cellular adhesion, the cell cycle, and connective tissue matrix remodeling.

The interpretation of antisense data tends to be more complex than that of small molecule inhibitors. This is due to the fact that the antisense mechanism of action involves inhibition of protein synthesis, but the rate of protein turnover is generally not influenced. The equation for inhibition of *c-myc* expression can be viewed as (1) below:

$$[MYC]^t = [MYC]_{ss} + \text{induced}[c\text{-}myc\text{mRNA}] - MYC_{\text{turnover}}. \quad (1)$$

Note: *MYC* all caps refers the protein, *c-myc* mRNA refers to the transcript and *c-myc* refers to the gene.

The magnitude of injury will determine the magnitude of new transcription of *c-myc* mRNA and the amount of *c-myc* mRNA is also a balance between the rate of synthesis and rate of decay.

The rate of *MYC* synthesis depends upon the concentration of *c-myc* mRNA. Equation (1) can be simplified to (2) as follows:

$$[MYC]^t = [MYC]_{ss} + \text{injury induced } MYC \text{ synthesis} - MYC_{\text{turnover}}. \quad (2)$$

Critical questions in the development of a delivery tool for the AVI-4126 involve: (1) what is the time course for *MYC* synthesis; (2) can AVI-4126 inhibit MYC expression in the appropriate cells in the vessel wall following appropriate magnitude and duration of injury, and (3) is the amount and duration of AVI-4126 delivered relative to injury to the vessel wall sufficient? The expected result of a study in which the initial synthesis of *MYC* is not inhibited is shown in Fig. 1.

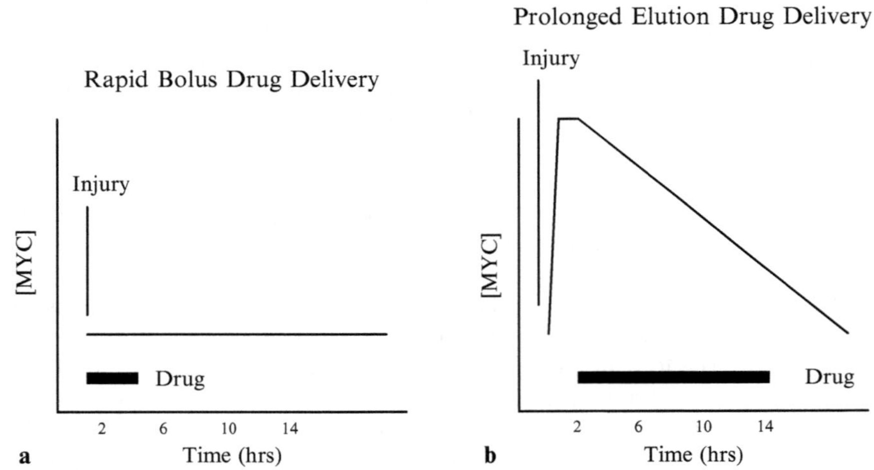

Fig. 1 Graphic cartoon of *MYC* expression as the dependent variable on the ordinate and time in hours as the independent variable on the abscissa. (**a**) The situation when AVI-4126 is delivered to the vessel wall at or near the time of vascular injury. The *MYC* concentration is efficiently prevented by inhibiting expression of new protein. (**b**) The situation when AVI-4126 is delivered to the vessel wall shortly after induced *MYC* protein synthesis has begun. The *MYC* protein level is elevated prior to AVI-4126 blockade. In this situation, the antisense inhibition is reflected only in creating a steeper slope in the decline in *MYC* concentrations in the vessel wall due to inhibition of lingering translation which occurs approximately 3 h after balloon injury

1.1 *MYC Expression Timecourse*

Background. Considerations important to the evaluation of *MYC* expression and antisense inhibition include the stability of the transcript and the translated protein. The half-life of the transcribed mRNA is between 30 min and 1 h (Dani et al. 1984). The half-life of the *MYC* protein is 20–50 min (Hann and Eisenman 1984; Waters et al. 1991). *MYC* expression is induced within 2–6 h of injury (Miano et al. 1991).

Interpretation. The injury caused by stent placement produces a rapid increase in *MYC* detected by western blot within 3 h. The peak *MYC* expression is followed by reduced but ongoing transcription and translation of *c-myc* mRNA which occurs simultaneously with the normal degradation of *MYC* protein. The expected result of a polymer-coated stent delivering AVI-4126 into the vascular wall will be to reduce the translation of *c-myc* mRNA. The rate of *MYC* degradation is not influenced so the result will be to observe an enhanced rate of loss of *MYC*.

Concern for lingering *MYC* expression influences duration for antisense present in the injured vessels. This combined with a report of *MYC* expression at seven days post injury prompted further evaluation at seven days. Swine coronary vessels were injured by balloon overstretch and then vessels were recovered 2 h later. The expression of *MYC* was determined by western blot and a ratio of blot intensity for *MYC* divided by intensity for β-actin, an internal control, to compensate for sample preparation variability (Table 1).

Conclusion. These data indicate that elevated *MYC* expression is shorter than 7 days after vessel injury. This is true for bare stents and polymer-coated stents. Therefore delivery of AVI-4126 does not require prolonged release (greater than 4 days).

2 Local Drug Delivery

Local drug delivery was designed to bring the antisense agent to the coronary artery during the period of time corresponding to peak injury response. The earliest attempts to deliver antisense agents for prevention of restenosis involved a rat carotid artery model using adventitial (Simons et al. 1992) or surgical application (Morishita et al. 1993). The initial clinically applicable devices were catheter-based providing local delivery as a bolus injection, followed by the subsequent withdrawal of the catheter. The combination of antisense targeting to *c-myc* with a catheter-based delivery to coronary arteries of pigs for prevention of restenosis began with the phosphorothioate oligonucleotides (Shi et al. 1994). The bolus

Table 1 Comparison of no injury with 7 days post-injury

No injury	7 days post-injury	*P* value	Interpretation
0.50 ± 0.11 (3)	0.42 ± 0.06 (12)	0.53; 13 d.f.	No difference

The numbers in parenthesis indicate the number of vessels evaluated injection of phosphorothioate oligomers produced reduced heart rate, blood pressure and cardiac output in primate models which in some cases were lethal (Cornish et al. 1993; Galbraith et al. 1994; Wallace et al. 1996; Henry et al. 1997; Iversen et al. 1999). The phosphorodiamidate morpholino oligomers (PMO) have been evaluated for similar effects after intravenous bolus injections in both primates [Good Laboratory Practice (GLP) studies by Sierra Biomedical] and man [Good Clinical Practice (GCP) studies at MDS Harris]. No alterations in heart rate, blood pressure or cardiac output have been observed. In summary, bolus injections of PMO by local catheter-based delivery devices are feasible.

2.1 *Transport Catheter Studies in Rabbit Iliac Vessels*

Twenty-five, male, New Zealand, white, atherosclerotic rabbits maintained on a diet of 0.25% cholesterol were anesthetized, a Transport Catheter™ inserted into the iliac artery and PTCA performed (8 atm for 30 s, three times). The endoluminal delivery of saline or 0.5 mg of AVI-4126 to the PTCA site was at 2 atm via the outer balloon for 2 min (Kipshidze et al. 2001). The area of the intima and media were determined by planimetry (Table 2). Quantitative angiography from these animals shows the maximum lumen diameter (MLD) at the time of harvest (60 days after PTCA) was significantly greater in the antisense-treated group than in the control animals. The morphometric analysis confirms the angiography in demonstrating significantly greater lumen area than in the control. The intimal area was also significantly smaller in the AVI-4126-treated animals. We also observed positive remodeling of the vessel. Vessel area was significantly greater ($P < 0.05$) in the treated animals.

2.2 *Infiltrator Catheter Studies in Swine*

We evaluated the long-term influence of intramural delivery of advanced c-myc antisense on neointimal hyperplasia following stenting in a pig model (Kipshidze et al. 2002). In acute experiments different doses (from 500 μg to 5 mg) Resten-NG

Table 2 Rapid bolus local delivery in rabbit iliac vessels

	Control	Treated
Maximum lumen diameter	0.8 ± 0.2	1.6 ± 0.3*
Late loss	1.8 ± 0.3	0.9 ± 0.2*
Lumen (mm^2)	0.62 ± 0.73	1.89 ± 0.35*
Intima (mm^2)	1.67 ± 0.44	0.82 ± 0.32*

*P value less than 0.05

Table 3 Rapid bolus local delivery in swine coronary vessels

	Control	Treated
Lumen area (mm^2)	3.26 ± 1.57	5.62 ± 1.40*
Intimal area (mm^2)	3.88 ± 1.04	1.95 ± 0.91*
IA/IS	4.08 ± 0.76	2.13 ± 0.55*

**P* value is less than 0.05

(n = 11) or saline (n = 14) were delivered prior to the stent implantation site with the Infiltrator™ delivery system. Animals were sacrificed at 2, 6 and 18 h after intervention, and excised vessels were analyzed for *c-myc* expression by western blot. In chronic experiments (n = 20) saline or 1, 5 and 10 mg of AVI-4126 were delivered in the same fashion and the animals were sacrificed at 28 days following intervention.

Western blot analysis demonstrated inhibition of *c-myc* expression and was dose dependent. Morphometry showed that the intimal area was significantly reduced relative to the control (Table 3). There was a statistically significant reduction of intimal areas in the 5- and 10-mg groups (2.01 ± 0.66 and 1.95 ± 0.91 respectively, $P < 0.001$, but no significant reduction in the 1-mg group (2.81 ± 0.56, $P > 0.5$) in comparison with the control. This study demonstrated that intramural delivery of advanced *c-myc* neutrally charged antisense morpholino compound completely inhibits *c-myc* expression and dramatically reduces neointimal formation in a dose-dependent fashion in a porcine coronary stent restenosis model while allowing for complete vascular healing.

2.3 *Coated Stents*

2.3.1 Phosphorylcholine (PC) Matrix for Drug Delivery

PC stents were loaded with AVI-4126 using soak-trap (ST) and dry-trap (DT) methods. Twelve pigs underwent AVI-4126 PC coronary stent implantation (3 stents/animal). At 2–6 h post-procedure, three pigs were sacrificed and stented segments were analyzed by western blot for *c-myc* expression. In chronic experiments, nine pigs (27 stent sites) were sacrificed at 28 days following intervention and vessels were perfusion-fixed. High performance liquid chromatography (HPLC) analysis of plasma showed minimal presence of the antisense oligomer. Western blot analysis involved determination of both *c-myc* and β-actin (an internal control protein) band intensities. The ratio of MYC to β-actin is 48% lower in the AVI-4126-treated vessels than in the untreated control vessels with stent implantation. The concentration of AVI-4126 in those vessels was 52 nM as determined by high performance liquid chromatography. Quantitative histologic morphometry (Table 4) showed that the neointimal area was significantly reduced (by 40%) in the ST group compared with control (2.2 ± 0.7 vs 3.7 ± 0.7 mm^2, respectively, $P = 0.0077$). Immunostaining

Table 4 Drug-eluting stent in swine coronary vessels

	Control	Treated
Maximum lumen diameter	2.3 ± 0.6	3.3 ± 0.6*
Late loss	1.1 ± 0.2	0.4 ± 0.5*
Lumen area (mm^2)	3.4 ± 0.9	5.2 ± 0.9*
Intima area (mm^2)	3.9 ± 0.8	2.3 ± 1.1*

and electron microscopy demonstrated complete endothelialization, without fibrin deposition, thrombosis, or necrosis in all groups.

Control arteries exhibited a substantial neointima consisting mostly of stellate and spindle-shaped cells, in a loose extracellular matrix. The neointima from treated arteries with antisense-loaded stent implantation was significantly smaller in size. Most importantly, there was no difference in the appearance of re-endothelialization. The transmission electron microscope (TEM) revealed a virtually normal appearance of the endothelium. A semi-quantitative histological grading system demonstrated similar smooth muscle cell (SMC) colonization in all groups and minimal residual fibrin deposition for the ST-eluting stents. However, Dry Trap and control polymer-coated (PC) stents had higher intimal fibrin scores.

We also observed less inflammation after implantation of the antisense-loaded stent. In general, the neointima of the ST- and DT-coated stents consisted of smooth muscle cells, matrix proteoglycans, and minimal focal regions of residual fibrin adjacent to the stent struts. Focal medial necrosis or intimal hemorrhage was an infrequent observation within any of the control or drug-coated stents. The antisense-loaded stent had a favorable influence on hyperplasia (reduction of intima by 40%) in the absence of endothelial toxicity and may represent an advantage over more destructive methods such as brachytherapy (Sheppard and Eisenberg 2001) or cytotoxic inhibitors (Herdeg et al. 2000). Recently, local antiproliferative strategies including pharmacological stent coatings (paclitaxel, rapamycin, etc.) have demonstrated inhibition of smooth muscle cell proliferation in vitro, reduced neointimal thickening in animal models of restenosis and produced promising results in the pilot human studies (Sousa et al. 2001). However, questions remain about the reendothelialization process after stent implantation with certain cytotoxic compounds which could put patients at risk for late stent thrombosis and cause late complications (Sousa et al. 2001). In contrast with other chemotherapeutic agents (paclitaxel, actinomycin D), Resten-NG (AVI-4126) is an antisense compound that inhibits the cell cycle in the G-1 by blocking *c-myc*, a regulatory gene that is the key factor in the cascade of effects that lead to restenosis in many angioplasty patients.

2.4 Experimental Matrices in Coated Stents

Some polymer coatings induce *MYC* to greater levels perhaps as a result of greater vessel wall injury. The bare stent vessel represents the best measure of the rate of the concentrations of *MYC* returning to steady state, which is approximately 4 days.

Table 5 shows data describing the rate of *MYC* loss in the injured vessels. The column referred to as d*x*/d*t* represents the difference in *MYC*: internal control ratio (both β-actin and glyceraldehyde-3-phosphate dehydrogenase (GAPDH) were utilized as internal controls) from the three-hour point to the 24-h point. This rate of MYC loss appears to be dose dependent.

If we assume the injury does not alter steady state *MYC* expression, then the time to recovery in the control group can be determined by (1.51–0.5)/(0.27/21) = 78.3 h. If synthesis of *MYC* were inhibited by 100%, then the *MYC* half-life would bring *MYC* to steady state in five half-lives or approximately 2.5 h. In the case of the low-dose stent we observe (1.68–0.5)/(0.86/21) = 28.8 h which represents a 63% reduction in the time to *MYC* steady state (IC_{63}). The high-dose stent would be (2.61–0.5)/(1.46/21) = 30.4 h or a 61.2% inhibition (IC_{61}). In these studies, the amount of AVI-4126 in the vessel wall was determined and the dose dependence of the rate of loss of *MYC* versus the vessel concentrations were measured at 24 h. These data are presented in Table 6. The studies demonstrate a remarkable agreement of inhibition of *MYC* with a resident amount of AVI-4126 in the vessel wall. Table 6 shows the concentrations of AVI-4126 in the vessels at the indicated times.

The observations in Table 6 reveal an excellent agreement between the published IC_{50} of 300 nM for Resten-NG in inhibiting *MYC* (Hudziak et al. 2000) and the observed (IC_{61}) at the 415 nM in the high-dose group. The inhibition by the low-dose stent was essentially equal to that of the high-dose stent. The concentration at 24 h was somewhat less, but was still in reasonable agreement given the differences between cell culture and in vivo blood vessels.

Limitations. The studies did not include a polymer-coated stent or scrambled PMO polymer-coated stent as controls. The polymer coating the stent tends to induce greater *MYC* expression observed at the 3-h time point over the bare stent.

Table 5 Rate of *MYC* loss in injured vessels

	MYC: β-actin and GAPDH ratios		
Group	3 h vessels	24 h vessels	d*x*/d*t*
Control	1.51 ± 0.31 (12)	1.24 ± 0.79 (12)	0.27
Low dose stent	1.68 ± 0.59 (6)	0.82 ± 0.45 (6)	0.86
High dose stent	2.61 ± 1.77 (5)	1.15 ± 0.47 (6)	1.46

GAPDH Glyceraldehyde-3-phosphate dehydrogenase

Table 6 Concentration of resten-NG in vessel walls (nM)

Group	24 h vessels
Control	63.4 ± 92.8 (12)
Low dose	233 ± 55.1 (6)
High dose	415 ± 204 (6)

The analysis of the rate of loss of *MYC* tends to minimize the shortfall in direct comparisons of observed *MYC* level at a given time point.

Conclusion: AVI-4126 inhibits *MYC* expression in the current polymer-coated stent format in proportion to the amount of AVI-4126 that remains resident in the vessel wall. The degree of inhibition is quantitatively in good agreement with earlier studies investigating *MYC* inhibition. The polymer-coated stent delivers micromolar concentrations into the vessel wall within 3 h of placement. The delivery would need to be faster in order to prevent initial expression of MYC demonstrated by the substantial *MYC* detected at 3 h. The polymer tends to add to vessel injury as measured by elevated *MYC* 3 h after placement. Ultimate success will require polymers capable of rapid elution of AVI-4126 with minimal capacity to inflame or otherwise additionally injure the vessel wall.

3 Perflourocarbon Gas Microbubble Carriers (PGMC) for Site-Specific Drug Delivery

Perfluorobutane gas microbubbles with a coating of dextrose and albumin efficiently bind antisense oligomers (Porter et al. 1996). These 0.3–10-μm particles bind to sites of vascular injury. Further, the perfluorobutane gas is an effective cell membrane fluidizer. The potential advantages of microbubble carrier delivery include minimal addition to vessel injury from delivery, no resident polymer to degrade leading to eventual inflammation, rapid bolus delivery, and repeated delivery is highly feasible. Further, the potential for PGMC to deliver to vessel regions both proximal and distal to stents in vessels suggests this mode of delivery will be excellent as an adjuvant to a variety of catheter and coated-stent delivery techniques.

3.1 PGMC Delivery Assisted by Diagnostic Ultrasound

Based on these advantages, 21 pigs received AVI-4217 (a pig version of AVI-4126) bound to PGMC, AVI-4217 alone or no antisense treatment after carotid balloon injury (Porter et al. 2000). The vessels were evaluated 30-days post injury. The results are expressed in Table 7.

Table 7 Perflourocarbon gas microbubble carrier delivery in swine carotid vessels

	Control	Treated
Maximal intimal thickness (mm^2)	0.34 ± 0.06	0.20 ± 0.06*
Lumen area (mm^2)	12.4 ± 6.6	21.2 ± 2.9*
Percent area stenosis	19 ± 8	8 ± 2*

3.2 *PGMC for Site-Specific Delivery of AVI-4217 in Porcine Coronary Vessels*

These results are impressive and additional studies have been conducted in coronary vessels with stents and did not involve ultrasound for site-specific delivery of the AVI-4217. Seven pigs underwent stent implantation (3 stents/animal). Four pigs received intravenous (IV) injection of PGMC and 1 mg of AVI-4217 and two served as controls. Four hours post-procedure, three pigs were sacrificed and stented segments analyzed by HPLC and western blot. In chronic experiments, four pigs were sacrificed at 28 days.

HPLC analysis of plasma samples of the treated animals showed minimal detected concentration of AVI-4217 but analysis of treated vessels demonstrated easily detected AVI-4217. Western blot analysis of the stented vessels demonstrated modest inhibition of *c-myc* with no alteration in expression of p21 or p27. Morphometry showed that the neointimal area was significantly reduced (Table 8). These data represent a limited number of vessels and studies that are underway to expand the use of PGMC for site-specific delivery of antisense agents.

4 Clinical Studies with AVI-4126

A Phase I study was conducted to evaluate the safety and pharmacokinetic properties of AVI-4126 at five dose levels (1, 3, 10, 30, and 90 mg) administered intravenously. Six subjects were tested at each dose level. Safety laboratory assessments (chemistry, hematology and urinalysis) were performed at baseline, and 24, 48, 72 h, 1 week and 2 weeks post-dose. Adverse experiences were collected on an ongoing basis from time of dosing to discharge from the study at follow-up week 2.

The most frequent adverse events reported included lower extremity aches and headache. The majority of adverse events were graded as mild in intensity and were generally self-limiting. Serum complement C3a was measured. Four subjects had elevated C3a greater than $2x$ the upper limit of normal (ULN) (normal = 0–400 ng ml^{-1}), including two subjects at the 3-mg dose level, one subject in the 10-mg cohort and one subject in the 90-mg dose group. Three of the four elevations

Table 8 Morphometric analysis of PGMC site-specific delivery of AVI-4217 in porcine coronary arteries

	Control	Treated
Lumen area (mm^2)	3.3 ± 0.7	6.1 ± 3.2
Percent area stenosis	57.8 ± 13.2	33.3 ± 24.6
Inflammatory score	0.7 ± 0.5	0.2 ± 0.3
Intimal vascularity	0.4 ± 0.5	0.2 ± 0.3
Intimal smooth muscle	3.0 ± 0.0	3.0 ± 0.0
Adventitial fibrosis	1.2 ± 0.8	0.7 ± 0.6

occurred at 24 h post-dose and one occurred at 0.5 h post-dose. In light of pharmacokinetic studies of the investigational compound, elevations of C3a at 24 h post-dose are unlikely to be related to administration of AVI-4126 but rather to spurious assay results. Further, there was no concurrent clinical symptomatology accompanying the elevated C3a, which is expected with elevated split complement levels.

4.1 Phase II Clinical Studies, AVAIL

The purpose of the AVAIL study was to investigate both safety and efficacy of AVI-4126 delivered locally via Infiltrator catheter after PCI in humans.

The AVAIL trial is prospective, evaluator blinded, randomized study including clinical follow-up at 30 days and 6 months after intervention and 6-months Angiographic and IVUS follow-up. Infiltrator catheter was advanced to target lesion and either drug was delivered (group A and B) or catheter was inflated (group C) after stent implantation in de novo lesions or PTCA in restenotic lesions. Primary endpoints include MACE, TVR, angiographic restenosis and IVUS at 6 months.

Forty four patients (pts) either with de novo lesions or restenosis were randomized into three groups: A (low dose): 3 mg (19 pts); B (high dose): 10 mg (15 pts); and C (control): (10 pts). Baseline angiographic characteristics did not differ between the groups (reference vessel diameter: 2.5–4 mm, lesion length <16 mm). Procedural success was 81–82% (unable to advance Infiltrator catheter to target lesion in 8 pts (5 pt group B and 3 pts from group C). There was no in-hospital or 30-day MACE recorded in any group. Clinical follow-up was available in 25 pts. At 6 months, 4 pts (50%) from control group ($n = 8$) and 7 (100%) pts from low dose A group ($n = 7$) required TVR.

In contrast in high dose group B ($n = 10$) only one patient (10%) needed TVR. Angiographic follow-up in 25 pts (8 pts group A, 7 pts group B and 10 pts group C) demonstrated Late loss of 1.4 ± 0.54, 0.8 ± 0.55 and 1.5 ± 0.65, respectively ($p = 0.025$). Binary restenosis was 38% in group C (control), 29% in group A (low dose) and 0% in group B (high dose).

We conclude that Local delivery of antisense is safe and feasible. These preliminary findings from the small cohort of patients require confirmation in a larger trial utilizing more sophisticated drug eluting technologies.

5 Conclusions

The most robust of the observations to date include the fact that AVI-4126 is safe and effective in reduction of restenosis in multiple species and conducted by multiple investigators. Three different methods for local and systemic delivery have been described, each with advantages and limitations. Efficacy in animal models is encouraging. Further, clinical trials with AVI-4126 indicate that the agent is very safe. The last remaining question: will AVI-4126 find a place in the future therapeutic regimen for the prevention of restenosis remains unanswered.

Acknowledgements The authors wish to thank the exceptional efforts of the AVI BioPharma chemistry group for the preparation of high purity and well-defined oligonucleotides, especially Dwight Weller, Doreen Weller, and Mary Martin. We also wish to thank the analytical group at AVI BioPharma for the high precision analysis of blood and tissue for AVI-4126. Finally, we wish to thank Dr. Thomas Porter and his laboratory at the University of Nebraska Medical Center for their efforts regarding perflourocarbon gas microbubble carriers.

References

Cornish KG, Iversen PL, Smith L et al (1993) Cardiovascular effects of a phosphorothioate oligonucleotide with sequence antisense t p53 in the conscious rhesus monkey. Pharmacol Commun 3:239–247

Dani C, Blanchard JM, Piechaczyk M et al (1984) Extreme instability of myc mRNA in normal and transformed human cells. Proc Natl Acad Sci USA 81:7046–7050

Galbraith WM, Hobson WC, Giclas PC et al (1994) Complement activation and hemodynamic changes following intravenous administration of phosphorothioate oligonucleotides in the monkey. Antisense Res Dev 4:201–206

Gruntzig AR, Senning A, Siegenthaler WE (1977) Nonoperative dilation of coronary artery stenosis: percutaneous transluminal coronary angioplasty. N Engl J Med 301:61–68

Hann SR, Eisenman RN (1984) Proteins encoded by the human c-myc oncogene: Differential expression in neoplastic cells. Mol Cell Biol 4:2486–2497

Henry SP, Bolte H, Auletta C et al (1997) Evaluation of the toxicity of ISIS 2302 a phosphorothioate oligonucleotide, in a four week study in cynamologous monkeys. Toxicology 120:145–155

Herdeg C, Oberhoff M, Baumbach A et al (2000) Local paclitaxel delivery for the prevention of restenosis: biological effects and efficacy in vivo. J Am Coll Cardiol 35:1969–1976

Hudziak RM, Summerton J, Weller DD et al (2000) Antiproliferative effects of steric blocking phosphorodiamidate morpholino antisense agents directed against c-myc. Antisense Nucleic Acid Drug Dev 10:163–176

Iversen PL, Cornish KG, Iversen LJ et al (1999) Bolus intravenous injection of phosphorothioate oligonucleotides causes hypotension by acting as a1-adrenergic receptor antagonists. Toxicol Appl Pharmacol 160:289–296

King S Br, Lembo NJ, Weintraub WS (1994) A randomized trial comparing coronary angioplasty with coronary bypass surgery. Emory Angioplasty versus Surgery Trial (EAST). N Engl J Med 331:1044–1050

Kipshidze N, Keane E, Stein, D et al (2001) Local delivery of c-myc neutrally charged antisense oligonucleotides with transport catheter inhibits myointimal hyperplasia and positively affects vascular remodeling in the rabbit balloon injury model. Catheter Cardiovasc Intervent 54:247–256

Kipshidze NN, Kim H-S, Iversen PL et al (2002) Intramural delivery of advanced antisense oligonucleotides with infiltrator catheter inhibits c-myc expression and intimal hyperplasia in the porcine. J Am Coll Cardiol 39:1686–1691

Miano JM, Vlasic N, Tota RR et al (1991) Smooth muscle cell immediate early gene and growth factor activation follows vascular injury. A putative mechanism for autocrine growth. Atheroscler Thromb 13:211–219

Morishita R, Gibbons GH, Ellison KE et al (1993) Single intraluminal delivery of antisense cdc2 kinase and proliferating-cell nuclear antigen oligonucleotides results in chronic inhibition of neointimal hyperplasia. Proc Natl Acad Sci USA 90:8474–8478

Parisi AF, Folland ED, Hartigan P (1992) A comparison of angioplasty with medical therapy in the treatment of single-vessel coronary artery disease. Veterans Affairs ACME Investigators. N Engl J Med 326:10–16

Porter TR, Iversen PL, Li S et al (1996) Interaction of diagnostic ultrasound with synthetic oligonucleotide-labeled perflourocarbon-exposed sonicated dextrose albumin microbubbles. J Ultrasound Med 15:577–584

Porter TR, Hiser W, Kricsfeld D et al (2000) Inhibition of carotid artery neointimal formation with intravenous microbubbles. Ultrasound Med Biol 27:259–265

Serruys P, Kutryk MJB, Bruining N et al (1998) Antisense oligonucleotide against c-myc administered with the Transport delivery catheter for the prevention of in-stent restenosis. Results of the randomized ITALICS trial. Circulation 98(Suppl):I-1–I-1016

Sheppard R, Eisenberg MJ (2001) Intracoronary radiotherapy for restenosis. N Engl J Med 344:295–297

Shi Y, Fard A, Galeo A et al (1994) Transcatheter delivery of c-myc antisense oligomers reduces neointimal formation in a porcine model of coronary artery balloon injury. Circulation 90:944–951

Simons M, Edelman ER, DeKeyser JL et al (1992) Antisense c-myb oligonucleotides inhibit intimal arterial smooth muscle cell accumulation in vivo. Nature 359:67–80

Sousa JE, Costa MA, Abizaid A et al (2001) Lack of neointimal proliferation after implantation of sirolimus-coated stents in human coronary arteries: a quantitative coronary angiography and three-dimensional intravascular ultrasound study. Circulation 103:192–195

Wallace TL, Bazemore SA, Kornburst DJ et al (1996) Single-dose hemodynamic toxicity and pharmacokinetics of a partial phosphorothioate anti-HIV oligonucleotide (AR177) after intravenous infusion to cynomolgus monkeys. J Pharmacol Exp Ther 278:1306–1312

Waters CM, Littlewood TD, Hancock DC et al (1991) c-myc protein expression in untransformed fibroblasts. Oncogene 6:797–805

Zalewski A, Shi Y, Mannion JD et al (1998) Synthetic DNA-based compounds for the prevention of coronary restenosis: current status and future challenges. In: Wickstrom E (ed) Clinical trials of genetic therapy with antisense DNA and DNA vectors. Dekker, New York, pp 363–393

Toxic RNA in Pathogenesis of Human Neuromuscular Disorders

D. Napierala and M. Napierala(✉)

Contents

Abstract In the past decade, substantial progress has been made in our understanding of the molecular pathogenesis of microsatellite expansion disorders. Some of these diseases, such as myotonic dystrophy, fragile X-associated tremor/ataxia syndrome, spinocerebellar ataxias 8, 10, 12 and Huntington's disease-like 2, are caused by an increased number of tri-, tetra- or pentanucleotide repeats in noncoding regions of the mutated genes. An RNA gain-of-function mechanism has been proposed to explain how mutations in noncoding fragments of genes lead to a disease phenotype. A number of studies demonstrated that RNA molecules harboring long repeating sequences affect the metabolism of other RNAs via deregulation of RNA processing proteins. This chapter presents the current status of knowledge related to mechanisms of pathogenesis, involvement of RNA structures and specific RNA binding proteins in seven neurodegenerative diseases associated with an RNA gain-of-function mechanism. Potential therapeutic strategies directed towards toxic RNAs are also reviewed.

M. Napierala
University of Texas M.D. Anderson Cancer Center, 1515 Holcombe Blvd., Houston, TX 77030, USA
e-mail: mnapiera@mdanderson.org

V.A. Erdmann et al. (eds.) *RNA Technologies in Cardiovascular Medicine and Research*, doi: 10.1007/978-3-540-78709-9,

1 Introduction

Expansions of short, tandemly repeated microsatellite sequences – most frequently trinucleotide but also tetra- and pentanucleotide repeats – are responsible for more than 20 human neurological and neuromuscular disorders (Orr and Zoghbi 2007). Microsatellite expansion diseases can be divided into two categories: loss-of-function diseases resulting from silencing of gene expression (e.g., fragile X syndrome or Friedreich's ataxia) and gain-of-function disorders resulting from various changes in the properties of the mutated gene products (e.g., Huntington's disease, myotonic dystrophies or spinocerebellar ataxias) (Ranum and Cooper 2006).

Two different types of gain-of-function mechanisms can be distinguished. First, a protein gain-of-function mechanism is related to the aberrant functions of mutated proteins harboring an expanded domain of homoaminoacids. Second, an RNA gain-of-function mechanism is associated with repeat expansions within noncoding regions of the genes. Synthesis and accumulation of toxic RNAs, containing long repeating tracts, lead to complex pathological changes in cellular metabolism.

The role of toxic RNAs was initially recognized in the pathogenesis of myotonic dystrophy type 1 and type 2 (DM1 and DM2). More recently, an RNA gain-of-function mechanism was demonstrated in the fragile X-associated tremor/ataxia syndrome (FXTAS). Several lines of evidence also suggest that spinocerebellar ataxias 8, 10 and 12 (SCA8, 10 and 12) as well as Huntington's disease-like 2 (HDL2) may result from expression of toxic RNAs (Table 1) (Osborne and Thornton 2006; Ranum and Cooper 2006).

Since the RNA and protein gain-of-function mechanisms are not mutually exclusive, a disease phenotype can also result from a combination of the two mechanisms.

In this chapter, we will review recent advances in the field of RNA-mediated microsatellite expansion disorders with special emphasis on the structure of repeat containing RNAs, the molecular mechanism of the RNA toxicity, and possible therapeutic strategies targeted towards pathologic RNAs.

Table 1 Diseases associated with the RNA gain-of-function mechanism

				Repeat size		
Disease	Gene	Repeat RNA	Locus	Normal	Pathological	Location
DM1	*DMPK*	CUG	19q13	5–37	50–>4,000	3′UTR
DM2	*ZNF9*	CCUG	3q21	<27	75–11,000	Intron
FXTAS[a]	*FMR1*	CGG	Xq27.3	5–45	55–200	5′UTR
	ASFMR1	CCG				?
SCA8[a]	*ATXN8OS*	CUG	13q21	16–34	71–>1,000	3′UTR/coding
	ATXN8	CAG				
SCA10	*ATXN10*	AUUCU	22q13	10–29	280–4,500	Intron
SCA12	*PPP2R2B*	CAG	5q32	7–32	55–78	5′UTR/intron/ coding?
HDL2	*JPH3*	CUG	16q24	6–28	40–59	3′UTR/coding?

[a] The role of bidirectional expression is postulated

2 Properties of Transcripts Containing Microsatellite Repeats

2.1 RNA Structure of Repeating Microsatellites

A small group of repeat sequences – including certain tri-, tetra- and pentanucleotide microsatellites – is associated with RNA gain-of-function disorders. RNAs containing CUG, CAG, CGG, CCG, CCUG and AUUCU repeats are implicated in the pathogenesis of DM1, DM2, SCA8, SCA10, SCA12, FXTAS, and HDL2 (Table 1). Generally, these repeat tracts are located in the noncoding sequences: introns and 5′ or 3′ untranslated regions (UTR). In the case of SCA12, depending upon the site of transcription initiation, the CAG tract is located either in the 5′UTR or upstream of the promoter region, or in the intron of the *PPP2R2B* gene (Holmes et al. 1999, 2001, 2006). In HDL2, the CUG repeats can be a part of the coding sequence or the 3′UTR, depending on the alternative splicing pattern of the *JPH3* gene (Margolis et al. 2006b).

Over the past ten years, the structural properties of transcripts containing microsatellite repeats have been analyzed using several biochemical and biophysical methods (Napierala and Krzyzosiak 1997; Michalowski et al. 1999; Tian et al. 2000; Handa et al. 2003; Sobczak et al. 2003; Sobczak and Krzyzosiak 2004b; Napierala et al. 2005; Zumwalt et al. 2007). Strikingly, all of these studies documented that each repeating sequence (tri-, tetra- and pentanucleotide) associated with the RNA gain-of-function mechanism formed a stable hairpin structure in vitro. Stability of repeat-containing hairpins depends on the length of the repeat tract and its sequence composition, i.e., repeating motif, presence of sequence interruptions disturbing the homogeneity of the repeat tract, and sequence context of the microsatellite region (Napierala and Krzyzosiak 1997; Handa et al. 2003; Sobczak et al. 2003; Sobczak and Krzyzosiak 2004a, 2005; Broda et al. 2005; Napierala et al. 2005; Zumwalt et al. 2007). Hairpins formed by trinucleotide repeats are stabilized by C-G and G-C pairs separated by single mismatches. NMR and thermodynamic analyses have demonstrated the existence of hydrogen bonds between N-N (N = A, C, G or U) mismatches (Broda et al. 2005). However, the overall thermodynamic stability of the hairpin stem does not depend significantly on the sequence composition of the mismatch.

Using biochemical methods and NMR spectroscopy, three independent groups demonstrated the formation of stable hairpins by CGG RNAs (Handa et al. 2003; Napierala et al. 2005; Zumwalt et al. 2007). AGG interruptions of the pure CGG tract, naturally occurring in human *FMR1* mRNA, influence conformation of the transcript and have an impact on the thermodynamic stability of the CGG repeat-containing RNAs (Napierala et al. 2005; Zumwalt et al. 2007). It should be pointed out that formation of four-stranded RNA structures (tetraplexes) by the CGG tracts has also been postulated (Khateb et al. 2007).

Crystal structure analysis of the 18-bp antiparallel RNA duplex containing CUG repeats showed a conformation similar to the A form of RNA with the exception of U-U mismatches (Mooers et al. 2005). Mismatched bases are an important element of the RNA recognition by specific binding proteins (Sect. 2.2).

The tetranucleotide CCUG repeats that are expanded in DM2 were demonstrated to adopt stable hairpins composed of C-G and G-C base pairs separated by C-U and U-C mispairs (Sobczak et al. 2003). CCUG repeat hairpin formation has important implications in the molecular pathogenesis of DM2. The apparent similarities between DM2 and DM1 result from the same type of structure adopted by their corresponding toxic RNAs.

Unexpectedly, even the U-rich pentanucleotide sequence, AUUCU, which is expanded in SCA10 patients, is capable of forming unusual RNA hairpins. Using NMR, CD and enzymatic probing, Handa et al. proved that tracts as short as 11 repeats exist in a hairpin conformation stabilized by an equal number of hydrogen bonded A-U and U-U pairs (Handa et al. 2005).

In summary, the ability to form hairpin structures is a common property of toxic RNAs. Although RNA conformation analyses were conducted in vitro, the existence of these structures in the cellular environment is supported by the demonstration of co-localization of the expanded transcripts with structure-specific RNA binding proteins (RBPs) (Fardaei et al. 2001, 2002). Formation of hairpins containing long and stable double-stranded stems may lead to a variety of biological effects, including direct inhibition of translation, aberrant interaction with specific proteins, enzyme activation, and regulation of gene expression via the RNA interference (RNAi) pathway. Some of these processes are indeed affected by toxic RNAs (Sects. 3–5).

2.2 Proteins Binding to the Repeated Sequences in RNAs

RNA molecules that contain repeating sequences as well as hairpin structures formed by microsatellites are recognized by RNA binding proteins (Table 2). Interactions between RBPs and transcripts containing expanded repeats are believed to be the major step in the molecular pathogenesis of the RNA gain-of-function disorders.

CUG repeat binding protein 1 (CUGBP1) was the first protein identified in gel retardation assays to bind to a $(CUG)_8$ probe (Timchenko et al. 1996a,b). This protein is a member of CELF (CUGBP1 and ETR-3 Like Factors, BRUNOL) family of RNA metabolism regulators. The CELF proteins (CELF1-6) participate in control and regulation of alternative splicing, translation and deadenylation (Ladd et al. 2001; Barreau et al. 2006; Thornton et al. 2006). CUGBP1 (CELF1) shows 76% sequence identity with CUGBP2 (CELF2, ETR-3), another protein which has been demonstrated to bind CUG repeats and play a role in the pathogenesis of DM1 (Lu et al. 1999). However, recent studies have questioned the binding of the CELF proteins, especially CUGBP1, to the CUG repeats. Experiments conducted using the SELEX procedure, the yeast three-hybrid system and surface plasmon resonance (SPR) showed that CUGBP1 interacts preferentially with UG/GU-containing transcripts (Kino et al. 2004; Marquis et al. 2006; Mori et al. 2008). More importantly, CUGBP1 demonstrated preference for single-stranded RNAs, which suggests that double-stranded CUG or CCUG hairpins are unlikely targets for its binding (Kino et al. 2004).

Table 2 Proteins interacting with microsatellite-containing transcripts

	Protein	Sequence preference[a]	Conformational preference[a]	Disease
MBNL family				
	MBNL1	CYG and CYYG	dsRNA with mismatches	DM1, 2
	MBNL2 (MBLL)	CCUG, CCCG	?	DM1, 2
CELF family				
	CUGBP1 (CELF1, BRUNOL2)	UG/GU	ssRNA	DM1, 2, FXTAS
	CUGBP2 (ETR3, BRUNOL3)	UGGU	?	DM1, 2
PKR (p20)		CUG, CAG	perfect dsRNA	DM1, 2
Purα		CGG, NGG	ssRNA	FXTAS
hnRNP A2/B1		CGG, telo-eric repeats	?	FXTAS
hnRNP H		CUG[b]	dsRNA	DM1, 2

[a] The studies did not include analyses of all possible repeating sequences
[b] Efficient binding requires the presence of specific non-repeating sequences

Despite the controversies related to the CUGBP1 binding specificity, CUGBP1 is certainly involved in molecular pathogenesis of DM1, DM2 and most likely other RNA gain-of-function diseases. It has been demonstrated that expression of expanded CUG repeats increases the half-life of CUGBP1 in cultured cells (Timchenko et al. 2001). Moreover, recent studies showed that the stabilization of CUGBP1 – and consequently its increased activity – are the result of increased phosphorylation of CUGBP1 in DM1 tissues (Kuyumcu-Martinez et al. 2007).

Further studies of proteins capable of binding CUG repeats revealed a family of muscleblind-like proteins (MBNL) (Fardaei et al. 2001, 2002; Ho et al. 2004). This group of zinc finger proteins, which are homologous to the *Drosophila* muscleblind proteins, consists of three members: MBNL1 (EXP), MBNL2 (MBLL) and MBNL3 (MBXL). Although all three proteins were shown to colocalize with RNA foci in DM1 and DM2 patient cells, the majority of RNA binding studies have been conducted with MBNL1 (Fardaei et al. 2002). This protein strongly prefers double-stranded, mismatch-containing RNA helices. The consensus MBNL1 binding sequences are CYYG and CYG (Y = C or U) (Kino et al. 2004; Warf and Berglund 2007). Perfect A-RNA duplexes (e.g., CAG-CUG) almost completely abolish the MBNL1 binding. Recently, Yuan et al. showed that MBNL1 forms oligomeric complexes which are capable of binding to double-stranded CUG repeats (Yuan et al. 2007). At higher protein to RNA ratios, multiple complexes bind to a single RNA molecule. This study provided a model explaining a mechanism of MBNL1 sequestration and formation of nuclear foci by expanded CUG/CCUG repeats.

Double-stranded, repeat-containing RNAs are also binding targets for the dsRNA-dependent protein kinase, PKR (p20) (Tian et al. 2000, 2005). PKR was demonstrated to bind expanded CUG repeats. Moreover, in vitro interactions between the CUG RNA and PKR lead to activation of PKR enzymatic activity (Tian et al. 2000, 2005). In vivo binding studies, using a yeast three-hybrid system, showed that dsRNAs lacking mismatches (e.g., CUG-CAG duplex) as well as CAG repeats interacted with PKR more efficiently than CUG tracts (Kino et al. 2004).

UV crosslinking experiments have led to the isolation of the heterogeneous nuclear ribonucleoprotein H (hnRNP H) associated with RNAs containing CUG repeats (Kim et al. 2005). This protein is known to participate in various processes related to RNA metabolism including splicing and polyadenylation (Chen et al. 1999; Arhin et al. 2002). HnRNP H is not a typical repeat binding protein. For efficient binding hnRNP H requires not only an expanded CUG tract (>46 repeats), but also a sequence containing a splice branch point. In DM1 cells upregulation of hnRNP H facilitates nuclear retention of the mutant *DMPK* transcript and stimulates aberrant splicing (Kim et al. 2005). Interestingly, hnRNP H was shown to interact in vivo with MBNL1 in an RNA-independent fashion and to form an RNA-dependent complex with the CUGBP1 (Paul et al. 2006).

None of the proteins described so far was capable of efficient binding to the CGG repeats expanded in patients with FXTAS. Recently, two groups reported the discovery of two proteins, purine-rich binding protein (Purα) and cytoplasmic fraction of hnRNP A2/B1, which interacted directly with CGG tracts, suggesting their role in molecular pathogenesis of FXTAS (Jin et al. 2007; Sofola et al. 2007b). Interestingly, CUGBP1 co-immunoprecipitated with the hnRNP A2/B1, which indicates an indirect association of the CUGBP1 with CGG repeats and a possible functional link between the molecular mechanism of DM and FXTAS (Sofola et al. 2007b).

The role of repeat binding proteins in the pathogenesis of RNA gain-of-function disorders will be presented later in this chapter (Sects. 3 and 4).

2.3 *Bidirectional Expression of Genes Coding for Toxic RNAs*

RNA toxicity originates from the expression of genes containing expanded CG-rich repeats. Alternatively, transcription of the gene encoded on the opposite DNA strand can lead to the accumulation of a completely different mRNA molecule containing a complementary repeating sequence. Taking into account that 61% of all human transcribed regions have a counterpart on the opposite DNA strand, the likelihood of antisense transcription is quite high (Cheng et al. 2005). Indeed, in the case of three RNA gain-of-function disorders such antisense transcripts have been detected and their putative role in molecular pathogenesis has been postulated.

In the case of DM1, an antisense transcript to the *DMPK* RNA was identified using sense-specific RT-PCR. It contained several open reading frames, but none of them encoded a polyglutamine tract. Cho et al. discovered that upon transcription of the antisense strand of the *DMPK* gene, short 21-mer RNAs can be isolated (Cho et al. 2005). These RNA fragments induced localized chromatin changes via histone H3-K9 methylation and recruitment of the heterochromatin protein 1 (HP1). The influence of the chromatin modifications at the CTG repeat region on the disease development is unclear and requires further investigation.

SCA8 represents a very interesting case of bidirectional transcription. Two genes spanning the trinucleotide CTG repeat tract are transcribed in opposite directions. Noncoding CUG repeats are part of the *ataxin 8 opposite strand* gene (*ATXN8OS*, originally *SCA8* or *KLHL1AS*), while in the recently discovered *ataxin 8* gene (*ATXN8*) CAG repeats encode a polyglutamine domain (Moseley et al. 2006). Pathological implications of this phenomenon are described in the Sect. 5.1.

Recently, an antisense transcript has been identified at the human *FMR1* locus. A novel gene, *ASFMR1* (*antisense transcript at the FMR1 locus*), overlaps with the CGG repeats of the *FMR1* gene in the antisense direction (Ladd et al. 2007). The antisense transcript is polyadenylated, has multiple splice forms, can be found in cytoplasm, and is widely expressed in human tissues. A putative open reading frame (ORF) contains a domain composed of polyproline. Since both CCG and CCU triplets code for proline, the presence of AGG interruptions in the CGG sequence (coding strand for *FMR1*) does not disrupt the purity of the putative homoaminoacid tract. Interestingly, in a pattern similar to the *FMR1* mRNA, expression of the *ASFMR1* transcript is elevated in cells harboring premutation alleles (55–200 repeats) and is silenced in the full mutation alleles (>200 repeats) (Ladd et al. 2007).

In summary, preliminary evidence suggests that bidirectional transcription may be an important aspect of toxic RNA metabolism.

3 Myotonic Dystrophy Type 1 and 2 (DM1 and DM2)

Myotonic dystrophy type 1 is the best characterized RNA gain-of-function-mediated disorder. Expansions of CTG repeats in the 3′UTR of the *DMPK* (*dystrophia myotonica protein kinase*) gene cause DM1, a multisystem disease that affects predominantly skeletal muscles, heart, eyes and endocrine system (Brook et al. 1992; Harper 2001). Like other microsatellite expansion disorders, the severity and age of onset of the DM1 correlate with the number of repeats. Shorter alleles containing 50–100 CTG repeats result in late onset cataract, alleles containing 100–500 repeats result in the classic form of DM1, while the largest expansions (over 500 CTG repeats) lead to a congenital form of the disease which manifests with infantile hypotonia and mental retardation. The classic form of DM1 is characterized by adult-onset myotonia, muscle weakness, cataract, cardiac conduction defects and endocrine abnormalities. Similar clinical features are characteristic for myotonic dystrophy type 2 (DM2), which is caused by expansions of the CCTG repeats in the first intron of the *ZNF9* (*zinc finger 9*) gene. Although the number of expanded repeats in DM2 is dramatically higher than in DM1, the general clinical presentation of DM2 is milder than DM1. Overlapping clinical features of DM1 and DM2 include myotonia, muscle weakness, cataract, cardiac arrhythmias, insulin resistance and testicular atrophy. Notably, DM2 does not present in a congenital form and is not associated with mental retardation.

3.1 *Molecular Pathogenesis of Myotonic Dystrophy*

How do mutations in the noncoding regions of two functionally unrelated genes lead to a similar clinical presentation? Although most of the myotonic dystrophy features can be explained by a gain-of-function effect of mRNA with expanded CUG or CCUG repeats, the molecular mechanism of myotonic dystrophy may be more complex.

3.1.1 Unlikely Role of *DMPK* and *ZNF9* Loss of Function

Both myotonic dystrophy type 1 and type 2 are inherited in a dominant mode, which can suggest a haploinsufficiency mechanism of a disease. The loss-of-function mechanism has also been suggested, based on the observation that in DM1 cells the expression of *DMPK*, as well as the adjacent *DMWD* and *SIX5* genes is decreased (Klesert et al. 1997; Thornton et al. 1997; Alwazzan et al. 1999; Eriksson et al. 1999). However, human data and mouse model studies indicate that loss-of-function is not the predominant molecular defect in myotonic dystrophy. First, no mutations in coding regions of either *DMPK* or *ZNF9* genes have been found in myotonic dystrophy patients. Secondly, mice deleted for the *Dmpk* gene do not reproduce the multisystem phenotype. Two research teams independently generated *Dmpk* knockout mice by targeted deletion of the first seven exons of the *Dmpk* gene (Jansen et al. 1996; Reddy et al. 1996). This mutation results in complete loss of the Dmpk protein in homozygous mice. *Dmpk* null mice demonstrated only mild cardiac conduction defects, late-onset myopathy (Jansen et al. 1996; Reddy et al. 1996) and insulin signaling defects (Llagostera et al. 2007). None of the *Dmpk* knockout mouse lines demonstrated the hallmark features of myotonic dystrophy such as myotonia, cataract and endocrine abnormalities. Furthermore, deficiency of the *Six5* gene in mice also failed to reproduce the features of myotonic dystrophy (Klesert et al. 2000; Sarkar et al. 2000).

ZNF9, the gene which is mutated in DM2 (Liquori et al. 2001), encodes a nucleic acid binding protein that is involved in cap-independent translation (Gerbasi and Link 2007). A recent report of $Znf9^{+/-}$ mice demonstrated that *Znf9* haploinsufficiency leads to heart, muscle and eye defects similar to these observed in myotonic dystrophy (Chen et al. 2007). While this is compelling evidence that a loss of the *Znf9* function leads to DM-like phenotype, human data provide evidence for a different conclusion. Analyses of *ZNF9* expression in myoblast cell lines from DM2 patients demonstrated that large CCUG expansions do not affect steady-state *ZNF9* mRNA levels (Margolis et al. 2006a), arguing against the *ZNF9* loss-of-function mechanism.

3.1.2 Evidence for the RNA Gain-of-Function Mechanism

Although deficiency of the *DMPK* or *ZNF9* genes, which are highly expressed in tissues affected in myotonic dystrophy, may contribute to the clinical manifestation of the disease, the major mechanism underlying the multisystem presentation is a toxic effect of mRNA molecules containing expanded CUG or CCUG repeats. The RNA gain-of-function mechanism had been postulated even before the identification of the *ZNF9* gene as the second gene mutated in some cases of myotonic dystrophy (Philips et al. 1998).

The first clues for a role of mRNAs in the pathogenesis came from analyses of the intracellular localization of *DMPK* mRNA in myoblastic and fibroblastic cells from DM1 patients. Using probes specific for expanded CUG repeats, Taneja et al. demonstrated that mutant *DMPK* transcripts accumulate in nuclei in multiple discrete foci (Taneja et al. 1995). Additional observations revealed that the localization of these foci is different from the site of *DMPK* transcription, and that the mutant *DMPK* mRNA is normally spliced and polyadenylated (Davis et al. 1997). These data suggested that the mRNA containing expanded CUG repeats fails to be exported from the nucleus. The presence of nuclear foci was also demonstrated in DM1 brain and DM2 muscle cells (Mankodi et al. 2001; Jiang et al. 2004). Similar to the mutant *DMPK* mRNA, the *ZNF9* pre-mRNA containing the expanded CCUG repeats was also correctly processed. In *DMPK* mRNA the CUG repeats are a part of the mature mRNA, while in *ZNF9* transcripts the CCUG repeats are located in intronic sequence which is spliced out. Interestingly, the mutant *ZNF9* mRNA was correctly spliced and exported to the cytoplasm (Margolis et al. 2006a). The *ZNF9* RNA that was found in the nuclear foci contained only expanded CCUG repeats. These observations support the hypothesis of a *trans*-dominant effect of mutant *DMPK* and *ZNF9* transcripts, whereby expanded CUG or CCUG repeats do not disrupt the processing of repeat-bearing pre-mRNA (*cis* effect), instead, the nuclear accumulation of repeat-containing RNAs affects cellular processes leading to the disease.

3.1.3 Aberrant Splicing as the Key Molecular Feature of DM

Currently, most of the data support the following model of the molecular pathogenesis of myotonic dystrophy: expanded CUG/CCUG repeats act in *trans* affecting the processing of other RNAs via dysregulation of specific RNA binding proteins. Consistent with this hypothesis, interactions of CUG/CCUG-containing RNAs with proteins regulating RNA metabolism have been demonstrated (discussed in Sect. 2.2) (Timchenko et al. 1996a; Miller et al. 2000). Altered nuclear and cytoplasmic activities of MBNL and CELF protein families' members have been identified and linked to molecular abnormalities observed in DM1 and DM2.

The CELF proteins regulate RNA metabolism at the level of splicing, editing, translation and stability, and their activity is up-regulated by repeat-containing mRNAs (Sect. 2.2). On the other hand, the activity of MBNL proteins is decreased due to their sequestration in nuclear foci that contain long CUG and CCUG repeats (Cardani et al. 2006). CELF and MBNL proteins have been demonstrated to regulate alternative splicing by binding to specific pre-mRNA sequences and other regulatory proteins. Analyses of DM tissues and mouse models of myotonic dystrophy revealed a number of pre-mRNAs that are splicing targets of both MBNL and CELF proteins. Interestingly, MBNL and CELF act antagonistically in a process of developmentally regulated alternative splicing events. Although both proteins are present in many tissues, CUGBP1 is mostly expressed during embryonic development, while MBNL1 is expressed postnatally (Lin et al. 2006). Therefore, CUGBP1 generates the "embryonic" mRNA isoforms, while the mRNA isoforms generated by MBNL1 are characteristic for postnatal development. As a consequence of loss of MBNL1 function due to sequestration in nuclear foci accompanied by gain of CUGBP1 function due to increased stabilization of the protein, changes of the splicing patterns of multiple mRNAs to the embryonic isoforms are observed in DM patients tissues and mouse models of myotonic dystrophy (Fig. 1).

These molecular abnormalities can be directly linked to the clinical findings in DM. Cardiac disturbances occur frequently in DM1. Conduction abnormalities range from first degree heart block to the more advanced forms including complete heart block. Hypertrophic cardiomyopathy and arrhythymias are also known to occur. Aberrant splicing of cardiac troponin T (cTNT), a protein that can regulate cardiac contractility, was the first identified example of splicing disruption in myotonic dystrophy (Philips et al. 1998). Splicing of cTNT is developmentally regulated in such a way that exon 5 is present in the fetal heart, but is excluded from the adult heart. Both CUGBP1 and MBNL1 can bind to cTNT pre-mRNA and regulate its splicing, generating either the embryonic or adult isoform (Philips et al. 1998; Ladd et al. 2005; Yuan et al. 2007). It has been demonstrated that due to increased CUGBP1 activity, an embryonic isoform of cTNT is expressed in DM1 muscles (Philips et al. 1998). While the embryonic isoform of cTNT is less sensitive to calcium, which can cause decreased contractility, the exact role of the aberrant splicing of cTNT in DM pathogenesis is not known. Missplicing of several other genes, such as *TNNT2*, *KCNAB1* and *ALP*, has also been found in DM1 hearts (Mankodi et al. 2005).

A characteristic feature of myotonic dystrophy is the occurrence of an unusual form of insulin-resistant diabetes that is due to defective insulin signaling in skeletal muscles. Misregulation of the insulin receptor (IR) splicing has been demonstrated in both DM1 and DM2 muscles (Savkur et al. 2001, 2004). It has been shown that overexpression of CUGBP1 alters the splicing in favor of a non-muscle isoform, which lacks exon 11, leading to insulin insensitivity in DM muscles. Alternatively, generation of a splice form containing exon 11 is facilitated by MBNL1. Therefore, the IR splicing events can be manipulated by changing a balance between CUGBP1 and MBNL1 activities (Dansithong et al. 2005; Ladd et al. 2005).

A classic feature of myotonic dystrophy, myotonia, is caused by aberrant splicing of the muscle-specific chloride channel 1 (CLC1) mRNA, leading to a premature termination of the message. As a consequence, loss of CLC1 protein in DM1 and DM2 skeletal muscles has been observed (Charlet et al. 2002; Mankodi et al. 2002). Reduced chloride conductance and a decreased number of functional Clc1 channels have been also demonstrated in mouse models of myotonic dystrophy (Mankodi et al. 2002; Lueck et al. 2007). Moreover, in vitro studies showed that overexpression of CUGBP1 in normal cells causes retention of intron 2 in the *Clc1* minigene, indicating that this intron is a target of CUGBP1 (Charlet et al. 2002). Subsequent studies

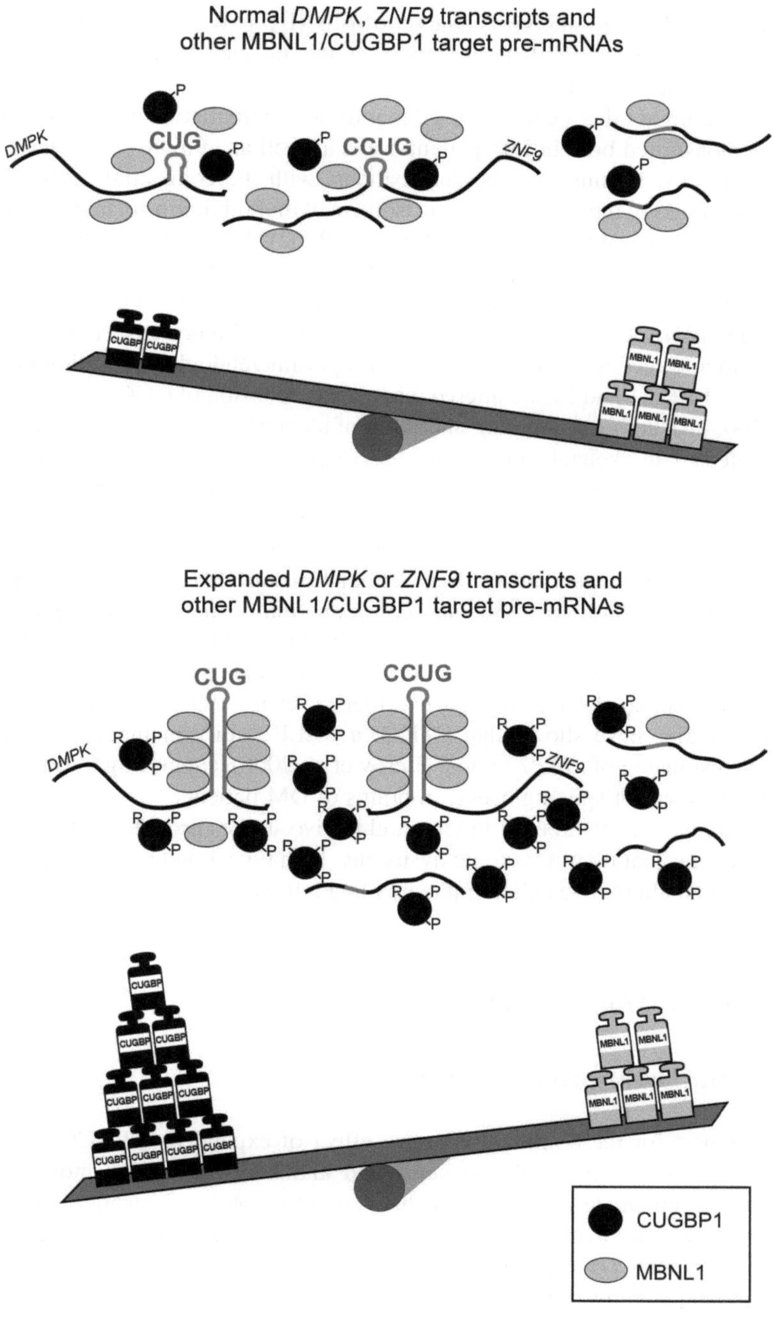

Fig. 1 Dominant negative effect of CUG/CCUG expansions. Sequestration of the *MBNL1* by expanded CUG and CCUG repeats accompanied by elevated activity (hyperphosphorylation and increased stability) of the *CUGBP1* are major steps in molecular pathogenesis of DM1 and DM2

revealed that Clc1 pre-mRNA is a target of MBNL1 as well (Kanadia et al. 2003b, 2006), again suggesting that disruption of the balance between CUGBP1 and MBNL activities leads to DM.

Missplicing of other genes involved in muscle differentiation and physiology has been demonstrated both in DM patients cells as well as in mouse models of myotonic dystrophy. In muscle cells from patients with the congenital form of DM1, decreased levels of the muscle-specific isoform of myotubularin-related 1 (MTMR1) mRNA have been observed (Buj-Bello et al. 2002). MTMR1 is a target of CELF proteins, and its abnormal splicing has also been shown in skeletal muscles and heart of transgenic mice overexpressing CUGBP1 (Ho et al. 2005). Splicing patterns and mRNA levels of proteins involved in calcium homeostasis in muscles, like the ryanodine receptor 1 (RYR1) and sarcoplasmic/endoplasmic reticulum Ca^{2+}-ATPase (SERCA1 and 2), are altered due to the expansion of CUG repeats (Kimura et al. 2005). In addition, splicing alteration of a cytoskeletal protein dystrophin has been detected in skeletal muscle and heart cells from DM1 patients. Since dystrophin is highly expressed in muscle cells, and mutations in this gene cause Duchenne/Becker's muscular dystrophy, its aberrant splicing may contribute to the muscle wasting seen in DM (Nakamori et al. 2007).

Neurofibrillary degeneration observed in brains of DM1 and DM2 patients may result, at least partially, from the altered production of the microtubule-associated protein tau isoforms (MAPT) (Maurage et al. 2005). Alteration of *MAPT* exon 2 splicing has been demonstrated in brains of DM patients as well as transgenic mice expressing expanded CUG repeats (Vermersch et al. 1996; Seznec et al. 2001). In vitro studies have shown that ETR-3, a CELF family member, promotes the selective exclusion of *MAPT* exon 2 (Leroy et al. 2006), suggesting a role of CELF proteins in aberrant splicing events in brains of DM patients.

To summarize, a number of in vitro and in vivo studies provide evidence that the key molecular feature of myotonic dystrophy, both type 1 and type 2, is expression of embryonic alternative splicing patterns in adult tissues.

3.2 Mouse Models of DM

3.2.1 Mice Overexpressing CUG Repeats

Clear support for the dominant-negative effect of expanded CUG/CCUG repeats comes from the analyses of transgenic and knockout mice. First mouse models generated to elucidate molecular mechanisms of myotonic dystrophy addressed a possible gene dose effect. However, neither loss of the *Dmpk* gene nor overexpression of normal human *DMPK* gene containing a $(CTG)_{11}$ repeat tract reproduced the characteristic features of myotonic dystrophy in young animals (Jansen et al. 1996). Interestingly, the older transgenic mice (11–15 months) with five to ten fold overexpression of the human *DMPK* gene developed myotonia and hypertrophic cardiomyopathy with dysrhythmia. Thus, the continuous overexpression of the wild type *DMPK* gene leads to accumulated distress which manifests phenotypically with age (O'Cochlain et al. 2004). To study effects of the CUG expansion in an untranslated region of a DM1-unrelated transcript, Mankodi et al. generated transgenic mouse expressing long CUG repeats within 3′UTR of actin mRNA. These mice developed early onset myotonia, suggesting that an increased number of CUG repeats alone is sufficient to induce a pathogenic effect (Mankodi et al. 2000). This notion was further confirmed by comparison of phenotypes of transgenic mice carrying the entire human *DMPK* gene with 350 CTG repeats versus mice carrying 20 CTG repeats (Seznec et al. 2001). Only mice expressing human *DMPK* mRNA with expanded repeats developed myotonia as well as histological and molecular abnormalities in skeletal muscles consistent with the myotonic dystrophy phenotype. Moreover, similar to data obtained from DM1 patients, brains of the transgenic mice bearing 350 CTG repeats demonstrated changes in the distribution of tau protein isoforms (Seznec et al. 2001).

Intriguing results were obtained using transgenic mice with inducible expression of CUG-containing mRNAs. Two groups demonstrated that severe cardiac abnormalities develop rapidly after induction of the expression of CUG-containing mRNAs in adult mice. In one of the inducible DM models, the heart-specific expression of the expanded (960 CUG) repeats resulted in 100% mortality within 2 weeks of induction (Wang et al. 2007). These mice developed physiological and molecular cardiac abnormalities similar to those observed in DM patients. Interestingly, the co-localization of MBNL1 with RNA in nuclear foci and induction CUGBP1 occurred within hours of induction, and preceded splicing changes. Surprisingly, the other mouse line developed DM-characteristic abnormalities upon induction of expression of the mRNA containing only five CUG repeats. In this case, neither MBNL1-containing nuclear foci, nor increased CUGBP1 activity were observed (Mahadevan et al. 2006).

To summarize, a number of in vivo studies conducted by different research groups provide evidence that the expression of expanded CUG/CCUG repeats, not mutations of the *DMPK* or *ZNF9* genes themselves, triggers a molecular response leading to the myotonic dystrophy phenotype.

3.2.2 CUGBP1 Transgenic and MBNL1 Knockout Mice

To test the hypothesis that increased CUGBP1 activity plays a role in the muscle pathology of myotonic dystrophy, two laboratories independently generated transgenic mice overexpressing CUGBP1 in cardiac and skeletal muscles. Consistent with the CUGBP1 gain-of-function mechanism, abnormal processing of CUGBP1-target mRNAs was observed in both transgenic mouse lines. The first group demonstrated characteristics for myotonic dystrophy splicing misregulation of the muscle-specific *CLC1*, *cTNT* and *MTMR1* mRNAs in the muscles and hearts of CUGBP1 transgenic mice (Ho et al. 2005). The second group reported that overexpression of CUGPB1 in mice increased the MEF2A and p21 proteins which resulted in delay of muscle development (Timchenko et al. 2004).

Targeted disruption of the mouse *Mbnl1* gene in such a way that only the isoforms interacting with CUG/CCUG-containing RNAs are not expressed, provided a tool to study the consequences of loss of Mbnl1's ability to process CUG/CCUG-containing RNAs. As predicted by the MBNL sequestration model, loss of this function was sufficient to cause myotonia, cataract and RNA splicing defects of cardiac troponin T and muscle-specific chloride channel similar to those seen in myotonic dystrophy (Kanadia et al. 2003a). Moreover, the overexpression of Mbnl1 in skeletal muscles of the DM mouse model carrying the human actin gene with insertion of 250 CTG repeats in 3′UTR rescued the myotonia and splicing defects (Kanadia et al. 2006). Thus, manipulations with CELF and MBNL protein activities in transgenic and knockout mice demonstrate that altered balance of opposite activities of the CUGBP1 and MBNL1 proteins underlie the aberrant splicing patterns of multiple mRNAs, which in turn leads to the multisystemic presentation of myotonic dystrophy.

Extensive studies of molecular abnormalities in DM patients' tissues and in mouse models of myotonic dystrophy provide solid evidence of the toxic RNA mechanism of the disease. Although the main stream of research is focused on the post-transcriptional RNA metabolism, there is evidence that expanded CUG/CCUG repeats also disrupt the process of transcription (Ebralidze et al. 2004). It has been demonstrated that long CUG repeats deplete basic transcription factors leading to the decreased expression of multiple genes, including *CLC1*. These unexpected findings suggest that not all features of long CUG/CCUG sequences have been revealed yet.

4 Fragile X-Associated Tremor/Ataxia Syndrome (FXTAS)

Fragile X syndrome (FXS) is caused by large CGG repeat expansion (>200 copies) leading to the DNA methylation and silencing of the *FMR1* gene (Fig. 2) (Fu et al. 1991; Pieretti et al. 1991; Orr and Zoghbi 2007). For a long time, it was believed that premutation carriers, with 55–200 CGG repeats, were simply an asymptomatic source of the full mutation alleles. However, recently, premutation alleles have been associated with two conditions: premature ovarian failure (POF) and fragile X tremor/ataxia syndrome (FXTAS) (Hagerman et al. 2001, 2004). FXTAS is a late onset neurodegenerative disorder characterized by tremor, gait instability, cerebellar dysfunction, cognitive decline and parkinsonism. It should be emphasized that the clinical characteristics of FXTAS are entirely different from fragile X syndrome, suggesting completely distinct molecular mechanisms for these disorders (Hagerman and Hagerman 2006). Cells of the premutation carriers express *FMR1* mRNA at levels two to ten times higher than cells harboring a normal (<55) number of CGG repeats, while the level of the FMR1 protein in the premutation carriers is not elevated. This suggests that the effects of the RNA level play a significant role in FXTAS (Tassone et al. 2000a,b).

Jin et al. demonstrated, in *Drosophila*, that expression of the premutation length $(CGG)_{90}$ RNA, even in the absence of the entire *FMR1* gene, leads to

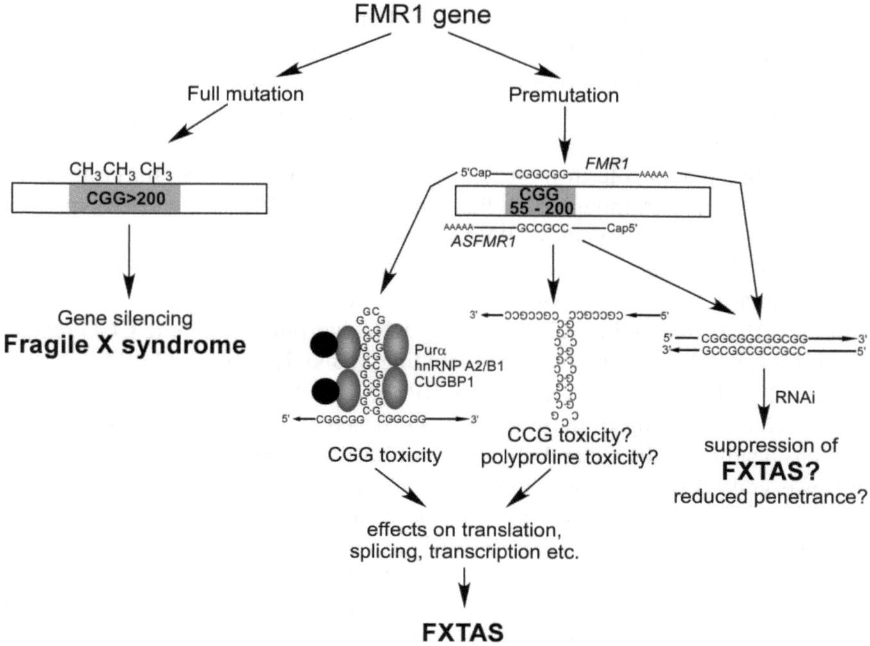

Fig. 2 Molecular pathogenesis of fragile X syndrome and FXTAS. Bidirectional transcription of the *FMR1* locus results in formation of transcripts containing CGG (*FMR1* mRNA) and CCG repeats (*ASFMR1* mRNA). Repeat regions in both types of RNAs can form hairpins and induce RNA toxicity. It is not known whether translation of the *ASFMR1* triggers protein toxicity. Toxic effects of CGG and CCG RNAs can potentially be eliminated by RNAi-mediated degradation of complementary transcripts

neurodegeneration (Jin et al. 2003). Neuropathologically – similar to the DM1 and DM2 – FXTAS is characterized by the presence of ubiquitin-positive, RNA-containing nuclear inclusions in both neurons and astrocytes (Greco et al. 2002). Inclusions were also present in the knock-in mice containing approximately 100 CGG repeats in the mouse *Fmr1* gene (Willemsen et al. 2003). The increase in number as well as in size of the inclusions was observed over the course of the lifetime of the mouse, corresponding to the progressive character of FXTAS. The inclusions isolated from post-mortem brains of FXTAS patients contain, in addition to the *FMR1* mRNA molecules, more than 20 different proteins, none of which contributed to more than 7% of total proteins detected in the inclusions (Tassone et al. 2004; Iwahashi et al. 2006). Two of the identified proteins, MBNL1 and hnRNP A2/B1, are well characterized RNA binding proteins, as discussed above (Sect. 2.2). The inclusions also contained intermediate filament (IF) proteins, which could potentially contribute to the RNA toxicity and neurodegeneration by regulating RNA synthesis and processing (Hutchison and Worman 2004; Iwahashi et al. 2006).

Recently, another RNA binding protein, Purα, has been detected in FXTAS inclusions (Jin et al. 2007). This protein, in concert with hnRNP A2/B1, is essential for the neurodegeneration observed in the *Drosophila* model of FXTAS. Overexpression of either of these proteins alleviates the neurodegenerative phenotype in the fly model providing evidence that sequestering and inhibition of Purα and hnRNP A2/B1 play a role in molecular pathogenesis of FXTAS (Jin et al. 2007; Sofola et al. 2007b). Although both Purα and hnRNP A2/B1 are widely expressed and involved in various aspects of RNA metabolism, their exact functions in development of FXTAS have not yet been elucidated.

In summary, substantial evidence has accumulated to support the following model for RNA toxicity in FXTAS (Fig. 2). RNA hairpins containing CGG repeats are targets for RNA binding proteins including MBNL1, Purα and hnRNP A2/B1. Additionally, the cellular level of CUGBP1 is affected by its interactions with hnRNP A2/B1. Binding of these proteins influences their downstream targets and functions, leading to the clinical phenotype of FXTAS. In parallel, expression of the antisense gene *ASFMR1* can stimulate RNA toxicity via CCG hairpins and/or protein toxicity via the putative polyproline-containing *ASFMR1* gene product as discussed in Sect. 2.3 (Fig. 2).

In a separate disease process, expanded CCG repeats (>200) located in the *FMR2* gene are associated with non-syndromic X-linked mental retardation (FRAXE) (Knight et al. 1993). Large expansion of these repeats induces transcriptional silencing of the *FMR2* gene. However, no neurodegenerative phenotype has been associated with premutation alleles of *FMR2*.

Interestingly, expression of premutation length CCG repeats (complementary to FXTAS CGG tract) can also trigger an RNA mediated neurodegenerative phenotype in a *Drosophila* model (Sofola et al. 2007a). Hence, independent expression of premutation size CGG or CCG transcripts leads to a neurodegenerative phenotype in transgenic *Drosophila*. Surprisingly, co-expression of both complementary RNAs in the fly model alleviates their independent neuronal toxicity (Sofola et al. 2007a). The rescue depends on the RNAi pathway suggesting a possible avenue for the future therapeutic intervention in toxic RNA-mediated disorders (Sect. 6). In humans, *FMR1* premutation carriers express high levels of both the *FMR1* mRNA and its antisense counterpart *ASFMR1* containing CCG repeats. Approximately one-third of the male premutation carriers will develop FXTAS (Jacquemont et al. 2004), implicating involvement of other factors, perhaps the expression of the complementary transcript, in the suppression of the disease (Fig. 2).

5 Other Disorders Associated with Toxic RNAs

5.1 Spinocerebellar Ataxias (SCAs)

Autosomal dominant spinocerebellar ataxias are a group of clinically and genetically heterogeneous neurodegenerative disorders. Approximately one-third of roughly

30 distinct SCAs are caused by expansion of microsatellite sequences (Duenas et al. 2006). The molecular pathogenesis of three diseases, SCA8, 10 and 12, is thought to be associated with RNA effects.

5.1.1 SCA8

The most characteristic features of SCA8 are dramatic instability of the expanded CTG repeats and extremely reduced penetrance (Moseley et al. 2000). Normal alleles harbor 16–34 repeats, while disease alleles contain from 71 to over 1,000 repeats in two bidirectionally transcribed genes: *ATXN8* and *ATXN8OS* (formerly *SCA8*) (Koob et al. 1999; Moseley et al. 2006; Ikeda et al. 2007). Since the discovery of bidirectional expression at the SCA8 disease locus, it has become apparent that the polymorphic CTG repeats are part of these two different genes that utilize opposite DNA strands as templates for transcription.

Initially, SCA8 was considered an RNA gain-of-function disease due to the CUG repeat expansion in the noncoding region of the *ATXN8OS* gene, making toxic RNA the most likely mechanism of the SCA8 pathogenesis. Expression of the expanded (112 CUG repeats) as well as normal (nine CUG repeats) transcripts in a *Drosophila* SCA8 model led to the neurodegeneration. Using this neurodegeneration phenotype in *Drosophila* retina, Mutsuddi et al. discovered four proteins that modify (enhance or suppress) the SCA8-induced pathogenesis (Mutsuddi et al. 2004). All of them encoded neuronally expressed RNA binding proteins: staufen, muscleblind, split ends and CG3249. The influence of these *Drosophila* SCA8 phenotype modifiers depended on the size of the CUG repeat (Mutsuddi et al. 2004).

The generally accepted RNA mechanism of SCA8 pathogenesis was re-evaluated upon the discovery of the *ATXN8* gene, which encodes a polyglutamine domain containing protein (Moseley et al. 2006). This gene is transcribed in the opposite direction to *ATXN8OS*. Translation of the *ATXN8* transcript results in formation of 1C2 positive intranuclear inclusions. This hallmark feature of the polyglutamine disorders was detected in cerebellar Purkinje and brainstem neurons in both SCA8 transgenic mice and human autopsy tissue.

In conclusion, simultaneous expression of the CUG (*ATXN8OS* gene) and CAG (*ATXN8* gene) expanded repeat-containing transcripts suggests that SCA8 may be the first known disease caused by both toxic RNA and protein gain-of-function.

5.1.2 SCA10

Spinocerebellar ataxia type 10 (SCA10) is a dominantly inherited ataxia caused by the expansion of the pentanucleotide repeat sequence ATTCT. SCA10 patients demonstrate progressive cerebellar dysfunction which manifests as limb and gait ataxia, ocular movement abnormalities, and dysarthria (Lin and Ashizawa 2005). The expansions occur within intron 9 of the *ATXN10* gene and are among the largest observed in the repeat expansion disorders, ranging from 280 to 4,500 repeats (Matsuura et al. 2000; Lin and Ashizawa 2005). ATTCT sequences can act as DNA unwinding elements (DUEs) – DNA sequences that are frequently associated with replication origins (Potaman et al. 2003, 2006). Aberrant initiation of DNA replication is likely to be responsible for dramatic expansions observed at the SCA10 locus (Liu et al. 2007).

In SCA10 patients, both normal and expanded alleles are expressed at a similar level thus arguing against the role of haploinsufficiency in the etiology of this disease (Ashizawa 2006). Thus far, two lines of evidence indicate a possibility of the involvement of toxic RNAs in the pathogenesis of SCA10. First, AUUCU transcripts can adopt hairpin structures in vitro (Handa et al. 2005). Secondly, overexpression of uninterrupted ATTCT repeats leads to the accumulation of the AUUCU transcript and the formation of RNA inclusions (Lin and Ashizawa 2005). However, proteins targeting AUUCU repeats and the downstream processes affected in the SCA10 cells have not been identified.

5.1.3 SCA12

Spinocerebellar ataxia type 12 is caused by a unique, among dominant ataxias, expansion of the CAG repeat in the noncoding region of the *PPP2R2B* gene. In unaffected individuals, the number of CAG repeats varies from 7 to 32, while pathogenic expanded alleles have 55–78 repeats (Holmes et al. 1999, 2001, 2006). The *PPP2R2B* gene encodes a regulatory subunit (Bβ) of a ubiquitous serine/threonine phosphatase PP2A (PP2), which is involved in the regulation of many cellular processes, including cell growth, apoptosis, differentiation, channel function and DNA replication (Price and Mumby 1999; Virshup 2000).

The structure of the *PPP2R2B* gene, and its splice variants, is quite complex. The predominant protein isoform, Bβ1, is derived from the exon 7 promoter with the CAG repeats located upstream of transcription start site or within the 5′UTR. Among several possible reading frames, only one putative variant, called Bβ7, predicts a polyserine tract at the N-terminus of the protein (Holmes et al. 2006). No evidence supporting polyglutamine toxicity in SCA12 has been discovered. Current data suggest that changes in the expression levels of one or more isoforms of the *PPP2R2B*, which result in alterations in PP2A activity, rather than RNA toxicity, are crucial in the development of SCA12 (Holmes et al. 2006). The role of toxic CAG RNAs cannot be completely ruled out since these repeats are capable of forming stable hairpin structures which are potential targets for MBNL1 and PKR binding (Sobczak et al. 2003; Kino et al. 2004; Michlewski and Krzyzosiak 2004; Sobczak and Krzyzosiak 2005).

5.2 Huntington's Disease-Like 2 (HDL2)

HDL2 is an autosomal dominant progressive neurodegenerative disease that is clinically and pathologically similar to Huntington's disease (HD). This disease is caused by expansion of the CTG repeat tract located in the variably spliced exon 2A of the *Junctophilin-3* (*JPH3*) gene on chromosome 16q24.3. Unaffected individuals carry 6–28 trinucleotide repeats while patients with HDL2 have 40–59 repeats. Depending on the alternative splicing, the CUG repeats may encode polyalanine or polyleucine tracts, or they may be a part of the 3′UTR (Margolis et al. 2006b).

Since no apparent evidence of protein toxicity could be detected in HDL2, an RNA gain-of-function mechanism of pathogenesis was considered. Rudnicki et al. detected RNA inclusions in the brains of HDL2 patients. These inclusions contained different variants of the *JPH3* mRNA, including pre-mRNA, and co-localized with the MBNL1 protein (Rudnicki et al. 2007). Additionally, overexpression of an untranslatable form of the *JPH3* gene, which contained an expanded CUG repeat, in HEK293 and HT22 cell lines led to the formation of inclusions and cellular toxicity. Importantly, abnormalities in splicing of microtubule-associated protein tau and amyloid-β precursor protein (APP) were found, suggesting direct involvement of the impaired function of MBNL1 in HDL2 development (Rudnicki et al. 2007). Although it would be premature to entirely exclude the possible contribution of a toxic protein (containing polyalanine, polyleucine or even polyglutamine, considering bidirectional transcription), results gathered so far strongly implicate the involvement of CUG RNA in the pathogenesis of HDL2.

6 Toxic RNAs as Therapeutic Targets

The first microsatellite sequence expansion responsible for a human disease was discovered in 1991 (La Spada et al. 1991). Although much progress has been made in understanding the molecular mechanisms of the repeat expansion disorders, effective treatments have not been developed thus far. Analyses of mechanisms of different genetic diseases suggest that the challenges faced in the case of toxic RNA-mediated disorders can be circumvented. Since in all of the noncoding repeat diseases structurally normal proteins are synthesized, the major therapeutic goal would be to eliminate the toxic RNA molecules.

In this section, we present current research in development of therapeutic strategies for RNA gain-of-function disorders with a focus on the toxic RNAs as targets for therapeutic intervention. Since the majority of studies related to RNA-targeted therapy have been conducted using DM1 as a model, we will concentrate predominantly on the strategies developed towards therapies for this disease. Selective inhibition of transcription of the allele harboring long repeating tract, blocking the toxic RNA interaction with RBPs, degradation-mediated removal of toxic RNAs by antisense, ribozymes and siRNAs can potentially alleviate the pathogenic effects of the alleles with expanded repeat tracts (Fig. 3).

Attempts have been made to target the mutant *DMPK* transcript using a retroviral vector expressing a 149-nucleotides-long antisense RNA that is complementary to 13 CUG repeats and 110 nucleotides of the unique *DMPK* sequence flanking the repeat region (Furling et al. 2003). DM1 myoblasts infected with this construct

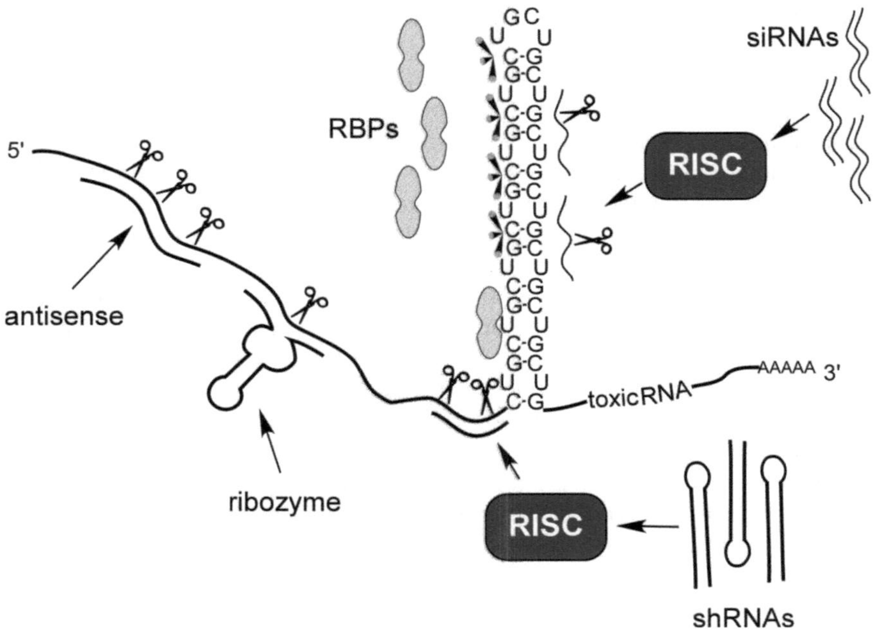

Fig. 3 Therapeutic approaches targeted directly towards toxic RNAs. RNA toxicity can be suppressed by eliminating RNA via ribozyme, antisense RNA or RNAi mediated cleavage. Both small interfering RNAs and short hairpin RNAs are processed by *RISC* complex and may be designed for unique RNA sequences as well as repeating regions. Interactions between RBPs and RNA can also be altered by specific small molecules

showed differential inhibition of expanded and normal transcripts at the levels of 80 and 50%, respectively. The expression of the antisense RNA effectively restored physiological functions of the DM1 myoblasts and lowered the intracellular levels of CUGBP1.

Another approach used to eliminate toxic repeat-containing RNAs was to target *DMPK* transcripts with ribozymes (Langlois et al. 2003). Ribozymes are small, catalytic RNA molecules capable of cleavage, in *trans*, of a specific phosphodiester bond in the RNA substrate. Transiently expressed hammerhead ribozyme, directed towards the 3′UTR of the *DMPK* mRNA downregulated the normal and expanded transcripts by 50–63%. The reduction of *DMPK* mRNA in human DM1 primary myoblasts was accompanied by significant reduction in the number and intensity of nuclear foci as well as partial restoration of expression of the insulin receptor isoform B (Langlois et al. 2003). The reasons for preferential reduction of the expanded *DMPK* transcript in both studies were not clear.

Recently, an adeno-associated virus expressing ribozyme was intramuscularly injected into the DM1 transgenic mice containing 350 CUG repeats. One month after the single injection into the tibialis anterior (TA) muscle, a significant, 55–60%, decrease of the *DMPK* mRNA level was detected, together with a 30% increase in the TA muscle volume (Doucet et al. 2007).

Downregulation of *DMPK* mRNAs, both normal and expanded, was also accomplished using lentivirus-delivered short hairpin RNAs (shRNAs) directed towards a non-repeating sequence of the 3′UTR (Langlois et al. 2005). These shRNAs were processed by the RNA interference pathway leading to specific knockdown of the *DMPK* transcripts due to their cleavage by Dicer. Surprisingly, this approach downregulated not only cytoplasmic, normal *DMPK* transcript containing 18 CUG repeats, but also the expanded transcripts, harboring approximately 3,200 repeats, which were retained in the nucleus of the DM1 cells. Although RNAi is thought to be primarily a cytoplasmic process, these results demonstrated that shRNAs can act both in the cytoplasm and in the nucleus. On the other hand, the cytoplasmic localization of the mutant *DMPK* transcript has also been reported, leaving the possibility that only the cytoplasmic fraction of the expanded *DMPK* mRNA was downregulated (Taneja et al. 1995).

Repeat regions of toxic RNAs adopt stable hairpin structures, but frequently short normal alleles, containing only a few repeats, are incapable of forming of stable hairpins. These structural differences proved to be useful in the design of allele-specific therapeutic strategies. Krol et al. demonstrated that the ribonuclease Dicer, involved in RNAi pathway, controls the level of expanded transcripts containing CUG or CAG repeats (Krol et al. 2007). Short repeat fragments cleaved out from long repeat hairpins subsequently act as siRNAs and further induce downstream silencing of the expanded transcripts. Similarly, transfection of short synthetic oligoribonucleotides containing $(CAG)_7$ or $(CUG)_7$ repeats into cells led to the silencing of the expanded transcripts. Interestingly, in the HeLa cells treated by the CUG or CAG siRNAs only longer transcripts, harboring 70 and 200 repeats, were affected (Krol et al. 2007). These results show that siRNA induced downregulation of CUG or CAG RNAs is, in fact, dependent on the length of the repeats thus establishing a rationale for the future allele-specific therapy.

Intriguingly, it is also possible that the RNAi pathway, induced by the dsRNA that results from bidirectional transcription, is involved in the natural regulation of the repeat-containing transcripts. This process may explain the reduced penetrance observed in SCA8 and FXTAS. In conclusion, the RNAi pathway becomes undoubtedly one of the most promising approaches for the future treatment of several incurable neurodegenerative disorders (Denovan-Wright and Davidson 2006; Paulson 2006; Rodriguez-Lebron and Paulson 2006).

Expanded RNA repeats, via sequestering some RNA binding proteins and activation of others, deregulate a delicate balance between these proteins and influence downstream processes leading to neurodegeneration. Restoration of this balance, by correcting the expression levels of appropriate proteins, can be therapeutically beneficial. Three groups, almost simultaneously, reported therapeutic effects of modifying the expression of MBNL1, CUGBP1 and hnRNP H in different models of DM1. Reversal of aberrant splicing and myotonia was observed in transgenic mice harboring 250 CUG repeats after AAV-mediated overexpression of *Mbnl1* (Kanadia et al. 2006). However, the normal structure of myofibers was not restored, most likely due to the low expression of *Mbnl1* relative to the CUG repeat-containing transcript. These results showed that either overexpression of MBNL1 (perhaps together with MBNL2) or suppression of interactions between CUG transcripts and the RBPs may represent a successful strategy for treatment of DM1 and DM2.

On the other hand, results of overexpression of the green fluorescent protein (*GFP*) gene containing a 3′UTR derived from the normal *DMPK* allele (five CUG repeats) suggested that binding of MBNL1 to the CUG repeats might be necessary for neutralizing repeat-induced toxicity (Mahadevan et al. 2006). In these circumstances, protein binding plays a protective role. Therefore, abolishing CUG RNA–MBNL1 interactions may be potentially deleterious to DM1 cells. An alternative therapeutic approach could involve the control of *DMPK* expression or a correction of the pathological effects of the increased CUGBP1 levels.

Recent data on the aberrant splicing of the insulin receptor as a model for DM1-specific spliceopathy demonstrated that siRNA-mediated knockdown of CUGBP1 and hnRNP H did not rescue aberrant IR splicing (Paul et al. 2006). Also, overexpression of MBNL1 only partially alleviated the splicing abnormalities. However, simultaneous increase of the MBNL1 and decrease of hnRNP H further stimulated correction of the IR splicing. Taken together, these results suggest that fine tuning of several elements is required to restore proper physiology of DM1 cells.

A significant disadvantage of therapeutic approaches that rely on expression of ribozymes, siRNAs, antisense RNAs or entire proteins is the necessity for the proper delivery of relatively large vectors into the specific cells. This task is further complicated by the necessity to cross the blood–brain barrier and evade the immunological response of the organism. Therefore, small molecules that specifically recognize and alter functions of toxic RNAs could present an alternative solution to these problems (Fig. 3). Future development of high-throughput screens (HTS), using both in vitro and cell-based assays, may result in identification of compounds capable of ameliorating the toxic activity of expanded RNAs. A cell-based assay, utilizing the PC12 neuronal cell line stably transfected with a luciferase gene containing 250 CUG repeats in the 3′UTR, was used to screen a library of 235 bioflavonoids and other chemicals structurally resembling flavonoids (Furuya et al. 2005). Some of the analyzed compounds, applied in micromolar concentrations, reduced CUG induced cytotoxicity, presumably by interfering with the affinity of protein binding to the toxic RNAs. Nevertheless, biochemical assays for specific inhibitors of RNA–protein interactions remain to be developed. Potentially, small molecules could also selectively target expression of the expanded allele by altering epigenetic modifications frequently associated with DNA regions containing long repeating sequences (Wang et al. 1996; Coffee et al. 2002; Herman et al. 2006).

In summary, although there has been tremendous progress in the understanding of the molecular mechanisms of RNA gain-of-function, many questions related to the pathogenesis of these disorders remain to be answered. Attempts of therapeutic intervention, such as those described above, indicate that RNA toxicity can be reversed.

Acknowledgments This work was supported by Grants from National Ataxia Foundation and Friedreich's Ataxia Research Alliance to MN. We thank Drs. Sandesh Chakravarthy, Daryl Scott, and Dustin Baldridge for reading the manuscript and for their helpful comments.

References

Alwazzan M, Newman E, Hamshere MG et al (1999) Myotonic dystrophy is associated with a reduced level of RNA from the DMWD allele adjacent to the expanded repeat. Hum Mol Genet 8:1491–1497

Arhin GK, Boots M, Bagga PS et al (2002) Downstream sequence elements with different affinities for the hnRNP H/H' protein influence the processing efficiency of mammalian polyadenylation signals. Nucleic Acids Res 30:1842–1850

Ashizawa T (2006) Spinocerebellar ataxia type 10: a disease caused by an expanded (ATTCT)n pentanucleotide repeat. In: Ashizawa T, Wells RD (eds) Genetic instabilities and neurological diseases. Elsevier–Academic, San Diego, pp 433–446

Barreau C, Paillard L, Mereau A et al (2006) Mammalian CELF/Bruno-like RNA-binding proteins: molecular characteristics and biological functions. Biochimie 88:515–525

Broda M, Kierzek E, Gdaniec Z et al (2005) Thermodynamic stability of RNA structures formed by CNG trinucleotide repeats. Implication for prediction of RNA structure. Biochemistry 44:10873–10882

Brook JD, McCurrach ME, Harley HG et al (1992) Molecular basis of myotonic dystrophy: expansion of a trinucleotide (CTG) repeat at the 3′ end of a transcript encoding a protein kinase family member. Cell 68:799–808

Buj-Bello A, Furling D, Tronchere H et al (2002) Muscle-specific alternative splicing of myotubularin-related 1 gene is impaired in DM1 muscle cells. Hum Mol Genet 11:2297–2307

Cardani R, Mancinelli E, Rotondo G et al (2006) Muscleblind-like protein 1 nuclear sequestration is a molecular pathology marker of DM1 and DM2. Eur J Histochem 50:177–182

Charlet BN, Savkur RS, Singh G et al (2002) Loss of the muscle-specific chloride channel in type 1 myotonic dystrophy due to misregulated alternative splicing. Mol Cell 10:45–53

Chen CD, Kobayashi R, Helfman DM (1999) Binding of hnRNP H to an exonic splicing silencer is involved in the regulation of alternative splicing of the rat beta-tropomyosin gene. Genes Dev 13:593–606

Chen W, Wang Y, Abe Y et al (2007) Haploinsuffciency for Znf9 in Znf9+/– mice is associated with multiorgan abnormalities resembling myotonic dystrophy. J Mol Biol 368:8–17

Cheng J, Kapranov P, Drenkow J et al (2005) Transcriptional maps of 10 human chromosomes at 5-nucleotide resolution. Science 308:1149–1154

Cho DH, Thienes CP, Mahoney SE et al (2005) Antisense transcription and heterochromatin at the DM1 CTG repeats are constrained by CTCF. Mol Cell 20:483–489

Coffee B, Zhang F, Ceman S et al (2002) Histone modifications depict an aberrantly heterochromatinized FMR1 gene in fragile X syndrome. Am J Hum Genet 71:923–932

Dansithong W, Paul S, Comai L et al (2005) MBNL1 is the primary determinant of focus formation and aberrant insulin receptor splicing in DM1. J Biol Chem 280:5773–5780

Davis BM, McCurrach ME, Taneja KL et al (1997) Expansion of a CUG trinucleotide repeat in the 3 untranslated region of myotonic dystrophy protein kinase transcripts results in nuclear retention of transcripts. Proc Natl Acad Sci USA 94:7388–7393

Denovan-Wright EM, Davidson BL (2006) RNAi: a potential therapy for the dominantly inherited nucleotide repeat diseases. Gene Ther 13:525–531

Doucet G, Vignaud A, Gourdon G et al (2007) Ribozyme-based gene therapy reverses muscle atrophy in a mouse model of myotonic dystrophy. Neuromuscul Disord 17:899

Duenas AM, Goold R, Giunti P (2006) Molecular pathogenesis of spinocerebellar ataxias. Brain 129:1357–1370

Ebralidze A, Wang Y, Petkova V et al (2004) RNA leaching of transcription factors disrupts transcription in myotonic dystrophy. Science 303:383–387
Eriksson M, Ansved T, Edstrom L et al (1999) Simultaneous analysis of expression of the three myotonic dystrophy locus genes in adult skeletal muscle samples: the CTG expansion correlates inversely with DMPK and 59 expression levels, but not DMAHP levels. Hum Mol Genet 8:1053–1060
Fardaei M, Larkin K, Brook JD et al (2001) In vivo co-localisation of MBNL protein with DMPK expanded-repeat transcripts. Nucleic Acids Res 29:2766–2771
Fardaei M, Rogers MT, Thorpe HM et al (2002) Three proteins, MBNL, MBLL and MBXL, colocalize in vivo with nuclear foci of expanded-repeat transcripts in DM1 and DM2 cells. Hum Mol Genet 11:805–814
Fu YH, Kuhl DP, Pizzuti A et al (1991) Variation of the CGG repeat at the fragile X site results in genetic instability: resolution of the Sherman paradox. Cell 67:1047–1058
Furling D, Doucet G, Langlois MA et al (2003) Viral vector producing antisense RNA restores myotonic dystrophy myoblast functions. Gene Ther 10:795–802
Furuya H, Shinnoh N, Ohyagi Y et al (2005) Some flavonoids and DHEA-S prevent the *cis*-effect of expanded CTG repeats in a stable PC12 cell transformant. Biochem Pharmacol 69:503–516
Gerbasi VR, Link AJ (2007) The myotonic dystrophy type 2 protein ZNF9 is part of an ITAF complex that promotes cap-independent translation. Mol Cell Proteomics 6:1049–1058
Greco CM, Hagerman RJ, Tassone F et al (2002) Neuronal intranuclear inclusions in a new cerebellar tremor/ataxia syndrome among fragile X carriers. Brain 125:1760–1771
Hagerman RJ, Hagerman PJ (2006) Fragile X syndrome and fragile X-associated tremor/ataxia syndrome. In: Ashizawa T, Wells RD (eds) Genetic instabilities and neurological diseases. Elsevier–Academic, San Diego, pp 165–174
Hagerman RJ, Leehey M, Heinrichs W et al (2001) Intention tremor, parkinsonism, and generalized brain atrophy in male carriers of fragile X. Neurology 57:127–130
Hagerman RJ, Leavitt BR, Farzin F et al (2004) Fragile-X-associated tremor/ataxia syndrome (FXTAS) in females with the FMR1 premutation. Am J Hum Genet 74:1051–1056
Handa V, Saha T, Usdin K (2003) The fragile X syndrome repeats form RNA hairpins that do not activate the interferon-inducible protein kinase, PKR, but are cut by Dicer. Nucleic Acids Res 31:6243–6248
Handa V, Yeh HJC, McPhie P et al (2005) The AUUCU repeats responsible for spinocerebellar type 10 form unusual RNA hairpins. J Biol Chem 280:29340–29345
Harper PS (2001) Myotonic dystrophy. Saunders, London
Herman D, Jenssen K, Burnett R et al (2006) Histone deacetylase inhibitors reverse gene silencing in Friedreich's ataxia. Nat Chem Biol 2:551–558
Ho TH, Charlet BN, Poulos MG et al (2004) Muscleblind proteins regulate alternative splicing. EMBO J 23:3103–3112
Ho TH, Bundman D, Armstrong DL et al (2005) Transgenic mice expressing CUG-BP1 reproduce splicing mis-regulation observed in myotonic dystrophy. Hum Mol Genet 14:1539–1547
Holmes SE, O'Hearn EE, McInnis MG et al (1999) Expansion of a novel CAG trinucleotide repeat in the 5′ region of PPP2R2B is associated with SCA12. Nat Genet 23:391–392
Holmes SE, Hearn EO, Ross CA, Margolis RL (2001) SCA12: an unusual mutation leads to an unusual spinocerebellar ataxia. Brain Res Bull 56:397–403
Holmes SE, O'Hearn E, Cortez-Apreza N et al (2006) Spinocerebellar ataxia type 12. In: Ashizawa T, Wells RD (eds) Genetic instabilities and neurological diseases. Elsevier–Academic, San Diego, pp 461–473
Hutchison CJ, Worman HJ (2004) A-type lamins: guardians of the soma? Nat Cell Biol 6:1062–1067
Ikeda Y, Daughters RS, Ranum LP (2007) Bidirectional expression of the SCA8 expansion mutation: One mutation, two genes. Cerebellum 5:1–9
Iwahashi CK, Yasui DH, An HJ et al (2006) Protein composition of the intranuclear inclusions of FXTAS. Brain 129:256–271

Jacquemont S, Hagerman RJ, Leehey MA et al (2004) Penetrance of the fragile X-associated tremor/ataxia syndrome in a premutation carrier population. JAMA 291:460–469
Jansen G, Groenen PJ, Bachner D et al (1996) Abnormal myotonic dystrophy protein kinase levels produce only mild myopathy in mice. Nat Genet 13:316–324
Jiang H, Mankodi A, Swanson MS et al (2004) Myotonic dystrophy type 1 is associated with nuclear foci of mutant RNA, sequestration of muscleblind proteins and deregulated alternative splicing in neurons. Hum Mol Genet 13:3079–3088
Jin P, Zarnescu DC, Zhang F et al (2003) RNA-mediated neurodegeneration caused by the fragile X premutation rCGG repeats in Drosophila. Neuron 39:739–747
Jin P, Duan R, Qurashi A et al (2007) Pur alpha binds to rCGG repeats and modulates repeat-mediated neurodegeneration in a Drosophila model of fragile X tremor/ataxia syndrome. Neuron 55:556–564
Kanadia RN, Johnstone KA, Mankodi A et al (2003a) A muscleblind knockout model for myotonic dystrophy. Science 302:1978–1980
Kanadia RN, Urbinati CR, Crusselle VJ et al (2003b) Developmental expression of mouse muscleblind genes Mbnl1, Mbnl2 and Mbnl3. Gene Expr Patterns 3:459–462
Kanadia RN, Shin J, Yuan Y et al (2006) Reversal of RNA missplicing and myotonia after muscleblind overexpression in a mouse poly(CUG) model for myotonic dystrophy. Proc Natl Acad Sci USA 103:11748–11753
Khateb S, Weisman-Shomer P, Hershco-Shani I et al (2007) The tetraplex (CGG)n destabilizing proteins hnRNP A2 and CBF-A enhance the in vivo translation of fragile X premutation mRNA. Nucleic Acids Res 35:5775–5788
Kim DH, Langlois MA, Lee KB et al (2005) HnRNP H inhibits nuclear export of mRNA containing expanded CUG repeats and a distal branch point sequence. Nucleic Acids Res 33:3866–3874
Kimura T, Nakamori M, Lueck JD et al (2005) Altered mRNA splicing of the skeletal muscle ryanodine receptor and sarcoplasmic/endoplasmic reticulum Ca2+-ATPase in myotonic dystrophy type 1. Hum Mol Genet 14:2189–2200
Kino Y, Mori D, Oma Y, Takeshita Y, Sasagawa N, Ishiura S (2004) Muscleblind protein, MBNL1/EXP, binds specifically to CHHG repeats. Hum Mol Genet 13:495–507
Klesert TR, Otten AD, Bird TD et al (1997) Trinucleotide repeat expansion at the myotonic dystrophy locus reduces expression of DMAHP. Nat Genet 16:402–406
Klesert TR, Cho DH, Clark JI et al (2000) Mice deficient in Six5 develop cataracts: implications for myotonic dystrophy. Nat Genet 25:105–109
Knight SJ, Flannery AV, Hirst MC et al (1993) Trinucleotide repeat amplification and hypermethylation of a CpG island in FRAXE mental retardation. Cell 74:127–134
Koob MD, Moseley ML, Schut LJ et al (1999) An untranslated CTG expansion causes a novel form of spinocerebellar ataxia (SCA8). Nat Genet 21:379–384
Krol J, Fiszer A, Mykowska A et al (2007) Ribonuclease dicer cleaves triplet repeat hairpins into shorter repeats that silence specific targets. Mol Cell 25:575–586
Kuyumcu-Martinez NM, Wang GS et al (2007) Increased steady-state levels of CUGBP1 in myotonic dystrophy 1 are due to PKC-mediated hyperphosphorylation. Mol Cell 28:68–78
Ladd AN, Charlet N, Cooper TA (2001) The CELF family of RNA binding proteins is implicated in cell-specific and developmentally regulated alternative splicing. Mol Cell Biol 21:1285–1296
Ladd AN, Stenberg MG, Swanson MS et al (2005) Dynamic balance between activation and repression regulates pre-mRNA alternative splicing during heart development. Dev Dyn 233:783–793
Ladd PD, Smith LE, Rabaia NA et al (2007) An antisense transcript spanning the CGG repeat region of FMR1 is upregulated in premutation carriers but silenced in full mutation individuals. Hum Mol Genet 16:3174–3187
Langlois MA, Lee NS, Rossi JJ et al (2003) Hammerhead ribozyme-mediated destruction of nuclear foci in myotonic dystrophy myoblasts. Mol Ther 7:670–680

Langlois MA, Boniface C, Wang G et al (2005) Cytoplasmic and nuclear retained DMPK mRNAs are targets for RNA interference in myotonic dystrophy cells. J Biol Chem 280:16949–16954
La Spada AR, Wilson EM, Lubahn DB et al (1991) Androgen receptor gene mutations in X-linked spinal and bulbar muscular atrophy. Nature 352:77–79
Leroy O, Dhaenens CM, Schraen-Maschke S et al (2006) ETR-3 represses Tau exons 2/3 inclusion, a splicing event abnormally enhanced in myotonic dystrophy type I. J Neurosci Res 84:852–859
Lin X, Ashizawa T (2005) Recent progress in spinocerebellar ataxia type-10 (SCA10). Cerebellum 4:37–42
Lin X, Miller JW, Mankodi A et al (2006) Failure of MBNL1-dependent post-natal splicing transitions in myotonic dystrophy. Hum Mol Genet 15:2087–2097
Liquori CL, Ricker K, Moseley ML et al (2001) Myotonic dystrophy type 2 caused by a CCTG expansion in intron 1 of ZNF9. Science 293:864–867
Liu G, Bissler JJ, Sinden RR et al (2007) Unstable spinocerebellar ataxia type 10 (ATTCT-AGAAT) repeats are associated with aberrant replication at the ATX10 locus and replication origin-dependent expansion at an ectopic site in human cells. Mol Cell Biol 27:7828–7838
Llagostera E, Catalucci D, Marti L et al (2007) Role of myotonic dystrophy protein kinase (DMPK) in glucose homeostasis and muscle insulin action. PLoS ONE 2:e1134
Lu X, Timchenko NA, Timchenko LT (1999) Cardiac elav-type RNA-binding protein (ETR-3) binds to RNA CUG repeats expanded in myotonic dystrophy. Hum Mol Genet 8:53–60
Lueck JD, Mankodi A, Swanson MS et al (2007) Muscle chloride channel dysfunction in two mouse models of myotonic dystrophy. J Gen Physiol 129:79–94
Mahadevan MS, Yadava RS, Yu Q et al (2006) Reversible model of RNA toxicity and cardiac conduction defects in myotonic dystrophy. Nat Genet 38:1066–1070
Mankodi A, Logigian E, Callahan L et al (2000) Myotonic dystrophy in transgenic mice expressing an expanded CUG repeat. Science 289:1769–1773
Mankodi A, Urbinati CR, Yuan QP et al (2001) Muscleblind localizes to nuclear foci of aberrant RNA in myotonic dystrophy types 1 and 2. Hum Mol Genet 10:2165–2170
Mankodi A, Takahashi MP, Jiang H et al (2002) Expanded CUG repeats trigger aberrant splicing of ClC-1 chloride channel pre-mRNA and hyperexcitability of skeletal muscle in myotonic dystrophy. Mol Cell 10:35–44
Mankodi A, Lin X, Blaxall BC et al (2005) Nuclear RNA foci in the heart in myotonic dystrophy. Circ Res 97:1152–1155
Margolis JM, Schoser BG, Moseley ML et al (2006a) DM2 intronic expansions: evidence for CCUG accumulation without flanking sequence or effects on ZNF9 mRNA processing or protein expression. Hum Mol Genet 15:1808–1815
Margolis RL, Holmes SE, Rudnicki DD et al (2006b) Huntington's Disease-like 2. In: Ashizawa T, Wells RD (eds) Genetic instabilities and neurological diseases. Elsevier–Academic, San Diego, pp 261–273
Marquis J, Paillard L, Audic Y et al (2006) CUG-BP1/CELF1 requires UGU-rich sequences for high-affinity binding. Biochem J 400:291–301
Matsuura T, Yamagata T, Burgess DL et al (2000) Large expansion of the ATTCT pentanucleotide repeat in spinocerebellar ataxia type 10. Nat Genet 26:191–194
Maurage CA, Udd B, Ruchoux MM et al (2005) Similar brain tau pathology in DM2/PROMM and DM1/Steinert disease. Neurology 65:1636–1638
Michalowski S, Miller JW, Urbinati CR et al (1999) Visualization of double-stranded RNAs from the myotonic dystrophy protein kinase gene and interactions with CUG-binding protein. Nucleic Acids Res 27:3534–3542
Michlewski G, Krzyzosiak WJ (2004) Molecular architecture of CAG repeats in human disease related transcripts. J Mol Biol 340:665–679
Miller JW, Urbinati CR, Teng-Umnuay P et al (2000) Recruitment of human muscleblind proteins to (CUG)(n) expansions associated with myotonic dystrophy. EMBO J 19:4439–4448

Mooers BH, Logue JS, Berglund JA (2005) The structural basis of myotonic dystrophy from the crystal structure of CUG repeats. Proc Natl Acad Sci USA 102:16626–16631
Mori D, Sasagawa N, Kino Y et al (2008) Quantitative analysis of CUG-BP1 binding to RNA repeats. J Biochem 143(3):377–383
Moseley ML, Schut LJ, Bird TD et al (2000) SCA8 CTG repeat: en masse contractions in sperm and intergenerational sequence changes may play a role in reduced penetrance. Hum Mol Genet 9:2125–2130
Moseley ML, Zu T, Ikeda Y et al (2006) Bidirectional expression of CUG and CAG expansion transcripts and intranuclear polyglutamine inclusions in spinocerebellar ataxia type 8. Nat Genet 38:758–769
Mutsuddi M, Marshall CM, Benzow KA et al (2004) The spinocerebellar ataxia 8 noncoding RNA causes neurodegeneration and associates with staufen in Drosophila. Curr Biol 14:302–308
Nakamori M, Kimura T, Fujimura H et al (2007) Altered mRNA splicing of dystrophin in type 1 myotonic dystrophy. Muscle Nerve 36:251–257
Napierala M, Krzyzosiak WJ (1997) CUG repeats present in myotonin kinase RNA form metastable "slippery" hairpins. J Biol Chem 272:31079–31085
Napierala M, Michalowski D, de Mezer M et al (2005) Facile FMR1 mRNA structure regulation by interruptions in CGG repeats. Nucleic Acids Res 33:451–463
O'Cochlain DF, Perez-Terzic C, Reyes S et al (2004) Transgenic overexpression of human DMPK accumulates into hypertrophic cardiomyopathy, myotonic myopathy and hypotension traits of myotonic dystrophy. Hum Mol Genet 13:2505–2518
Orr HT, Zoghbi HY (2007) Trinucleotide Repeat Disorders. Annu Rev Neurosci 30:575–621
Osborne RJ, Thornton CA (2006) RNA-dominant diseases. Hum Mol Genet 15:162–169
Paul S, Dansithong W, Kim D et al (2006) Interaction of muscleblind, CUG-BP1 and hnRNP H proteins in DM1-associated aberrant IR splicing. EMBO J 25:4271–4283
Paulson H (2006) RNA interference as potential therapy for neurodegenerative disease: applications to inclusion-body myositis? Neurology 66:114–117
Philips AV, Timchenko LT, Cooper TA (1998) Disruption of splicing regulated by a CUG-binding protein in myotonic dystrophy. Science 280:737–741
Pieretti M, Zhang FP, Fu YH et al (1991) Absence of expression of the FMR-1 gene in fragile X syndrome. Cell 66:817–822
Potaman VN, Bissler JJ, Hashem V et al (2003) Unpaired structures in SCA10 (ATTCT)n.(AGAAT)n repeats. J Mol Biol 326:1095–1111
Potaman VN, Pytlos MJ, Hashem VI et al (2006) DNA structures and genetic instabilities associated with spinocerebellar ataxia type 10 suggest a DNA amplification model for repeat expansion. In: Ashizawa T, Wells RD (eds) Genetic instabilities and neurological diseases. Elsevier–Academic, San Diego, pp 447–460
Price NE, Mumby MC (1999) Brain protein serine/threonine phosphatases. Curr Opin Neurobiol 9:336–342
Ranum LP, Cooper TA (2006) RNA-mediated neuromuscular disorders. Annu Rev Neurosci 29:259–277
Reddy S, Smith DB, Rich MM et al (1996) Mice lacking the myotonic dystrophy protein kinase develop a late onset progressive myopathy. Nat Genet 13:325–335
Rodriguez-Lebron E, Paulson HL (2006) Allele-specific RNA interference for neurological disease. Gene Ther 13:576–581
Rudnicki DD, Holmes SE, Lin MW et al (2007) Huntington's disease-like 2 is associated with CUG repeat-containing RNA foci. Ann Neurol 61:272–282
Sarkar PS, Appukuttan B, Han J et al (2000) Heterozygous loss of Six5 in mice is sufficient to cause ocular cataracts. Nat Genet 25:110–114
Savkur RS, Philips AV, Cooper TA (2001) Aberrant regulation of insulin receptor alternative splicing is associated with insulin resistance in myotonic dystrophy. Nat Genet 29:40–47
Savkur RS, Philips AV, Cooper TA et al (2004) Insulin receptor splicing alteration in myotonic dystrophy type 2. Am J Hum Genet 74:1309–1313

Seznec H, Agbulut O, Sergeant N et al (2001) Mice transgenic for the human myotonic dystrophy region with expanded CTG repeats display muscular and brain abnormalities. Hum Mol Genet 10:2717–2726

Sobczak K, Krzyzosiak WJ (2004a) Imperfect CAG repeats form diverse structures in SCA1 transcripts. J Biol Chem 279:41563–41572

Sobczak K, Krzyzosiak WJ (2004b) Patterns of CAG repeat interruptions in SCA1 and SCA2 genes in relation to repeat instability. Hum Mutat 24:236–247

Sobczak K, Krzyzosiak WJ (2005) CAG repeats containing CAA interruptions form branched hairpin structures in spinocerebellar ataxia type 2 transcripts. J Biol Chem 280:3898–3910

Sobczak K, de Mezer M, Michlewski G et al (2003) RNA structure of trinucleotide repeats associated with human neurological diseases. Nucleic Acids Res 31:5469–5482

Sofola OA, Jin P, Botas J et al (2007a) Argonaute-2-dependent rescue of a Drosophila model of FXTAS by FRAXE premutation repeat. Hum Mol Genet 16:2326–2332

Sofola OA, Jin P, Qin Y et al (2007b) RNA-binding proteins hnRNP A2/B1 and CUGBP1 suppress fragile X CGG premutation repeat-induced neurodegeneration in a Drosophila model of FXTAS. Neuron 55:565–571

Taneja KL, McCurrach M, Schalling M et al (1995) Foci of trinucleotide repeat transcripts in nuclei of myotonic dystrophy cells and tissues. J Cell Biol 128:995–1002

Tassone F, Hagerman RJ, Chamberlain WD et al (2000a) Transcription of the FMR1 gene in individuals with fragile X syndrome. Am J Med Genet 97:195–203

Tassone F, Hagerman RJ, Taylor AK et al (2000b) Elevated levels of FMR1 mRNA in carrier males: a new mechanism of involvement in the fragile-X syndrome. Am J Hum Genet 66:6–15

Tassone F, Iwahashi C, Hagerman PJ (2004) FMR1 RNA within the intranuclear inclusions of fragile X-associated tremor/ataxia syndrome (FXTAS). RNA Biol 1:103–105

Thornton CA, Wymer JP, Simmons Z et al (1997) Expansion of the myotonic dystrophy CTG repeat reduces expression of the flanking DMAHP gene. Nat Genet 16:407–409

Thornton C, Swanson MS, Cooper TA (2006) The RNA-mediated disease process in myotonic dystrophy. In: Ashizawa T, Wells RD (eds) Genetic instabilities and neurological diseases. Elsevier–Academic, San Diego, pp 37–54

Tian B, White RJ, Xia T, Welle S et al (2000) Expanded CUG repeat RNAs form hairpins that activate the double-stranded RNA-dependent protein kinase PKR. RNA 6:79–87

Tian B, Mukhopadhyay R, Mathews MB (2005) Polymorphic CUG repeats in human mRNAs and their effects on gene expression. RNA Biol 2:149–156

Timchenko LT, Miller JW, Timchenko NA et al (1996a) Identification of a (CUG)n triplet repeat RNA-binding protein and its expression in myotonic dystrophy. Nucleic Acids Res 24:4407–4414

Timchenko LT, Timchenko NA, Caskey CT et al (1996b) Novel proteins with binding specificity for DNA CTG repeats and RNA CUG repeats: implications for myotonic dystrophy. Hum Mol Genet 5:115–121

Timchenko NA, Cai ZJ, Welm AL et al (2001) RNA CUG repeats sequester CUGBP1 and alter protein levels and activity of CUGBP1. J Biol Chem 276:7820–7826

Timchenko NA, Patel R, Iakova P et al (2004) Overexpression of CUG triplet repeat-binding protein, CUGBP1, in mice inhibits myogenesis. J Biol Chem 279:13129–13139

Vermersch P, Sergeant N, Ruchoux MM et al (1996) Specific tau variants in the brains of patients with myotonic dystrophy. Neurology 47:711–717

Virshup DM (2000) Protein phosphatase 2A: a panoply of enzymes. Curr Opin Cell Biol 12:180–185

Wang YH, Gellibolian R, Shimizu M et al (1996) Long CCG triplet repeat blocks exclude nucleosomes: a possible mechanism for the nature of fragile sites in chromosomes. J Mol Biol 263:511–516

Wang GS, Kearney DL, De Biasi M et al (2007) Elevation of RNA-binding protein CUGBP1 is an early event in an inducible heart-specific mouse model of myotonic dystrophy. J Clin Invest 117:2802–2811

Warf MB, Berglund JA (2007) MBNL binds similar RNA structures in the CUG repeats of myotonic dystrophy and its pre-mRNA substrate cardiac troponin T. RNA 13:2238–2251

Willemsen R, Hoogeveen-Westerveld M, Reis S et al (2003) The FMR1 CGG repeat mouse displays ubiquitin-positive intranuclear neuronal inclusions; implications for the cerebellar tremor/ataxia syndrome. Hum Mol Genet 12:949–959

Yuan Y, Compton SA, Sobczak K et al (2007) Muscleblind-like 1 interacts with RNA hairpins in splicing target and pathogenic RNAs. Nucleic Acids Res 35:5474–5486

Zumwalt M, Ludwig A, Hagerman PJ et al (2007) Secondary structure and dynamics of the r(CGG) repeat in the mRNA of the fragile X mental retardation 1 (FMR1) gene. RNA Biol 4:93–100

Index

MIX
Papier aus verantwortungsvollen Quellen
Paper from responsible sources
FSC® C105338

If you have any concerns about our products,
you can contact us on
ProductSafety@springernature.com

In case Publisher is established outside the EU,
the EU authorized representative is:
Springer Nature Customer Service Center GmbH
Europaplatz 3, 69115 Heidelberg, Germany

Printed by Libri Plureos GmbH
in Hamburg, Germany